# Microelectronic
# Switched-Capacitor Filters

# Microelectronic Switched-Capacitor Filters

*with ISICAP: a computer-aided-design package*

## H. Baher

**JOHN WILEY & SONS**

Chichester • New York • Brisbane • Toronto • Singapore

*National*      01243 779777
*International*  (+ 44) 1243 779777

*Other Wiley Editorial Offices*

John Wiley & Sons, Inc., 605 Third Avenue,
New York, NY 10158-0012, USA

Jacaranda Wiley Ltd, 33 Park Road, Milton,
Queensland 4064, Australia

John Wiley & Sons (Canada) Ltd, 22 Worcester Road,
Rexdale, Ontario M9W 1L1, Canada

John Wiley & Sons (SEA) Pte Ltd, 37 Jalan Pemimpin #05-04,
Block B, Union Industrial Building, Singapore 2057

*Library of Congress Cataloging-in-Publication Data*

Baher, H.
     Microelectronic switched capacitor filters/H. Baher.
          p.   cm.
     Includes bibliographical references and index.
     ISBN 0 471 95404 7
     1. Switched capacitor circuits.   2. Electric filters.   3. Metal
     oxide semiconductors.   4. Integrated circuits.   I. Title.
     TK7868.S88B35 1996
     621.3815′324—dc20                                    95–38186
                                                             CIP

*British Library Cataloguing in Publication Data*

A catalogue record for this book is available from the British Library

ISBN 0 471 95404 7

Typeset in 10½/12pt Times by Aarontype, Bristol
Printed and bound in Great Britain by Bookcraft (Bath) Ltd
This book is printed on acid-free paper responsibly manufactured from sustainable forestation,
for which at least two trees are planted for each one used for paper production.

*The universities are thus reverting to a position more analogous to that which they occupied in the Middle Ages; they are becoming training schools for the professions . . . If pure learning is to survive as one of the purposes of universities, it will have to be brought into relation with the life of the community as a whole not only with the refined delights of a few gentlemen of leisure. I regard disinterested learning as a matter of great importance, and I should wish to see its place in academic life increased, not diminished. Both in England and in America, the main force tending to its diminution has been the desire to get endowments from ignorant millionaires. The cure lies in the creation of an educated democracy, willing to spend public money on objects which our captains of industry are unable to appreciate. This is by no means impossible, but it demands a general raising of the intellectual level'.*

Bertrand Russell

*On Education*

# Contents

# Preface

*O wüsst ich doch den Weg zurück,*
*Den lieben Weg zum Kinderland!*
*O warum sucht'ich nach dem Glück*
*Und liess der Mutter Hand?*
*O wie ich selnet auszuruhn,*
*von keinem Streben aufgeweckt,*
*Die müden Augen zuzutun,*
*von Liebe sanft bedeckt!*
*Und nichts zu forschen, nichts zu späln,*
*Und nur zu träumen leicht und lind;*
*Der Zeiten Wandel nicht zu sehn,*
*Zum zweiten Mal ein Kind!*

Klaus Groth
*A Lied by Brahms*

This is a book on the design of switched-capacitor filters and the associated analog metal-oxide-semiconductor (MOS) integrated circuits. The latter form the basis for placing switched-capacitor filters firmly in the very large scale integration (VLSI) field. The book is accompanied by a diskette containing **ISICAP**: a comprehensive computer-aided design package. The source code of the program is in the directory **ISICAP** while the executable version of the program is also included on the diskette in the directory **ISICAPS**.

The book also provides some background material which is, on the one hand essential for the understanding of switched-capacitor filters, while on the other hand there is some reason to believe that some readers may not be fully familiar with this information. So, the book starts at a fairly introductory level, but proceeds in a gradual manner to develop the design techniques of switched-capacitor filters, reaching a highly sophisticated level. Thus, the presentation in the book is, to a large extent, self-contained. As a consequence of this approach, the book covers the topics which are of essential importance to all types of filter not just switched-capacitor ones.

Switched-capacitor filters have now reached a sufficient degree of maturity to justify a book devoted entirely to the design techniques of this important category of signal processing subsystem. At present there is no comprehensive text-book which concentrates on switched-capacitor filters. This book fills this gap in the literature by putting together the design techniques of switched-capacitor filters from two complementary stand-points: those of the filter design specialist and the integrated circuit design engineer. There is a slight bias, however, in favour of the former since there are many accounts of integrated circuit design available in the literature. But the sharply focused exposition of switched-capacitor filters, appears here for the first time. The inclusion of computer-aided design programs is also a unique feature, thought highly desirable because of the rather heavy computations involved in the design procedure of some categories of switched-capacitor filters.

Integrated circuits for telecommunications applications usually contain both analog and digital circuits on the same chip. The term *analog VLSI*, therefore, was originally used to denote the analog part of the VLSI chip even though the digital circuits constituted the larger portion. On the other hand, there have recently been efforts to design VLSI chips that contain mainly analog circuits leading to a more appropriate use of the term. At any rate, in either of its guises, analog VLSI has benefitted to a considerable measure from the use of switched-capacitor techniques. The quest for MOS building blocks with low power consumption, high precision linear performance, insensitivity to parasitics, and leading to area-efficient modular designs, has been the subject of intensive and successful efforts in the design of switched-capacitor filters. But these are precisely the requirements of analog VLSI circuits, and consequently switched-capacitor research has always opened many avenues for analog VLSI design in general. This is in addition to the fact that the most successful approach to analog VLSI has, arguably, been that of switched-capacitor techniques.

The book is divided into three parts in addition to the diskette containing **ISICAP** and **ISICAPS**.

**Part I** is a general introduction including one chapter. Chapter 1 sets the scene for the exposition in the book and puts switched-capacitor filters in the proper perspective relative to analog integrated filters and VLSI circuits. The main features and attributes of signal processing using switched-capacitor techniques are highlighted. Several applications of these filters are indicated, particularly in the telecommunications field. Also the typical frequency limits, power consumption and the prevalent technology are discussed.

**Part II** contains five chapters dealing with MOS integrated circuits which form the building blocks of switched-capacitor filters. **Chapter 2** gives an account of the MOS transistor operation and fabrication and concludes with a discussion of the subject of noise in MOS integrated circuits. **Chapter 3** deals with elementary circuits such as amplifiers, current mirrors, and active loads. **Chapter 4** gives the details of the design of two-stage CMOS operational amplifiers together with complete design examples and a summary of the design procedure. In **Chapter 5** high performance operational amplifiers are presented with particular emphasis on designs suitable for low noise and high frequency applications in the design of switched-capacitor filters. **Chapter 6** deals with the other two building blocks of switched-capacitor filters, namely: the capacitor and switch, with the associated non-ideal effects such as capacitor ratio errors and clock feed-through.

**Part III** contains five chapters dealing with the design techniques of switched-capacitor filters using the building blocks discussed in Part II. **Chapter 7** gives the fundamentals of filter design in general and concentrates on the continuous-time passive filter models which are used as reference designs from which switched-capacitor filters are derived. **Chapter 8** gives the methods of description of switched-capacitor filters and explains how switched-capacitor filter transfer functions may be derived from continuous-time prototypes. In **Chapter 9**, the lossless discrete integrator is introduced and used to construct the amplitude-oriented filters which have achieved remarkable popularity due to their simplicity and low sensitivity properties. It is shown that these filters have no lumped-element equivalents, and the design techniques are given in rigorous form; then it is shown how these may be implemented with considerable ease using the programs contained in **ISICAP** and **ISICAPS**. **Chapter 10** gives the amplitude-oriented design of ladder filters based on lumped-element prototypes. Then the cascade method of realization is also given which realizes the same transfer functions. Again, the computer implementation of the design techniques is presented relying on **ISICAP** and **ISICAPS**. **Chapter 11** gives the design techniques of selective linear phase and data-transmission filters. These are of prime importance in numerous applications including high frequency filters and pulse transmission as well as video signal processing. **Chapter 12** is the final chapter, dealing with several practical considerations and special techniques. These include minimization of clock feedthrough, scaling for maximum dynamic range and minimum capacitance, fully differential balanced designs, high-frequency filters, programmable filters, pre-filtering and post-filtering, and layout considerations.

There are design-oriented problems at the ends of all chapters, except those which are mainly descriptive.

The **disk** accompanying this book contains the source codes, written in FOR-TRAN of the package in the directory **ISICAP**, in the form of Appendices, each containing a program that is carefully and clearly documented. For use, a FOR-TRAN compiler is needed on the computer. The reader may load the entire source program, then compile all or some of the various parts according to need. This was provided for maximum flexibility and to allow the user to introduce any modifications producing different programs for special purposes. The programs also allow greater depth of understanding of the finer points of the design techniques, and as such provide a valuable educational tool. Furthermore, many of the supporting subroutines are useful for general filter design problems; these include the subroutines for generation of Chebyshev polynomials, the factorization and squaring of polynomials, and many other general purpose programs.

The diskette also contains *four* executable programs in the directory **ISICAPS** which may be loaded and used directly. Each is menu driven and may be run by typing the appropriate name of the program then pressing return and following the menu instructions. These are as follows:

(a) The main programs in ISICAP are in **ISICAP1**. The executable "contracted" version ISCAP1 gives the design of (1) lossless-discrete-integrator (LDI) filters for an arbitrary transfer function, (2) Chebyshev low-pass LDI filters, (3) low-pass filters with finite zeros of transmission (such as elliptic filters) and (4) band-pass filters with finite zeros of transmission. In (2) the program produces odd-degree ladder filters and it gives an error message if the given specifications require an even degree filter. The user then repeats entering the specifications while increasing the stopband attenuation (or reducing the stopband edge relative to the sampling frequency) while holding all other specifications the same until the *next* higher odd-degree filter is produced by the program. Alternatively, the following program in (b) below may be used.

(b)  This is called **ISICAP1M** which is similar to ISICAP1 except that it allows the user to obtain the even-degree Chebyshev LDI case or choose the higher odd-degree case. For this purpose, the user should only use item (1) on the menu of ISICAP1M ignoring the rest so that for other designs ISICAP1 should always be used.

(c)  In the same directory **ISICAPS**, there is also an executable program called **ISICAP2** which gives the complete design of the generalized Chebyshev filter. This is a new filter design that may replace the elliptic filter with a number of advantages that are explained in the text. It obviates the need for using the tables of elliptic filter prototypes and gives the complete design from the specifications directly.

(d)  The picture is completed by another program **ISICAPC**. This gives the cascade designs for all types of amplitude-oriented filters: maximally-flat, Chebyshev, elliptic, and generalized chebyshev for all cases of low-pass, high-pass, band-pass and band-stop bilinear designs.

There are also sample data files on the diskette to illustrate the manner in which the input and output to the programs are presented. Before using the programs in ISICAPS, the reader should type and read the file README1 for important information and guidance.

The first three letters of the package constitute the name of the ancient Egyptian goddess ISI(S): the prototype of true love and fidelity, towards whom I feel great affection and whose spirit still echoes in Mozart's *Magic Flute* and Beethoven's *Fidelio*.

The references at the end of the book are intended to be representative and no attempt was made at completeness; the switched-capacitor filter design literature being too vast for such an attempt.

The computational work in this book owes much to contributions from my former research students, notably Emad Afifi, Jacques Beneat, Songxin Zhuang and Mark O'Malley. The support of KFUPM is also gratefully acknowledged.

I believe the book is suitable for incorporation in senior undergraduate courses and as a graduate level or professional course on the subject as well as being of direct benefit to electronic circuit design engineers in industry. The design package alone should be quite useful to many circuit designers.

# Part I
## GENERAL INTRODUCTION

*'There is a tide in the affairs of men*
*Which, taken at the flood, leads on to fortune;*
*Omitted, all the voyage of their life*
*Is bound in shallows and in miseries*
*On such a full sea are we now afloat,*
*And we must take the current when it serves,*
*Or lose our ventures'.*

Shakespeare

*Julius Caesar*

## OUTLINE

Switched-capacitor filters are placed in their proper perspective in relation to analog integrated filters and the analog VLSI area. The advantages and main features of signal processing using switched-capacitor techniques are discussed. Some examples of the applications of switched-capacitor filters in the telecommunications field are indicated. Thus, a degree of motivation for the exposition in the book is provided.

# 1

# Perspective

## 1.1  ANALOG INTEGRATED FILTERS

The earliest filters to be firmly established in the electronic circuit design repertoire were lumped passive ones, employing inductors, capacitors, and transformers [1], [2]. These have become reference designs against which any other category of filter is measured and compared. This is due to a number of reasons. First, these passive structures have been shown to be capable of satisfying the most stringent specifications that may be encountered in all disciplines of electrical engineering design. Secondly, they have been found to possess low-sensitivity properties with respect to variations in element values; a highly desirable attribute from the practical view-point. Finally, they do not consume power.

However, as the operating frequencies fall into and below the audio range, the required inductors develop severe limitations and disadvantages: they become large-valued, bulky, with low quality factors, and expensive to construct. For this reason, active filters [3]–[5] were introduced with the primary objective of overcoming the practical disadvantages of inductors at the lower end of the frequency spectrum. These employ resistors, capacitors and active devices such as transistors or operational amplifiers. Active structures which imitate the low-sensitivity properties of passive models were also introduced.

With the technological advances in integrated circuit design, the rather obvious step and natural tendency to implement analog active filters as monolithic integrated circuits, were frustrated by practical factors. To appreciate this, consider Figure 1.1 showing an integrator. Its transfer function $V_{out}(s)/V_{in}(s)$ is given by

$$T(s) = -\frac{1}{RCs} \tag{1.1}$$

where $s$ is the complex frequency.

Now, the basic difficulty in implementing such a transfer function in integrated circuit form is that one would have to realize the RC product with high precision which, in turn, requires the realization of the absolute value of each component with an even greater precision. Since the errors in realizing an integrated resistor and a

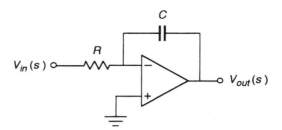

**Figure 1.1**    A continuous-time integrator.

cacpacitor could be as large as 20% for each element, the error in the RC product may be as large as 40%, which is, of course, unacceptable in any application. Thus the implementation may be achieved in hybrid form, implying that the active parts of the circuit are realized as an *integrated circuit* while the passive parts are *discrete* components using thin and thick film techniques to allow for accurate trimming. It is well known that any filter transfer function can, in principle, be realized using building blocks of the type shown in Figure 1.1, hence the inherent errors constitute a strong argument against the realization of active RC filters in monolithic integrated circuit form. Furthermore, the required absolute values of the elements can be too large, requiring a large area on the integrated circuit. In the event of implementing the resistors using transistors, one has to contend with the inherent non-linearities. There are, however, schemes for cancellation of these effects [6], but in most cases they require additional circuits which may be more complex than the filter itself, something like chasing a fly with a sledgehammer (let alone killing it !)

## 1.2   SWITCHED-CAPACITOR FILTERS

The state of affairs described in the above section changed dramatically with the advent of switched-capacitor techniques [7], [8]. These made possible, for the first time, the implementation of analog filters in monolithic integrated circuit form using metal-oxide semiconductor (MOS) technology; the same technology that had matured in the design of digital circuits. The key idea is deceptively simple. Consider the basic building block of Figure 1.2(a) which typifies the entire category of switched-capcitor filters. It is made up of an operational amplifier, two capacitors and analog switches. The switches are actuated periodically by a clock as shown in Figure 1.2(b) so that the input voltage is sampled then switched to the input of the operational amplifier half a period later. Under the given clocking scheme, the transfer function $V_{out}/V_{in}$ of this circuit is given by [1]

$$T(s) = \frac{e^{-Ts/2}}{2(C_b/C_a)\sinh Ts/2} \tag{1.2}$$

The basic difference between this circuit and its continuous-time counterpart of Figure 1.1 is that the transfer function in (1.2) is determined by the *capacitor ratio* $C_b/C_a$ and not by absolute values as the function in (1.1). Hence a circuit composed of building blocks with this key property, will have a response which is also determined

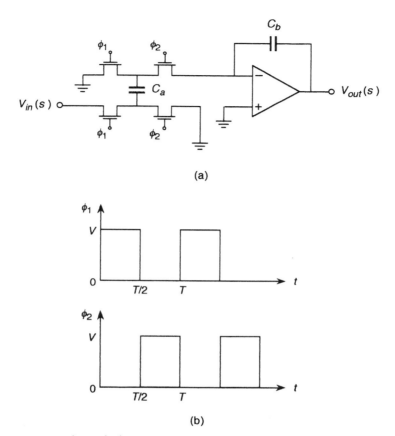

(a)

(b)

**Figure 1.2**    (a) A typical switched-capacitor circuit; (b) the bi-phase clock driving the switches.

by capacitor ratios, not by absolute element values. In integrated circuit implementations, it is possible to realize capacitor ratios with high accuracy down to 0.1% and
this may improve further in the future. Furthermore, since the absolute capacitor
values are of no consequence, these can be reduced at will to values that are limited
only by the practical lower limits of the fabrication process. This allows the reduction
of the overall area occupied by the capacitors on the integrated circuit.

In addition to the above key advantages, circuits of the generic type in Figure 1.2(a)
are largely insensitive to the parasitic capacitances inherent in the integrated circuit
manufacturing process.

To sum up, the main features of switched-capacitor filters are the following:

1. The building blocks are operational amplifiers, capacitors, and analog switches.

2. The circuits are of the analog sampled-data type: they operate directly on analog
signals and produce outputs that are essentially analog; the filtering process is
accomplished on samples of the signals. Thus, no coding or quantization are
needed as in the case of digital filters which require the process of analog-to-
digital (A/D) conversion, and subsequent digital-to-analog (D/A) conversion if
the filter is to produce analog outputs.

3. The performance of the filter depends on capacitor ratios, not on absolute values.

4. The filters can be made programmable.

5. The circuit structures can be made insensitive to parasitic capacitances.

6. The filters can be designed to imitate the excellent low sensitivity properties of reference passive designs.

7. High quality precision filters are possible for telecommunications applications.

8. The filters can be fabricated using standard MOS technology, the same one used in digital circuit design. Hence these can be placed on the same chip with digital circuits.

9. Switched-capacitor filters designed using complementary MOS (CMOS) circuits have much simpler structures and consume less power than a digital counterpart performing the same signal processing function.

Standard power consumption is typically 1 mW, or better, per filter pole [8]. Micro-power designs are also possible down to 1 $\mu$W per filter pole . For example, a fifth-order filter with 350 $\mu$W, a power supply of 2.5 V and clock frequency of 128 kHz has been reported [9].

Operation with switching frequencies of 1 MHz is routine, while operation with a switching frequency of 30 MHz is possible with CMOS and special circuit techniques. For this technology, a theoretical maximum limit of 130 MHz has been predicted [10].

## 1.3   RELATION TO THE ANALOG VLSI PROBLEM

Until recently, digital circuits dominated the very large scale integration (VLSI) area. In order for a new category of circuits to be a viable alternative to digital circuits, it must be shown capable of realizing high precision signal processing systems and in particular the design of filters. Such a criterion could not be met by analog RC active filters, although some recent advances in this area have shown reasonable but limited promise. It is really with the use of switched-capacitor techniques that monolithic precision filters were made possible, thus penetrating the VLSI area. Prior to that, analog filters could only be realized in hybrid integrated form, with a very uncertain future for this dubious approach dictated by the obvious limitations and advocated by those whose research interests have, unfortunately, assumed a 'tongue-in-cheek' complexion.

Another reason for the success of switched-capacitor techniques is that they are particularly amenable to implementation using MOS devices in the same technology used for digital circuits, thus allowing the analog part of a system to be integrated with the digital one on the same chip. This is in line with the modern trend of both digital and analog signal processors complementing one another.

The term *analog VLSI* had, for some time, meant an integrated circuit 90% of which contained digital circuits while the remaining 10% were analog, whose function was mainly as peripherals and interface circuits with the analog world. On the other hand, the digital part required 10% of the design time while the analog part required 90% of the time. Thus the term *analog VLSI* derived from the association with digital VLSI. This situation is gradually changing and a major reason for the change is the establishment of switched-capacitor techniques with their inherent compatibility with MOS integrated circuit design methodology. One may envisage a real analog VLSI chip as one composed mainly of analog circuits while a digital part is used only for memory and control. In any event, the situation that exists at present is that a VLSI chip contains both digital and switched- capacitor circuits in a complementary rather than a competing manner, which is perhaps the constructive attitude with which the analog VLSI problem should be approached. The question is not whether the chip should be entirely digital or entirely analog, but rather how the various tasks required are to be allocated to the analog and digital parts. However, at present, analog circuit design, by contrast with digital design, is knowledge intensive, and requires longer design time, and gives lower yield. Analog circuit design experts are very few, and the area does not really lend itself to the extreme simplification that has allowed digital circuit design to be approachable by non-experts.

In addition to the most important application in linear filtering of signals, switched-capacitor circuits can be used as oscillators, modulators, A/D converters, speech processors in speech synthesis, rectifiers, detectors, and comparators [8], [11]. However, it is with filters that this book is concerned, and as such there have been many demanding applications where switched-capacitor filters have excelled. These include the transmit and receive filters for speech codecs, data modems, programmable filters, and tone receiver filters [8]. More recently, the frequency range of operation has been extended to higher frequencies well beyond the audio range and includes filters for applications in FM as well as video systems [12], [13], [14]. Oversampling techniques [15] (sigma-delta) for A/D and D/A conversion have proliferatd the areas of ISDN and digital music recording and reproduction (compact disc recording and reproduction systems).

Furthermore, Gallium Arsenide (Ga As) technology [16] has been used instead of MOS to investigate the possibility of extending the frequency limits to the radio frequency and lower microwave range. The results look quite promising. With GaAs, operation with a 100 MHz switching frequency is possible to contemplate with a predicted theoretical maximum of 500 MHz for certain design techniques.

The above merits notwithstanding, switched-capacitor circuits have limitations, a point which is shared by any other signal processing subsystem. These limitations are a result of the non-ideal effects in the integrated circuit building blocks used to design the filter. Therefore, the discussion of these limitations will only be undertaken at the appropriate point in the book where a sufficient level of understanding of the filter design techniques has been reached.

## 1.4 APPLICATIONS IN TELECOMMUNICATIONS

The most demanding specifications on the performance of filters are to be found in the telecommunications area [17]. A number of the applications of switched capacitor filters in this field have helped to establish their success.

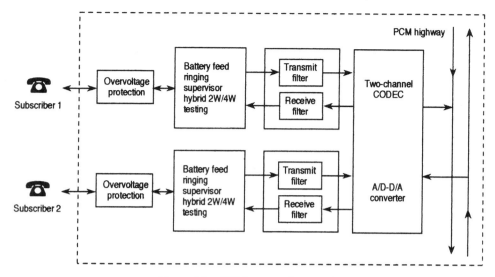

**Figure 1.3**  PCM system for speech.

Pulse-code-modulation (PCM) is used to transmit speech over telephone channels according to the scheme shown in Figure 1.3. Prior to being sampled at a rate of 8 kHz, the speech signal must be bandlimited to 3.4 kHz. It is also desirable to remove the 50 Hz or 60 Hz power line frequency. These functions are performed by the transmit filter which is a switched-capacitor band-pass type. At the receiver end, a low-pass switched-capacitor filter is used. The *idealized* responses of the transmit and receive filters are shown in Figure 1.4, which have to be approximated subject to a tolerance scheme. There are two approaches to the design of the integrated coder-decoder (CODEC). The first is to implement the coder and decoder on one chip, with the transmit and receive filters on a second chip. Alternatively, the coder and transmit filter could be on one chip while the decoder and receive filter are on a second chip. The separation of the two filters in the second approach leads to reduced cross-talk and noise when the CODEC operates asynchronously. The implementation of the CODEC chip is one of the major success stories of switched-capacitor techniques [18].

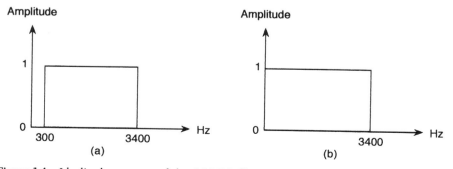

**Figure 1.4**   Idealized responses of the CODEC filters: (a) transmit filter; (b) receive filter.

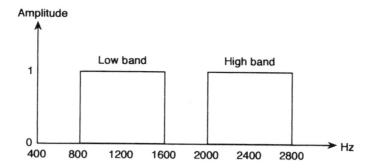

**Figure 1.5**   Idealized responses of full duplex data MODEM filters.

Switched-capacitor techniques have also been a very powerful tool for the implementation of the two band-pass filters needed for data modulators-demodulators (MODEMs).They are used to transmit and receive data simultaneously on telephone channels. These filters have the idealized amplitude responses shown in Figure 1.5. They must also approximate a linear-phase characteristic in their passbands, and have specific time responses. The set of two filters is available on a single chip from many manufacturers [19], [20].

Switched-capacitor filters have also been very successful in the design of integrated tone receivers [21], [22]. Tone signalling is used in a number of systems in telephony. Examples are the multifrequency detection systems used for signalling between switching machines, and the dual-tone multifrequency detection system used for signalling between the subscriber and central exchange that is incorporated in touch-tone receivers. The detection of these signals is accomplished by switched-capacitor filters. A typical single-chip receiver contains 37 operational amplifiers.

The use of switched-capcitor filters for speech processing and recognition has also been successful [23], [24]. More demanding high-frequency designs have been implemented for use in video processing [14].

Finally, switched-capacitor filters can easily be made programmable by varying the clock frequency to produce a varying cutoff filter. The clock can be programmed digitally. An alternative to clock programmability is mask programmability which is similar to the method used in digital circuits.

## 1.5   COMPUTER-AIDED ANALYSIS AND DESIGN TOOLS

The above applications are only a few examples of the success of switched-capacitor filters in the telecommunication field. The filters can be used in any application, conventional or otherwise, where the specifications are within the tolerance limits on the fabrication process. We must bear in mind, however, that switched-capacitor filters are ultimately realized as integrated circuits and offer no advantages if used in discrete form. Hence they fit very nicely into the VLSI circuit design field.

Since they are integrated filters, switched-capacitor filters demand the use of layout tools such as MAGIC [25] prior to submitting the design for fabrication. Computer simulation tools such as CAzM [26] can also be used for testing and fine-tuning the design. Dedicated *analysis* programs have also been introduced, notably SWITCAP [27], [28] and SCANAL [29]. The comprehensive *design* package **ISICAP** supplied with this book covers a very wide range of filter design techniques which are made available for the first time.

# Part II
## ANALOG MOS INTEGRATED CIRCUIT BUILDING BLOCKS

*'Not that the story need be long,*
*but it will take a long while to make it short'.*

H. D. Thoreau
*Letter*

## OUTLINE

The building blocks of analog MOS integrated circuits are studied with particular emphasis on those employed in the design of switched-capacitor filters. These are the operational amplifier, the analog switch and the capacitor as well as the necessary subsidiary circuits. These are discussed from the view-point of the integrated circuit design engineer and with the ultimate objective of placing switched-capcitor filters into the context of analog VLSI circuits. Particular attention is paid to the non-ideal effects inherent in the practical implementation of these circuits. This is due to the fact that all the limitations on the performance of switched-capacitor filters have their origin in the integrated circuit design stage, particularly in the fabrication process. Therefore, the discussion begins at the most basic level of the single transistor operation and fabrication, then proceeds to the construction of complex circuits. High performance operational amplifier design is also included, together with methods of minimmization of other non-ideal effects in switches and capacitors.

# 2
# MOS Transistor Operation and Fabrication

## 2.1  INTRODUCTION

This chapter gives an account of the operation of the metal-oxide-semiconductor (MOS) transistor together with the technological processes employed for the fabrication of MOS transistors and integrated circuits. Knowledge of device physics is assumed, including the operation of the pn junction and basic transistor operation [30]–[32]. The chapter begins with a review of the MOS transistor operation and basic equations, and introduces the complementary (CMOS) circuits. A description is next given of the fabrication of MOS devices which is necessary for understanding the performance and limitations of MOS integrated circuits in general, and switched-capacitor filters in particular. Layout rules and area requirements of integrated circuits are also described. The chapter concludes by a discussion of the subject of noise in MOSFETs.

## 2.2  THE MOS TRANSISTOR

Figure 2.1 shows the physical structure of an n-channel enhancement-type MOSFET. The device is fabricated on a p-type substrate consisting of a single-crystal silicon wafer. The $n^+$ regions are heavily-doped n-type silicon, constituting the source and drain regions. A thin silicon dioxide ($SiO_2$) layer is grown on the substrate, extending over the area between the source and drain. For the electrodes, metal can be used as contacts to the gate, source, drain and substrate. The gate electrode can also be made from poly-crystalline silicon (polysilicon) in the modern process of silicon-gate technology. The oxide layer results in the current in the gate terminal being very small ($\approx 10^{-15}$ A).

The normal operation of the MOS transistor requires the pn junctions formed between the substrate and each of the drain and source to be reverse-biased. Generally the drain voltage is higher than that of the source, and the two pn junctions referred to above would be reverse-biased if the substrate is connected to the source.

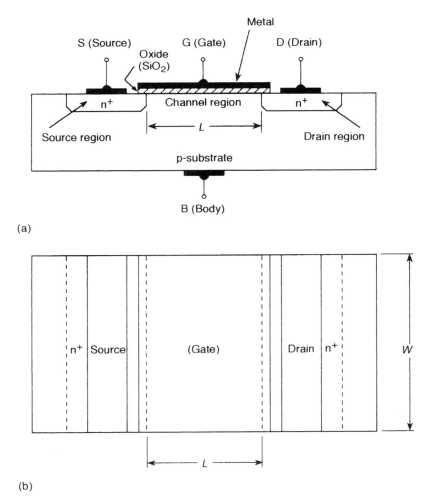

**Figure 2.1**  The enhancement-type MOSFET: (a) cross-section; (b) top view.

We shall assume this to be the case in the following analysis, then examine the effect of the substrate shortly afterwards.

### 2.2.1  Operation

1.  For $v_{GS} = 0$, the drain-substrate and source-substrate pn junctions form two diodes back-to-back. Thus no current flows between the source and drain, since the path between these regions has a resistance of the order of $10^{12}\ \Omega$.

2.  With $v_{GS} > 0$, the free holes in the p-substrate region near the gate are repelled into the substrate, thus creating a *depletion region*, as illustrated in Figure 2.2,

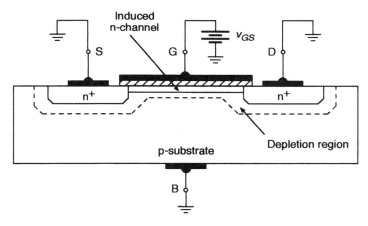

**Figure 2.2** Enhancement NMOS device showing the induced *n*-channel as a result of applying a positive $v_{GS} > V_t$.

which contains bound negative charges. In addition, the positive $v_{GS}$ attracts electrons from the source and drain $n^+$ regions upwards towards the gate. These conditions result in the formation of a negatively charged *induced channel* connecting the source and drain regions. Consequently, if a voltage is applied between the drain and source, a current flows between these regions via the induced channel. This n-type channel, obviously, results from the *inversion* of the p-type region near the gate into an n-type region. Hence the channel is called an *inversion layer* and the resulting structure is called an n-channel MOSFET or an NMOS transistor.

Now, there is a critical value of $v_{GS}$ at which a conducting channel is formed. This value is called the *threshold voltage* $V_t$, which has typical values in the range of 1–3 V, determined during the fabrication process.

3. With $v_{GS} > V_t$ the n-channel is formed and applying a voltage $v_{DS}$ results in the $i_D$ against $v_{DS}$ characteristic shown in Figure 2.3 for values of $v_{DS}$ in the range 0–0.2 V. Thus, the induced channel is *enhanced* by taking $v_{GS} > V_t$ and this gives the device its name of *enhancement-type* MOSFET.

4. Again, with $v_{GS} > V_t$, $v_{DS}$ is increased further. Clearly there is a voltage gradient across the n-channel since the voltage between the source and drain varies from $v_{GS}$ at the source to $v_{GS} - v_{DS}$ at the drain. Thus, the channel depth varies in a tapered shape as shown in Figure 2.4. At a value $v_{DS} = (v_{GS} - V_t)$, the channel depth at the drain becomes zero and the channel is said to be *pinched-off*. Thus, the channel shape is unaffected by a further increase in $v_{DS}$. Beyond pinch-off the device enters *saturation* resulting in the $i_D$ against $v_{DS}$ characteristic shown in Figure 2.5.

Figure 2.6 shows the circuit symbols for the NMOS device.

We now consider quantitatively the various regions of operation of the MOS transistor.

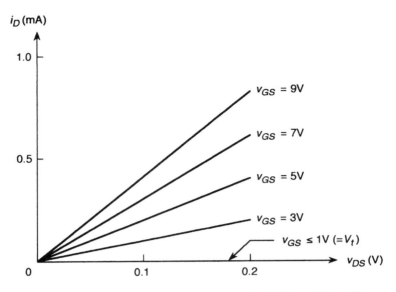

**Figure 2.3**   The $i_D - v_{DS}$ characteristics of a MOSFET with $V_t = 1$ V for different $v_{GS}$.

(a) The device is *cut off* when

$$v_{GS} < V_t \qquad (2.1)$$

(b) To operate in the *triode* region we must have

$$v_{GS} \geqslant V_t \qquad (2.2)$$

and keep $v_{DS}$ small enough for the channel to remain continuous, i.e.

$$v_{GD} > V_t \qquad (2.3)$$

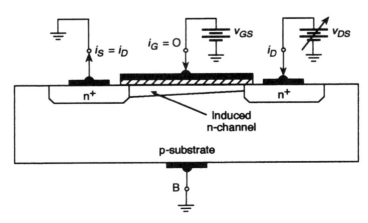

**Figure 2.4**   Effect of increasing $v_{DS}$ on the shape of the induced channel.

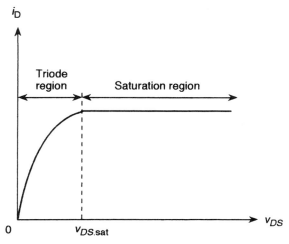

**Figure 2.5**  Typical $i_D$, $v_{DS}$ characteristic for enhancement NMOS transistor with $v_{GS} > V_t$.

or, noting that

$$v_{GD} = v_{GS} + v_{SD} = v_{GS} - v_{DS} \tag{2.4}$$

we have

$$v_{DS} < v_{GS} - V_t \tag{2.5}$$

In this region of operation, an approximate relation between $i_D$ and $v_{DS}$ is given by [30]–[32]

$$i_D = K\left[2(v_{GS} - V_t)v_{DS} - v_{DS}^2\right] \tag{2.6}$$

with

$$K = \frac{1}{2}\mu_n C_{ox}\left(\frac{W}{L}\right) \tag{2.7}$$

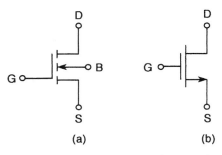

(a)                                  (b)

**Figure 2.6**  Symbols of the enhancement NMOS: (a) Showing the substrate; (b) simplified symbol for B is connected to S.

where $\mu_n$ is the electron mobility in the induced n-channel, $C_{ox}$ is the oxide capacitance per unit area of the gate to body capacitor,

$$C_{ox} = \frac{\epsilon_{ox}.}{t_{ox}} \tag{2.8}$$

with $\epsilon_{ox}$ being the permittivity of the oxide and $t_{ox}$ is its thickness. $L$ is the length of the channel and $W$ is its width. Usually the quantity $\frac{1}{2}\mu_n C_{ox}$ is determined by the fabrication process, and is of the order of $10\,\mu A/V^2$ for an NMOS process with a $0.1\,\mu m$ oxide thickness. Thus, $K$ is determined by the *aspect ratio* $(W/L)$ of the device. It is often convenient to work with the *transconductance parameter* $K'$ defined as

$$K' = \mu C_{ox} \tag{2.9}$$

so that

$$K = K'(W/L)/2 \tag{2.10}$$

Near the origin, for very small $v_{DS}$, we can neglect $v_{DS}^2$ and (2.6) becomes

$$i_D \cong 2K(v_{GS} - V_t)v_{DS} \tag{2.11}$$

This allows the use of the transistor as a voltage-controlled linear resistor of value

$$r_{DS} = v_{DS}/i_D = 1/[2K(v_{GS} - V_t)] \tag{2.12}$$

which is controlled by $v_{GS}$.

(c) In the *saturation* region we must have

$$v_{GS} \geqslant V_t \tag{2.13}$$

and

$$v_{GD} \leqslant V_t \tag{2.14}$$

or

$$v_{DS} \geqslant v_{GS} - V_t \tag{2.15}$$

The three regions of operation (a)–(c) are shown in Figure 2.7 for differnent $v_{GS}$ values.

From (2.5) and (2.15) the boundary between the triode and saturation regions is defined by

$$v_{DS} = v_{GS} - V_t \tag{2.16}$$

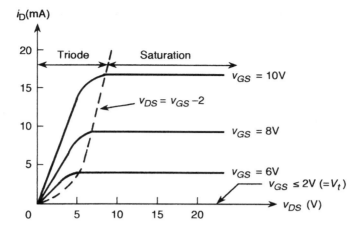

**Figure 2.7** Typical $i_D$, $v_{DS}$ characteristics of an NMOSFET with varying $v_{GS}$, for which $V_t = 2\,V$.

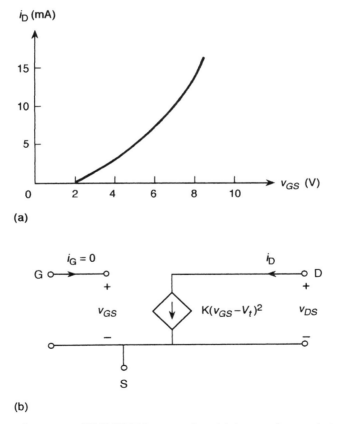

(a)

(b)

**Figure 2.8** The enhancement NMOSFET in saturation: (a) $i_D$, $v_{DS}$ characteristic; (b) large signal equivalent circuit.

which, upon substitution in (2.6) gives the saturation value

$$i_D = K(v_{GS} - V_t)^2 \qquad (2.17)$$

which is independent of $v_{DS}$. This relationship is shown in Figure 2.8, which implies that in saturation, the MOSFET behaves as an ideal voltage-controlled current source whose strength is determined by $v_{GS}$ in the non-linear fashion shown in Figure 2.8(a). Figure 2.8(b) shows the large signal model of the n-channel device in saturation, based on the foregoing analysis.


## 2.2.2  The Transconductance

For a MOSFET in saturation we can use (2.17) with $v_{GS}$ having the dc component $V_{GS}$ (corresponding to the bias point) and the signal component $v_{gs}$ to write:

$$i_D = K(V_{GS} + v_{gs} - V_t)^2$$
$$= K(V_{GS} - V_t)^2 + 2K(V_{GS} - V_t)v_{gs} + K v_{gs}^2 \qquad (2.18)$$

and for small signal operation, the last term can be neglected. The first term is a dc or quiescent current. Therefore the signal current is

$$i_d = 2K(V_{GS} - V_t)v_{gs} \qquad (2.19)$$

The transconductance of a MOSFET device is given by

$$g_m = \frac{i_d}{v_{gs}}$$
$$= 2K(V_{GS} - V_t) \qquad (2.20)$$

so that substitution for $K$ from (2.7) gives

$$g_m = (\mu_n C_{ox})(W/L)(V_{GS} - V_t) \qquad (2.21)$$

More specifically,

$$g_m = \left.\frac{\partial i_D}{\partial v_{GS}}\right|_{v_{GS}=V_{GS}} \qquad (2.22)$$

For the dc value $I_D$ we also have

$$I_D = K(V_{GS} - V_t)^2 \qquad (2.23)$$

which upon use in (2.20) and (2.21) gives the alternative expressions for the trans-conductance as

$$g_m = \sqrt{2\mu_n C_{ox}(W/L)I_D}$$

$$= \sqrt{2K'(W/L)I_D} \qquad (2.24)$$

$$= 2\sqrt{KI_D}$$

### 2.2.3   Channel Length Modulation

The idealized behaviour of the MOSFET in saturation shown in Figure 2.7 assumes that an increase of $v_{DS}$ beyond pinch off does not affect the channel shape. This leads to the horizontal portion of the characteristic which implies that the output resistance in saturation

$$r_{o\,sat} = \frac{\partial v_{DS}}{\partial i_D} \qquad (2.25)$$

is infinite. However, in practice, increasing $v_{DS}$ beyond $v_{DS,\,sat}$, has an effect on the channel length, which actually decreases. This is called a *channel-length* modulation phenomenon. But from (2.7) $K$ is inversely proportional to the channel length, and therefore, $i_D$ increases with $v_{DS}$ resulting in the modified set of curves shown in Figure 2.9. The linear dependence of $i_D$ on $v_{DS}$ is taken into account by multiplying (2.17) by the factor $(1 + \lambda v_{DS})$ to give

$$i_D = K(v_{GS} - V_t)^2 (1 + \lambda v_{DS}) \qquad (2.26)$$

where, by reference to Figure 2.9

$$\lambda = 1/V_A \qquad (2.27)$$

which is a device parameter called the *channel length modulation parameter*, and $V_A$ is in the range 30–200 V, while $\lambda \approx 0.005 - 0.03 \ \mathrm{V}^{-1}$.

As a consequence of the relationship in (2.26), the output resistance of the device in saturation is now finite and given by

$$r_0 = \left[\frac{\partial i_D}{\partial v_{DS}}\right]^{-1}, \ v_{GS} = \text{constant}$$

$$= 1/\lambda K(V_{GS} - V_t)^2 \qquad (2.28)$$

or

$$r_0 \cong 1/\lambda I_D$$

$$\cong V_A/I_D \qquad (2.29)$$

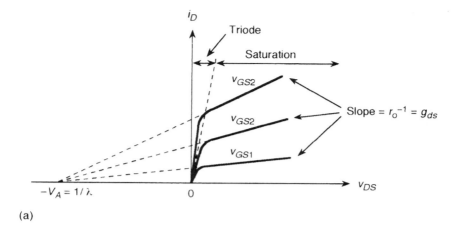

(a)

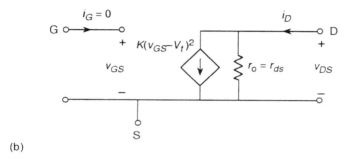

(b)

**Figure 2.9** Effect of finite output resistance of the MOSFET in saturation: (a) $i_D$, $v_{DS}$ characteristics, $v_{GS1} < v_{GS2} < v_{GS3}$; (b) large signal equivalent circuit.

with $I_D$ as the drain current for a specific $V_{GS}$. An alternative notation is

$$g_{ds} = g_o = 1/r_o$$

$$= \frac{I_D\lambda}{1 + \lambda V_{DS}} \qquad (2.30)$$

$$\cong \lambda I_D$$

which is also called the *small signal channel conductance*

## 2.2.4   PMOS Transistors and CMOS Circuits

A MOS transistor can be fabricated with an n-type substrate, $p^+$-type drain and source regions. The created channel becomes of the p-type. In this case the operation is similar to the NMOS case but $V_t$, $v_{GS}$, and $V_A$ are negative. The symbol of the

PMOS device is shown in Figure 2.10 and the description of the various regions of operation is as follows

   (a) *Cut-off*

$$v_{GS} > V_t \tag{2.31}$$

   (b) *Triode*

$$v_{GS} < V_t \tag{2.32}$$

$$v_{DS} \geqslant v_{GS} - V_t \tag{2.33}$$

    and

$$i_D = K\,[2(v_{GS} - V_t)v_{DS} - v_{DS}^2] \tag{2.34}$$

   (c) *Saturation*

$$v_{GS} < V_t \tag{2.35}$$

$$v_{DS} \leqslant v_{GS} - V_t \tag{2.36}$$

    and

$$i_D = K(v_{GS} - V_t)^2(1 + \lambda v_{DS}) \tag{2.37}$$

   Also,

$$r_0 = |V_A|/I_D \tag{2.38}$$

$$K = \frac{1}{2}\mu_p C_{ox}\left(\frac{W}{L}\right) \tag{2.39}$$

From our standpoint the most important use of the p-channel or PMOS transistor lies in employing it together with an NMOS device in a complementary manner,

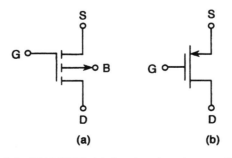

**(a)**                        **(b)**

**Figure 2.10**  Symbols of the PMOSFET: (a) showing the substrate; (b) simplified symbol when B is connected to S.

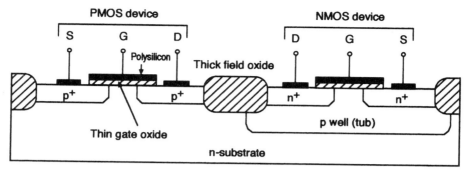

Figure 2.11   Cross-section of a CMOS circuit.

resulting in a CMOS circuit. Figure 2.11 shows a CMOS integrated circuit containing both an NMOS and a PMOS transistor. The PMOS device is fabricated directly on an n-type substrate while a p-well (or p-tub) is created for the construction of the NMOS device. The two transistors are isolated by a thick oxide region.

### 2.2.5   The Depletion-type MOSFET

The structure of this device is identical to that of the enhancement-type with one important difference: it has a physically implanted channel, and there is no need to create one. In this case, a positive $v_{GS}$ depletes the channel from its charge carriers; hence the name *depletion-mode*. As $v_{GS}$ becomes more negative, a value is reached where the n-channel is completely depleted of charge carriers and $i_D$ is reduced to zero even for a non-zero value of $v_{DS}$. This negative value of $v_{GS}$ is the threshold voltage of the device.

Naturally, a depletion-type MOSFET can be operated in the enhancement mode by keeping $v_{GS}$ positive. The current-voltage characteristics of the device are similar to those of the enhancement type except that $V_t$ is negative. The circuit symbol of the depletion-type n-channel MOSFET is shown in Figure 2.12.

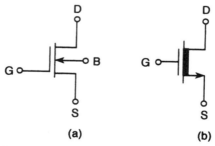

Figure 2.12   Symbols of the depletion-type NMOSFET: (a) showing the substrate; (b) simplified symbol when B is connected to S.

## 2.3   INTEGRATED CIRCUIT FABRICATION

All the non-ideal effects in integrated circuits have their origins in the fabrication process of the devices. Therefore, a reasonable understanding of this process is quite helpful in predicting and, possibly, reducing these non-ideal effects. In this section an outline of the fabrication process of MOS integrated circuits is given.

The basic processing steps in the production of MOS transistors, and subsequently larger circuits composed of a number of transistors, are the following:

(a) Diffusion and ion implantation.

(b) Oxidation.

(c) Photolithography.

(d) Chemical vapour deposition.

(e) Metallization.

These steps are applied to a silicon crystal which must be produced and prepared first. This is outlined next before the processing itself is discussed.

### 2.3.1   Wafer Preparation

The starting point in silicon processing according to the above steps is a single crystal silicon wafer of appropriate conductivity and doping. The starting material for crystal growth is very pure polycrystalline silicon referred to as *semiconductor-grade* silicon corresponding to an impurity concentration of less than 1 part per billion silicon atoms. The number of silicon atoms is usually $5 \times 10^{22}$ atoms per cm$^3$. The resistivity corresponding to the number of impurities, if they are of the acceptor boron type, is about $300\,\Omega\,cm$. Actually, polycrystalline silicon with impurity concentrations down to 0.1 parts per billion is also available.

This semiconductor-grade silicon is used to grow single crystals in the form of ingots, each having a diameter of about 10–15 cm and length of the order of 1 m. A crystallographic orientation flat is also ground along the length of the ingot. Then the extreme top and bottom of the ingot are cut off and the ingot surface is ground to produce a constant and precise diameter. The ingot is next sliced producing circular slices or *wafers* about 0.5–1.0 mm thick. Then, the wafer is subjected to a polishing and cleaning process in order to remove the silicon damaged from the slicing operation, and to produce a highly planar flat surface which is necessary for fine line device geometries and for improving the parallelism of the two surfaces in preparation for photolithography. Usually one side of the wafer is given a mirror-smooth finish while the other (backside) is only treated for an acceptable degree of flatness.

### 2.3.2   Diffusion and Ion Implantation

Diffusion is a process by means of which dopants are introduced into the surface of the silicon wafer. The most commonly used diffusants are *substitutional* dopants.

These have atoms which are too large to fit in the interstices between the silicon atoms and therefore the only way they can enter the silicon crystal structure is by replacing silicon atoms. Those of the donor type are phosphorus, arsenic, and antimony with phosphorus being used most often. Boron is practically the only acceptor type dopant used.

For the process of diffusion to occur, the presence of vacancies in non-ideal crystals is necessary. The vacancy density can be increased by raising the temperature. The vacancies can be generated at the crystal surface or interior to it. The diffusion process relies on the flow of atoms caused by a *concentration gradient*. The particle flow or flux $F$ is proportional to the concentration gradient, i.e.

$$F = -D \frac{\partial N}{\partial y} \tag{2.40}$$

where $F$ is the rate of flow of atoms per unit area (per second), and the derivative indicates the concentration gradient. $D$ is the diffusion coefficient. Also the diffusion rate obeys the *continuity equation*

$$\frac{\partial N}{\partial t} = -\frac{\partial F}{\partial y} \tag{2.41}$$

Combining (2.40) with (2.41) we obtain

$$\frac{\partial N(y, t)}{\partial t} = D \frac{\partial^2 N(y, t)}{\partial y^2} \tag{2.42}$$

The solution of the above equations, subject to the boundary conditions, gives the profile of the distribution of the diffusant particles $N(y, t)$. This is considered below, according to the type of diffusion employed.

There are two types of diffusion. The first occurs under constant surface concentration or infinite source conditions and is referred to as *deposition diffusion*. The diffusion concentration at the surface of the silicon crystal $(y = 0)$ is assumed constant so that $N(0, t) = N_o$ a constant. The other boundary condition is that the concentration should tend to zero as $y \rightarrow \infty$. The solution to equation (2.42) under these conditions is given by

$$N(y, t) = N_o \operatorname{erf} c \left( \frac{y}{2\sqrt{Dt}} \right) \tag{2.43}$$

where erf $c$ is the complementary error function. Figure 2.13 shows an example of the diffusion profile of p-type boron diffusion into an n-type (phosphorous-doped) substrate.

Deposition diffusion of the type described above is usually followed by a second diffusion process called *drive-in diffusion* in which the external source is removed. There are no more external dopants entering the silicon, but those already there move further inside and are redistributed. The impurity profile is obtained in this case by applying the boundary condition that the total diffusant density (atoms per unit area)

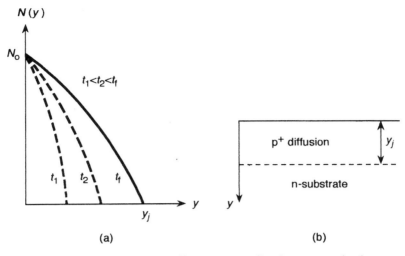

**Figure 2.13** Deposition diffusion: (a) profile; (b) junction depth.

remains constant. Alternatively, this implies that the net flow of diffusant atoms in or out of the silicon is zero at the surface $(y = 0)$. Thus, solution of (2.41) and (2.42) with the boundary condition that (2.40) is zero at $y = 0$ gives

$$N(y, t) = \frac{Q}{\sqrt{\pi Dt}} \exp\left(-\frac{y^2}{4Dt}\right) \qquad (2.44)$$

which is a Gaussian distribution with typical behaviour as shown in Figure 2.14.

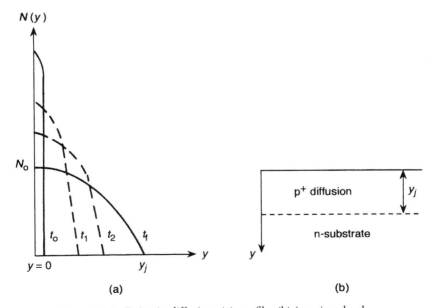

**Figure 2.14** Drive-in diffusion: (a) profile; (b) junction depth.

## Sheet resistance

Two main parameters are used to characterize diffusion layers, these are the junction depth $y_j$ and the sheet resistance. The former has been described, while the latter is given by $R = \rho\ell/A = \rho\ell/wt$ where $\rho$ is the resistivity of the layer, $\ell$ its length, $t$ its thickness and $w$ its width. In the case of a square shape, $\ell = w$ and we have

$$R = \rho/t \qquad (2.45)$$

which is independent of the lateral dimensions. This is referred to as the *sheet resistance* $R_s$ expressed in $\Omega/\square$. Thus, the resistance of a layer is expressed as

$$R = R_s\ell/w \qquad (2.46)$$

Sheet resistances of diffused layers are generally in the range of $1$–$1000\,\Omega/\square$.

## Ion implantation

This can be used as an alternative to deposition diffusion to produce a shallow region of dopant atoms. It has the advantage of more precise control over the sheet resistance and is therefore more commonly used than deposition diffusion. In this technique, silicon wafers are placed in vacuum and scanned by a beam of high energy dopant atoms which have been accelerated by high voltages. The atoms impinge on the surface of the wafer and penetrate into a small layer. The depth of penetration which is called the *projected range* is of the order of $0.1$–$1.0\,\mu m$. For boron ions, this is typically $0.067\mu m$ for an acceleration voltage of $20\,kV$, which can be increased to $0.3\,\mu m$ by increasing the voltage to $100\,kV$. For phosphorus ions in silicon, the range is $0.026\,\mu m$ at $20\,kV$ and $0.123\,\mu m$ at $100\,kV$.

The distribution of the implanted ions is Gaussian, and a typical profile is shown in Figure 2.15.

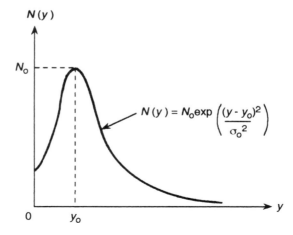

**Figure 2.15**   Ion implantation impurity profile, $y_o$ = projected range with $\sigma_o$ its standard deviation and $N_o$ the peak concentration.

Ion implantation is definitely more costly than deposition diffusion, but it offers greater control over the sheet resistance and allows very small values of dosage $<10^{14}\,cm^3$, hence very large sheet resistance values ($>1000\,\Omega/\square$) become possible. This is particularly useful for obtaining large resistor values ($>5\,k\Omega$). Very low dosage ion implantation can also be used for the adjustment of the threshold voltage of a MOSFET.

An ion implantation pattern on a wafer is defined by a mask made up of $SiO_2$ or $Si_3N_4$ and photoresist for low energy ions ($<100\,kV$). Heavy metals such as gold or tantalum deposited on $SiO_2$ layer are used as masks with high energy ions.

### 2.3.3   Oxidation

A *native oxide* layer about 20–30 Å thick would naturally form on a silicon surface if exposed to air. But this will inhibit further growth of oxide to a desired thickness, which is of the order of 5000–10 000 Å. This is used as either a diffusion or implantation mask or as a passivation layer for the protection of the devices. For this purpose, thermal oxidation is used, which consists in heating the wafers to about 1000°C and simultaneously exposing them to a gas containing $O_2$ or $H_2O$. The diffusion rate of the gas increases with temperature so that oxidation will continue until the desired thickness is reached. In this thermal oxide growth, some of the silicon will be consumed, and if the resulting oxide thickness is $t_{ox}$ then the thickness of the silicon consumed will be $0.44t_{ox}$.

### *Oxide masking*

Oxide layers are used for producing patterns on a wafer by creating openings or *windows* through which ion implantation or diffusion may be applied while protecting the other parts of the wafer from the process. Thus, the dopant pattern coincides with that of the windows in the oxide layers. This technique is essential for the production of microscopic size devices. The oxide mask thickness for this purpose is usually about 5000 Å which is found sufficient for all diffusion processes while it is also adequate for all but the exceptionally high energy ion implantation processes.

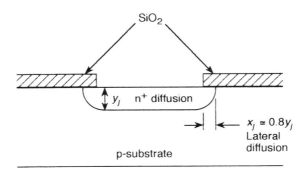

**Figure 2.16**   The use of an oxide layer for diffusion masking.

Figure 2.16 shows a pn junction produced by diffusion through an oxide window. Diffusion will occur vertically and also laterally as shown. The junction intersects the silicon surface well below the protective thermally grown oxide layer. Thus, this layer protects the junction against the surrounding effects and is called a *passivation junction*. The production of the oxide mask requires the use of *photolithography* which is discussed next.

### 2.3.4   Photolithography

This is the process by which a pattern is created on the wafer defining the locations of devices and circuits to be produced. Device dimensions of the order of 1.5 $\mu$m are possible with conventional ultraviolet light exposure techniques. Using electron-beam or X-ray methods, submicron dimensions are achievable. Photolithography involves a number of steps which are illustrated in Figure 2.17. First, a photoresist which is a light-sensitive liquid, is applied to the surface of the wafer. A thin layer about 5000 to 100 000 Å is produced by placing a drop of photoresist at the centre of the wafer then spinning it at a high speed, from 3000 rpm to 7000 rpm for 30 to 60 seconds. The wafer is then "prebaked" by heating it at 80°C for 30–60 minutes to expel solvents from the photoresist and harden it to a semi-solid fluid. The coated wafer is then aligned with a *photomask*. This is a glass plate about 125 mm square and 2 mm thick which has a photographic emulsion or thin film metal pattern on one side. The pattern consists of clear and opaque areas. The photomask is normally produced by computer-aided design tools which define the pattern by a computer connected to the mask- making machine. This pattern is produced such that the wafer will be divided into a number of chips each containing an integrated circuit made up of a large number of transistors. Then follows the process of the development of the photoresist by ultraviolet (UV) radiation. There are two types of photoresist, negative and positive. If the former is used, the areas exposed to UV light become polymerized, hardened virtually insoluble in the developer solution. This will develop into a copy of the mask patterns as shown in Figure 2.17(c) where the clear areas on the mask coincide with the areas where the photoresist remains on the wafer. If positive photoresist is used, the opposite situation results in which exposure to UV radiation depolymerizes the photoresist making the exposed areas soluble in the developer solution while the unexposed ones are rendered virtually insoluble. In this case, the clear areas in the photomask will coincide with the areas where the photoresist was removed. After development, the wafer is postbaked at about 150°C for 30–60 minutes to harden the photoresist for better adhesion to the wafer and increase its resistance to the acids used for the etching of the oxide. This oxide etching process is next, and can be either wet or dry. In the former method, the wafer is exposed to an etching solution to remove the oxide layer in the areas where there is no photoresist as shown in Figure 2.17(e). The outcome is the pattern of windows coinciding with the corresponding photomask. In dry etching, gaseous plasma is used instead of the chemical solution. The plasma is produced by an RF (radio frequency) field , and this type of etching results in the ability to produce smaller windows in the neighbourhood of 1 $\mu$m. The final step in photolithography is the removal or stripping of the photoresist. Positive photoresists can be easily removed by organic solvents such as acetone. Negative ones

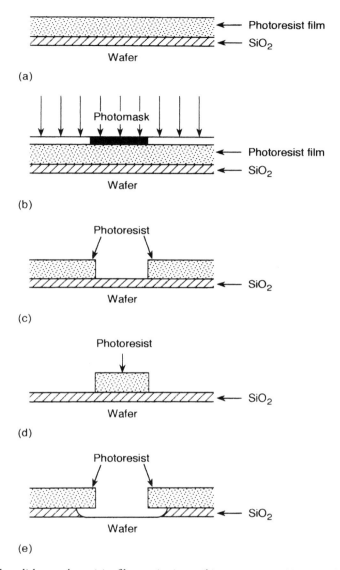

**Figure 2.17** Photolithography: (a) film spinning; (b) exposure; (c) negative photoresist development; (d) positive photoresist development; (d) oxide etching.

require a more elaborate process such as immersion in sulphuric acid together with mechanical scrubbing.

## 2.3.5 Chemical Vapour Deposition (CVD)

This is the process by means of which thin films of material are deposited on a substrate. Examples of materials deposited in IC processing are silicon dioxide, silicon nitride, silicon epitaxial layers and silicon hetero-epitaxial films. The material to be deposited is in gaseous form and is placed in a reaction chamber with the substrate where the chemical reaction causes the atoms that are produced to be deposited on the substrate.

$SiO_2$ CVD can be achieved at relatively lower temperatures (450–600°C) by comparison with the thermal oxide growth discussed earlier. The process is also much faster, being completed in a few minutes, but the resulting oxide film is generally of lower quality giving lower dielectric strength than a thermally grown one. The oxide, however, can be deposited on a wafer for post-metallization passivation without seriously affecting the existing condition, while providing a protective layer after all the processes, including metallization, have taken place. CVD oxide can also be used for isolating metallization levels.

Silicon nitride can also be deposited using CVD for device protection and passivation against penetration of certain contaminants against which $SiO_2$ is less effective. It can also be used as a diffusion or ion implantation mask.

### 2.3.6   Metallization

This is the final step in the wafer processing sequence, and consists in producing a thin metal layer serving as the conducting medium linking the devices and circuits on a chip. It is also used to produce bonding pads around the periphery of the chip which are needed to bond the wire leads from an IC package to the chip. These bonding wires are usually $25\,\mu m$ diameter gold . The pads are $100\,\mu m \times 100\,\mu m$ squares to allow for the flattened ends of the wires and for some placement errors.

The metallization process normally uses aluminium of thickness $1\mu m$ and widths of $2$–$25\,\mu m$. First, the thin film is deposited on the wafer using vacuum evaporation. Then the required pattern of the metal is created using photolithography. Etching uses phosphoric acid or plasma etching may also be used. Alternatively, metallization can be accomplished using a lift-off process. A positive photoresist is deposited and patterned using standard photolithography, then the metal film is deposited on the remaining photoresist. Next the photoresist is dissolved and is lifted off the wafer, taking those parts of the metal on its top with it. This process is capable of producing very fine line width patterns even for a larger film thickness.

### 2.3.7   MOSFET Processing Steps

Based on the previous description of the fabrication process, we may now state the sequence of processing steps necessary to produce an n-channel aluminium-gate MOSFET of the type shown in Figure 2.1.

1. The starting material is p-type silicon with resistivity of $10\,\Omega\,cm$.

2. A thermal oxide layer of about $10\,000\,\text{Å}$ is grown on the silicon.

3. A first photolithography process produces windows for the source and drain diffusions.

4. $n^+$ phosphorous diffusion is used to produce the source and drain regions.

5. A second photolithography process is used to remove the oxide from the channel region between the source and drain.

6. A very thin oxide layer (about 300–800 Å) is grown over the channel region.

7. A third photolithography step produces the contact windows.

8. Metallization is employed using an aluminium thin film.

9. A fourth photolithography step produces metallization patterns for the gate electrode, as well as the source and drain contact areas.

10. Contact sintering is applied followed by backside metallization.

### 2.3.8  Self-aligned Gate Structures

As described in Section 2.2, the inversion layer producing the channel must cover the entire region between the source and drain. Hence, the gate electrode must extend throughout this region. To allow for possible mask errors, the gate electrode is designed to overlap the source and drain regions by small distances of the order of $5\,\mu$m. This will create small *overlap capacitances* $C_{gs}$ between the gate and source, and $C_{gd}$ between the gate and drain. These are of the order of 1–3 pF. $C_{gd}$ represents a feedback capacitance from drain to gate and will affect the frequency response of the transistor adding to the Miller effect.

To reduce the overlap capacitances a *self-aligned gate* structure is used as shown in Figure 2.18 .This is produced using the same steps as a p-channel device, except that the source and drain regions do not extend all the way under the gate. Boron ion implantation is used to produce p-type extension of the source and drain regions up to the edge of the gate. The high-energy boron ions penetrate the thin gate oxide but are blocked by the thick field oxide and by the gate. Therefore, the gate electrode itself also acts as an implantation mask and the source and drain regions effectively end just under the gate edges. This reduces the overlap capacitances.

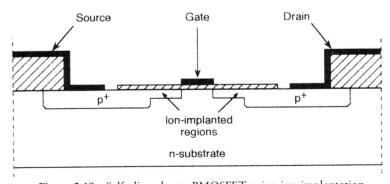

**Figure 2.18**  Self-aligned gate PMOSFET using ion implantation.

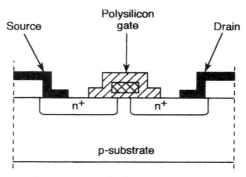

Figure 2.19  Polysilicon gate MOSFET.

An alternative self-aligned gate structure is shown in Figure 2.19, which uses polysilicon as the gate . This can withstand the high temperatures of the diffusion process and as a consequence the gate can also serve as a diffusion mask. Naturally, there will still be lateral diffusion under the gate, but the overlap capacitances are still much smaller than those created in the conventional structure ions

## 2.4  LAYOUT AND AREA CONSIDERATIONS FOR IC MOSFETS

In integrated circuits, minimum clearance values are required between the various diffused regions, contact windows,and metallized contacts to allow for mask registration errors and the minimum line resolution determined by the photolithographic process. The minimum dimension resolution $\lambda$ (not to be confused with the channel length modulation parameter) is typically one micrometer or less. The minimum clearances and dimensions are usually given as multiples of $\lambda$ resulting in a set of dimensions called the *design rules*.

MOSFETs produced on the same IC have two important attributes which allow high density of devices to be produced on the same chip. The first is the simple geometry and the second is the self- isolating property. Figure 2.20 shows PMOS-FETS sharing the same n-type substrate. The oxide thickness under the gate is about

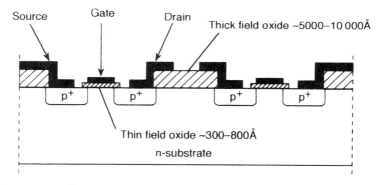

Figure 2.20  Illustrating the self-isolating properties of integrated MOSFETs.

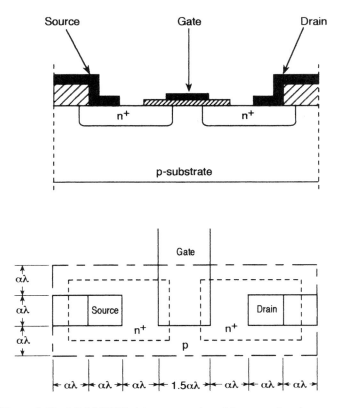

**Figure 2.21** NMOSFET: (a) cross section; (b) top-surface layout.

300–800 Å which is less than one-tenth of the thickness of the field oxide. The channel is created at the threshold voltage $V_t$ which is much smaller than that necessary to produce an inversion layer under the thick field oxide. Therefore, no n-channels will be created between adjacent MOSFETS, and they are, therefore, self-isolating requiring no special isolation regions.

Consider the NMOS and its layout shown in Figure 2.21. For illustration we take a uniform design rule dimension of $10\lambda$ (corresponding to $\alpha = 10$ in Figure 2.21) assumed for all clearances and spacings. The overall dimensions of the PMOS transistor are approximately $75\lambda$, and $30\lambda$ so that the area is $2250\lambda^2$ per transistor. For $\lambda = 1\,\mu$m the area is $0.00225\,\text{mm}^2$ yielding 444 MOSFETS per $\text{mm}^2$. For design rules of $5\lambda$ ($\alpha = 5$ in Figure 2.21) the figure becomes 1800 MOSFETS/$\text{mm}^2$ so that a 1cmx1cm chip can accommodate 180 000 MOSFETS.

Generally, the design rules are specified by the wafer manufacturer and must be observed by the designer in the layout of the integrated circuit. This serves as a standardization methodology of the design and in reality makes the layout accessible to system designers without the need to acquire deep knowledge of integrated circuits. Computer layout tools such as MAGIC [25] are indispensable in this regard, so are simulation programs such as CAzM [26] and SPICE.

## 2.5  NOISE IN MOSFETs

MOS components suffer from noise generated by small fluctuations of analog signals within the components. The various types of this noise are now discussed.

### 2.5.1  Types of Noise

#### Shot noise

This results from the dc current flowing across a pn junction. It has the mean square value:

$$\langle i^2 \rangle = 2qI_D\Delta f \tag{2.47}$$

where $q$ is the electronic charge, $I_D$ is the average dc current of the pn junction, and $\Delta f$ is the bandwidth. The noise current spectral density is given by

$$\langle i^2 \rangle / \Delta f = 2qI_D \tag{2.48}$$

#### Thermal noise

This is a result of random electron motion, with a mean square value having the typical form

$$\langle v^2 \rangle = 4kTR\Delta f \tag{2.49}$$

where $R$ is the equivalent resistance of the noise source, $T$ is the absolute temperature and $k$ is Boltzmann's constant. For the device in saturation, the channel is tapered and $R$ can be approximated by $R \approx 3/2g_m$.

#### Flicker (1/f) noise

This results from charge carrier traps which capture and release carriers in a random manner. The time dependence of this phenomenon results in noise which, for a specific device and process, has a spectral density of the form

$$\langle i^2 \rangle / \Delta f = K_f(2K'/C_{ox}L^2)[I^a/f] \tag{2.50}$$

where $K_f$ is a constant with a typical value of $10^{-24}$ V$^2$. Farads and $a = 0.5$–$2.0$, taken here to be $=1$.

### 2.5.2  Modelling of Noise

In evaluating the performance of CMOS circuits, taking into account the above types of noise, they can be incorporated in the equivalent circuits as sources. Both thermal and $1/f$

noise can be modelled by a current source in parallel with $i_D$ in the large signal model of the MOS transistor of Figure 2.8(b). The mean square current noise source value is

$$\langle(i_N)^2\rangle = [(8/3)kTg_m + (2K'K_fI_D)/fC_{ox}L^2]\Delta f \tag{2.51}$$

This current can be referred to the gate of the MOS transistor, as shown in Figure 2.22(a) by dividing the expression in (2.51) by $g_m^2(=4KI_D)$, which leads to

$$\langle(v_{eq})^2\rangle = [(8/3g_m)kT + (K_f)/fC_{ox}WL]\Delta f \tag{2.52}$$

For frequencies below 1 kHz, the $1/f$ noise is the dominant source and for many practical cases, the above expression reduces to

$$\langle(v_{eq})^2\rangle = [(K_f)/fC_{ox}WL]\Delta f \tag{2.53}$$

Figure 2.22(b) shows the noise spectrum of a typical MOSFET.

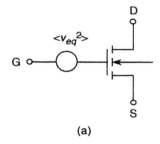

(a)

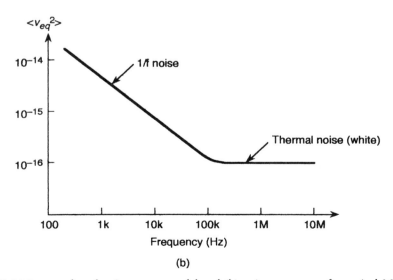

(b)

**Figure 2.22** (a) Input-referred noise source model and (b) noise spectrum of a typical MOSFET.

## CONCLUSION

In this chapter, the basic operation of MOSFETS was reviewed. The fabrication process of integrated circuits employing MOSFETS was described with emphasis on the aspects which are of particular help in the understanding of switched-capacitor circuits. The concepts discussed in this chapter will be used to discuss the design of composite circuits which make up the basic building blocks of switched-capacitor filters. The interaction between circuit designers and integrated circuit fabrication specialists has become so close in the case of analog circuits, that each party has to acquire familiarity with the other's methods to guarantee a successful outcome of the complete design process; hence the need for this chapter from the view-point of the circuit designer.

## PROBLEMS

2.1   An NMOS transistor has a drain current of 6.5 mA at $V_{GS} = V_{DS} = 10$ V. The drain current decreases to 2 mA for $V_{GS} = V_{DS} = 6$ V. Calculate the values of $K$ and $V_t$ for this transistor.

2.2   In a certain fabrication process, the transconductance parameter $K' = 20 \, \mu A/V^2$ and the threshold voltage $V_t = 1$ V. It is required to operate an NMOS transistor device over a range in which $v_{GS} = v_{DS} = 5$ V and producing a drain current of 1 mA for the device having a minimum length of 10 $\mu$m. Find the required value of channel width.

2.3   A MOSFET has $V_t = 1$ V and $K = 500 \, \mu A/V^2$. If the device is to operate in saturation with $i_D = 10$ mA, calculate the required value of $v_{GS}$ and the minimum required value of $v_{DS}$.

2.4   A MOSFET operating in saturation at a constant $v_{GS}$ has $i_D = 1$ mA and $v_{DS} = 2$ V. When $v_{DS} = 7$ V, $i_D$ becomes 1.1 mA. Find the corresponding values of $r_o$, $V_A$ and $\lambda$.

2.5   An enhancement type NMOSFET has $V_t = 1$V and $K = 200 \, \mu A/V^2$. If it is to be used as a voltage-controlled linear resistor, find the required range of $v_{GS}$ to obtain a resistance range of 1–4 k$\Omega$.

2.6   A depletion-type NMOSFET operates in the triode region with $v_{DS} = 0.2$ V. It carries a drain current of 2 mA at $v_{GS} = -2$ V and 10.8 mA at $v_{GS} = 2$ V. Calculate $V_t$ and $K$ for the device.

2.7   An enhancement-type NMOS transistor has $K' = 20 \, \mu A/V^2$ and an aspect ratio $(W/L) = 4$. It operates in saturation with $I_D = 20$ mA. Calculate the transconductance of the device.

2.8   An enhancement-type PMOS transistor has its gate grounded and its source connected to 5 V. The device has $K = 60 \, \mu A/V^2$ $V_t = -1$ V and $\lambda = -0.03$ V$^{-1}$. Calculate the drain current for each of the values of $v_D = 3$ V, 0.

2.9   Consider the circuit shown in Figure P2.9. The PMOS device has $V_t = -1$ V, $K' = 10 \, \mu A/V^2$, $L = 5 \, \mu$m and $\lambda = 0$. Find the required values of $R$ and channel width $W$ to obtain a drain current of 0.2 mA.

5V

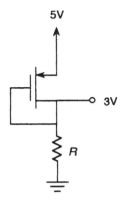

3V

R

Figure P2.9

2.10 Consider the circuit shown in Figure P2.10. Each transistor has $K' = 10\,\mu A/V^2$, $V_t = 1\,V$, $\lambda = 0$, and $L = 10\,\mu m$. Find the values of channel widths of the devices which are required to obtain the current and voltage values shown on the circuit diagram.

5V

1mA

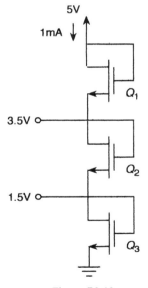

$Q_1$

3.5V

$Q_2$

1.5V

$Q_3$

Figure P2.10

# 3

# Basic Integrated Circuit Building Blocks

## 3.1 INTRODUCTION

This chapter gives a concise presentation of the elementary circuits [30]–[32], employing the metal-oxide-semiconductor (MOS) structure, which form the basis for the design of the more composite building blocks of switched-capacitor filters. The use of MOS transistors as load devices is first introduced followed by the design principles of MOS amplifiers. Next, parasitic capacitances are discussed due to their importance to the operation of switched-capacitor filters especially at high frequencies. In this context, the cascode amplifier is introduced with the objective of reducing the Miller effect. Finally, the current mirror is introduced followed by a discussion of the CMOS amplifier.

## 3.2 MOS ACTIVE RESISTORS AND LOAD DEVICES

An active resistor is used instead of a passive polysilicon or diffused resistor to generate a dc voltage drop, or to form a small signal resistance that is linear over a small range. This requires a much smaller area than a corresponding polysilicon resistor, but at the expense of linearity.

Active loads in MOS technology can be constructed from either enhancement-type or depletion-type devices. In the former case, the drain is connected to the gate, while in the latter the source is connected to the gate.

Figure 3.1 shows an NMOS diode-connected enhancement-type transistor together with its $v - i$ characteristic, which is defined by

$$i = K(v - V_t)^2 \tag{3.1}$$

and it always operates in saturation. For a transistor biased at a voltage $V$, we can write

$$i = K(V + v_{gs} - V_t)^2$$
$$= K(V - V_t)^2 + 2K(V - V_t)v_{gs} + K v_{gs}^2 \tag{3.2}$$

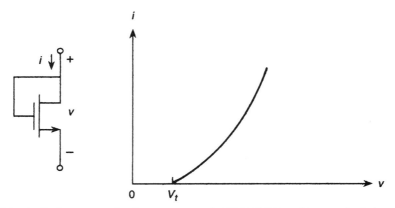

**Figure 3.1** A diode-connected enhancement-mode NMOSFET and its terminal characteristic.

and for small signal operation the last term can be neglected. The first term is a dc or quiescent current. Therefore the signal current is

$$i = 2K(V - V_t)v_{gs} \tag{3.3}$$

But, as we have seen, the transconductance $g_m$ of the device is given by $i/v_{gs}$. Therefore, the incremental resistance of the diode-connected transistor is $1/g_m$.

Similarly, the diode-connected depletion-type MOS transistor shown in Figure 3.2 can be used as an active load. For the device to operate in saturation, the voltage across the two terminals must exceed $-V_{iD}$ (the threshold voltage). In this case,

$$i \cong KV_{tD}^2\left(1 + \frac{v}{V_A}\right) \tag{3.4}$$

and the device can be used to provide large resistance load values as usually required for high gain amplifiers.

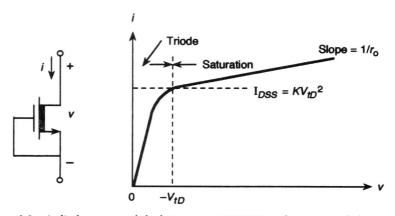

**Figure 3.2** A diode-connected depletion-type MOSFET and its terminal characteristic.

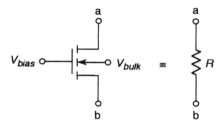

**Figure 3.3**   An MOSFET as an active resistor.

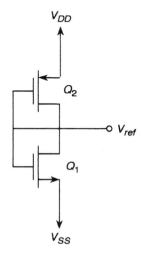

**Figure 3.4**   Active resistors for voltage division.

An alternative use of the MOS transistor as a resistor is shown in Figure 3.3. In the vicinity of $v_{DS} \approx 0$, the resistance of the device is given by $r_{ds} = r_o$.

A number of diode-connected transistors can be stacked to form simple CMOS voltage references as shown in Figure 3.4 for two devices. Analysis of the circuit gives

$$V_{ref} = \frac{V_{SS} + V_{tn} + \sqrt{K_2/K_1}(V_{DD} - |V_{tp}|)}{1 + \sqrt{K_2/K_1}} \tag{3.5}$$

## 3.3   MOS AMPLIFIERS

### 3.3.1   NMOS Amplifier with Enhancement Load

The simplest amplifier circuit employing a diode-connected enhancement-type NMOS transistor as an active load is shown in Figure 3.5 together with its small signal equivalent circuit. Assume that $Q_1$ and $Q_2$ have infinite resistances in saturation, and identical threshold voltages $V_t$ but different $K$-values. Then, when $Q_1$ is in

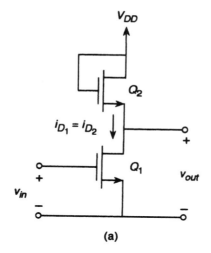

(a)

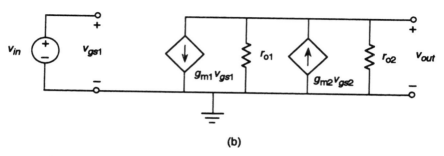

(b)

Figure 3.5   Enhancement-load amplifier: (a) circuit; (b) equivalent circuit.

saturation,

$$i_{D1} = K_1(v_{GS1} - V_t)^2 \tag{3.6}$$

But $i_{D1} = i_{D2} = i_D$, $v_{GS1} = v_{in}$; then

$$i_D = K_1(v_{in} - V_t)^2 \tag{3.7}$$

Also $Q_2$ is such that

$$i_D = K_2(v_{GS2} - V_t)^2 \tag{3.8}$$

But

$$v_{GS2} = V_{DD} - v_{out} \tag{3.9}$$

then,

$$i_D = K_2(V_{DD} - v_{out} - V_t)^2 \tag{3.10}$$

Thus, (3.7)–(3.10) yield

$$v_{out} = \left(V_{DD} - V_t + \sqrt{\frac{K_1}{K_2}}V_t\right) - \sqrt{\frac{K_1}{K_2}}v_{in} \qquad (3.11)$$

signifying a linear transfer characteristic. The large signal gain is, therefore,

$$A_V = -\sqrt{K_1/K_2}$$
$$= -\sqrt{(W_1/L_1)/(W_2/L_2)} \qquad (3.12)$$

In order to determine the small signal gain of the amplifier, we use the equivalent circuit shown in Figure 3.5(b). This gives

$$v_{out} = -g_{m1}v_{gs1}[(1/g_{m1})\|r_{01}\|r_{02}] \qquad (3.13)$$

and with $v_{gs1} = v_{in}$, we have

$$A_v = \frac{v_{out}}{v_{in}} = \frac{-g_{m1}}{g_{m2} + (1/r_{01}) + (1/r_{02})} \qquad (3.14)$$

Usually $r_{01}, r_{02} \gg 1/g_{m2}$ and the above expression simplifies to

$$A_v = -g_{m1}/g_{m2} \qquad (3.15)$$

which can be easily put in the form (3.12) using (2.7) and (2.39).

### 3.3.2 Effect of the Substrate

In the above analysis, we have tacitly assumed that the substrate of each transistor is connected to the source. However, in integrated circuit realizations, $Q_1$ and $Q_2$ share the same substrate, which is grounded. Thus, for $Q_2$ the substrate is grounded while the source is not, and a signal voltage $v_{bs}$ develops between the body and source, giving rise to a drain current component given by [31]

$$i_{db} = g_{mb}v_{bs} \qquad (3.16)$$

where

$$g_{mb} = \frac{\partial i_D}{\partial v_{BS}}\bigg|_{\substack{v_{GS}=\text{constant} \\ v_{DS}=\text{constant}}} \qquad (3.17)$$

is the *body transconductance*. But $i_D$ depends on $v_{BS}$ via the dependence of $V_t$ on $V_{BS}$ as given by [31]

$$V_t = V_{to} + \gamma(\sqrt{2\phi_f + V_{SB}} - \sqrt{2\phi_f}) \qquad (3.18)$$

where $\phi_f$ is a parameter $\approx 0.3\,V$, $V_{to}$ is the threshold voltage for $V_{BS} = 0$ and $\gamma$ is a process parameter with a typical value of $0.5\,V^{1/2}$.

Therefore we can write

$$g_{mb} = \chi g_m \qquad (3.19)$$

with

$$\chi = \frac{\partial V_t}{\partial V_{SB}} \qquad (3.20)$$

$$= \frac{\gamma}{2\sqrt{2\phi_f + V_{SB}}}$$

where $\chi$ is the *body factor* with typical values in the range 0.1–0.3.

The NMOSFET model incorporating the controlled current source $g_{mb}v_{bs}$ used to model the body effect is shown in Figure 3.6, which can always be used in situations where the substrate is not connected to the source. Thus the enhancement-load amplifier incorporating the body effect as shown in Figure 3.7(a) has the equivalent circuit shown in Figure 3.7(b). Direct analysis shows that the voltage gain is given by

$$A_v = \frac{-g_{m1}}{g_{m2} + g_{mb2} + (1/r_{01}) + (1/r_{02})} \qquad (3.21)$$

Usually $r_{01}$, $r_{02} \gg 1/g_{m2}$, and we have

$$A_v \cong \frac{-g_{m1}}{g_{m2} + g_{mb2}} \qquad (3.22)$$

which, upon use of (3.19), gives

$$A_v = -\frac{g_{m1}}{g_{m2}} \frac{1}{1 + \chi} \qquad (3.23)$$

Thus, the body effect reduces the gain by a factor of $1/(1 + \chi)$.

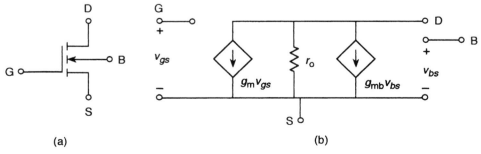

(a)                                        (b)

**Figure 3.6**   Effect of the substrate: (a) MOSFET with B not connected to S; (b) small signal equivalent circuit.

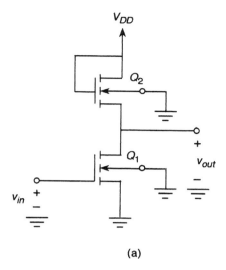

(a)

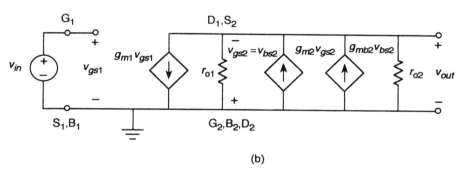

(b)

**Figure 3.7** Amplifier with the effect of the substrate: (a) circuit; (b) equivalent circuit.

We note that the enhancement-load amplifier has a limited output signal swing which cannot exceed $V_{DD} - V_t$.

### 3.3.3 NMOS Amplifier with Depletion Load

Figure 3.8(a) shows an amplifier using an enhancement transistor $Q_1$ together with a depletion-type diode-connected load transistor $Q_2$. The $i$ against $v$ characteristic of the load is similar to that in Figure 3.2. We shall now show that this amplifier has a performance superior to that of the enhancement-load type discussed earlier. It is stipulated that the amplifier is biased to operate in the region where both $Q_1$ and $Q_2$ are in saturation. The small-signal gain is then given by

$$A_v = v_{out}/v_{in} = -g_{m1}[r_{01}\|r_{02}]$$ (3.24)

neglecting the body effect.

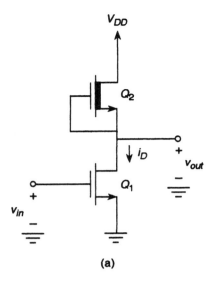

(a)

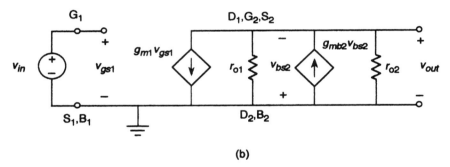

(b)

**Figure 3.8**   (a) Depletion-load amplifier; (b) equivalent circuit.

To incorporate the body effect into the analysis, the equivalent circuit of Figure 3.8(b) is used. This gives

$$v_{out} = -g_{m1}v_{gs1}\left[\frac{1}{g_{mb2}}\|r_{01}\|r_{02}\right] \tag{3.25}$$

But $v_{gs1} = v_{in}$ and we have

$$A_v = \frac{v_{out}}{v_{in}} = -g_{m1}\left[\left(\frac{1}{g_{mb2}}\right)\|r_{01}\|r_{02}\right] \tag{3.26}$$

and usually $r_{01}, r_{02} \gg 1/g_{mb2}$ which gives

$$A_v \cong \frac{-g_{m1}}{g_{mb2}}$$

$$\cong \frac{-g_{m1}}{g_{m2}}\left(\frac{1}{\chi}\right) \tag{3.27}$$

or

$$A_v \cong -\sqrt{\frac{(W_1/L_1)}{(W_2/L_2)}}\left(\frac{1}{\chi}\right) \qquad (3.28)$$

By comparison with the gain of the enhancement-load amplifier given by (3.23) it is seen that the depletion-load amplifier provides a higher gain by a factor of $(1+\chi)/\chi$.

It is also observed that the bias current of this amplifier is $\approx I_{DSS}$ of the depletion load which is $K_D V_{tD}^2$; implying that it is fixed by the technology and device geometry.

### 3.3.4   The Source Follower

The basic amplifiers discussed above are of the *common-source* type, since a signal ground is established at the source. They provide a high input impedance, a large negative voltage gain, and a large output resistance. The latter is not a desirable property for voltage amplifiers. Therefore, it is often required to design an output buffer stage for obtaining a low output resistance without affecting the gain of the previous stage. This can be achieved by using the *common-drain* or *source follower* configuration shown in Figure 3.9. The output voltage appears across the parallel combination of $r_0$ and $1/g_{mb}$. The voltage gain is obtained as

$$\frac{v_{out}}{v_{in}} = \frac{[(1/g_{mb})\|r_0]}{(1/g_m) + [(1/g_{mb})\|r_0]} \qquad (3.29)$$

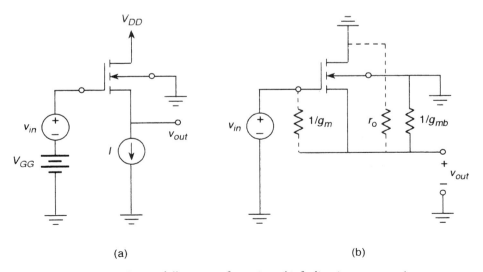

(a)                                          (b)

**Figure 3.9**   (a) Source-follower configuration; (b) finding its output resistance.

For $r_0 \gg 1/g_m$ we have

$$\frac{v_{out}}{v_{in}} \cong \frac{g_m}{g_m + g_{mb}} \tag{3.30}$$

and with

$$g_{mb} = \chi g_m \tag{3.31}$$

$$\frac{v_{out}}{v_{in}} = \frac{1}{1 + \chi} \tag{3.32}$$

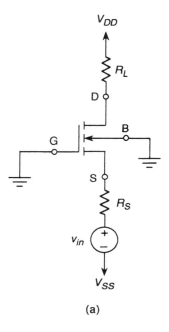

(a)

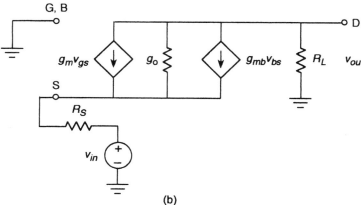

(b)

**Figure 3.10**  The common-gate amplifier: (a) circuit with the substrate shown;
(b) equivalent circuit.

which is almost unity if the body effect is neglected. This is the no-load (open-circuit) voltage gain. The output resistance of the source follower is given, from Figure 3.9(b), by

$$R_0 = (1/g_m)\|(1/g_{mb})\|r_0 \tag{3.33}$$

For example, with a transistor with $(W/L) = 10$, $\mu_n C_{ox} = 100\,\mu\text{A/V}^2$, $V_t = 1\,\text{V}$, $V_A = 100\,\text{V}$, $I = 20\,\text{mA}$, $\chi = 0.1$, we have the voltage gain as 0.9 and the output resistance as $140\,\Omega$.

### 3.3.5  The Common Gate Amplifier

This configuration is shown in Figure 3.10 and gives high input and output conductances. Using the equivalent circuit of Figure 3.6(b) incorporating the body effect we obtain:

$$A_v = \frac{g_m(1+\chi)R_L}{1 + g_m(1+\chi)R_S} \tag{3.34}$$

$$g_{in} = g_m(1+\chi) \tag{3.35}$$

$$g_{out} = \frac{g_o}{1 + g_m(1+\chi)R_S} \tag{3.36}$$

It follows that unlike other configurations, the body effect does not degrade the performance of the amplifier; on the contrary it increases the effective transconductance. The main advantage of this configuration, however, lies in its wider bandwidth as we shall see in the next section.

## 3.4  HIGH-FREQUENCY CONSIDERATIONS

### 3.4.1  Parasitic Capacitances

Figure 3.11 shows the high-frequency equivalent circuit of a MOSFET, containing the intrinsic components of the terminal capacitances as well as the extrinsic ones. Those of the former type are associated with reverse-biased p-n junctions, channel and depletion regions. These are strongly dependent on the region of operation of the devices. Those of the latter type (extrinsic) are made up of components which are largely constant, and are due to layout parasitics and overlapping regions. In saturation, the most significant capacitances are the following:

1. $C_{gd}$: Gate-to-drain capacitance. This is a thin oxide capacitance due to the overlap of the gate and drain diffusion, and as such can be assumed to be voltage-independent.

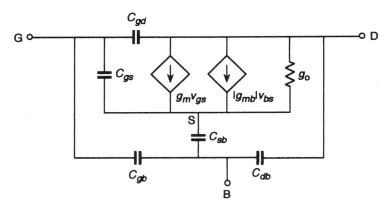

**Figure 3.11**  High-frequency equivalent circuit of MOSFET.

2. $C_{gs}$: Gate-to-source capacitance, which has two components:

(a) $C_{gs1}$ which is a thin oxide capacitance due to the gate-to-source overlap.

(b) $C_{gs2}$ which is the gate-to-channel capacitance. In saturation, this is around $\frac{2}{3}C_{ox}$ here $C_{ox}$ is the total thin oxide capacitance between the gate and the surface of the substrate.

In saturation, $C_{gs}$ is almost voltage-independent.

3. $C_{sb}$: Source-to-substrate capacitance. This has two components:

(a) $C_{sbpn}$ which is the p-n junction capacitance between the source diffusion and substrate

(b) $C'_{sb}$ which is about $\frac{2}{3}$ of the capacitance of the depletion region below the channel.

$C_{sb}$ has a voltage-dependence similar to that of an abrupt p-n junction.

4. $C_{db}$: drain-to-substrate capacitance, which is a voltage-dependent capacitance of a p-n junction.

5. $C_{gb}$: Gate-to-substrate capacitance, which is normally small in saturation, typically around $0.1C_{ox}$.

Now, for the common source amplifier of Figure 3.12(a), the capacitances discussed above come into play at high frequencies, and dictate the frequency response of the amplifier. Figure 3.12(b) shows the equivalent circuit of such an amplifier including the parasitic capacitances, when the amplifier is loaded with $C_L$, i.e. a capacitive load. This equivalent circuit simplifies to that of Figure 3.12(c), by straightforward addition of capacitors and sources, in which

$$G_{Leq} = g_{d1} + g_{d2} + g_{m2} + |g_{mb2}|$$

$$C_{Leq} = C_{db1} + C_{gs2} + C_{sb2} + C_L$$

(3.37)

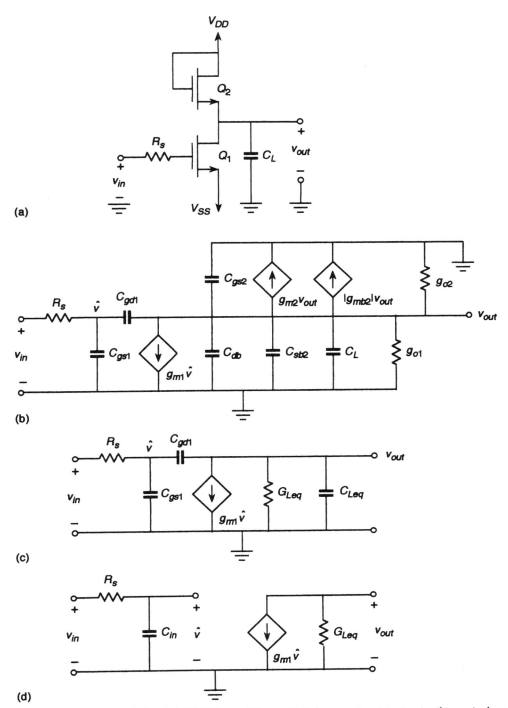

**Figure 3.12**   Capacitively-loaded NMOS amplifier at high frequencies: (a) circuit; (b) equivalent circuit; (c) simplified equivalent circuit; (d) approximate equivalent circuit.

Direct analysis of this circuit gives for the gain

$$A_v(s) = \frac{V_{out}(s)}{V_{in}(s)}$$

$$= \frac{G_s(sC_{gd1} - g_{m1})}{[s(C_{gs1} + C_{gd1}) + G_s][s(C_{gd1} + C_{Leq}) + G_{Leq}] - sC_{gd1}(sC_{gd1} - g_{m1})}$$

(3.38)

At $s = j\omega$, and for moderate frequencies we can take

$$g_{m1} \gg \omega C_{gd1}$$

$$G_{Leq} \gg \omega(C_{gd1} + C_{Leq})$$

(3.39)

and the gain

$$A_v(j\omega) = \frac{-g_{m1}G_s}{G_s G_{Leq} + j\omega[G_{Leq}(C_{gs1} + C_{gd1}) + g_{m1}C_{gd1}]}$$

$$= \frac{-g_{m1}/G_{Leq}}{1 + j\omega R_s[C_{gs1} + C_{gd1}(1 + g_{m1}/G_{Leq})]}$$

$$= \frac{A_v(0)}{1 + j\omega R_s C_{in}}$$

(3.40)

where

$$A_v(0) = -g_{m1}/G_{Leq}$$

(3.41)

and

$$C_{in} = C_{gs1} + C_{gd1}(1 + g_{m1}/G_{Leq})$$

$$= C_{gs1} + C_{gd1}[1 + |A_v(0)|]$$

(3.42)

Therefore $A_v(j\omega)$ in (3.40) is seen to be the gain of the approximate equivalent circuit of Figure 3.12(d). In particular the input capacitor $C_{in}$ results from the gate-to-source capacitance $C_{gs1}$ plus the gate-to-drain capacitor $C_{gd1}$ magnified by the factor $[1 + |A_V(0)|]$. The latter is the familiar Miller effect, and since $|A_V(0)| \gg 1$, the effect results in a serious reduction in the bandwidth.

### 3.4.2   The Cascode Amplifier

The Miller effect can be eliminated or at least reduced by using a common gate MOSFET $Q_2$ together with the common source amplifier resulting in the cascode configuration shown in Figure 3.13(a). $Q_2$ isolates the input and output nodes. It provides a low input resistance $1/g_{m2}$ at its source, and a high one at its drain to drive $Q_3$.

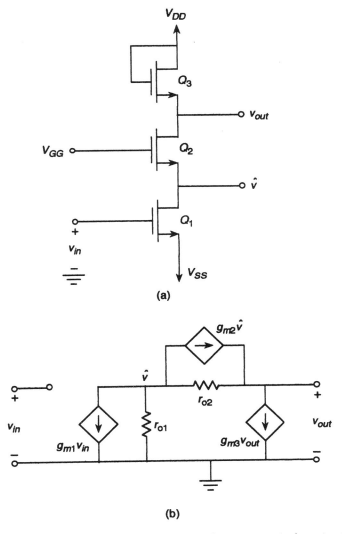

**Figure 3.13**   (a) Cascode amplifier; (b) low-frequency equivalent circuit.

The low-frequency small signal equivalent circuit is shown in Figure 3.13(b) which gives

$$g_{m1}v_{in} = -g_{m2}\hat{v} = -g_{m3}v_{out} \tag{3.43}$$

or

$$\hat{v} \cong -\frac{g_{m1}}{g_{m2}}v_{in} \tag{3.44}$$

and

$$v_{out} \cong \frac{g_{m2}}{g_{m3}} \hat{v} \cong -\frac{g_{m1}}{g_{m3}} v_{in} \tag{3.45}$$

Hence the gate-to-drain gain of $Q_1$ is $-g_{m1}/g_{m2}$ and $C_{gd1}$ of the driver transistor is multiplied by $(1 + g_{m1}/g_{m2})$. If we choose $g_{m1} = g_{m2}$ then this factor is 2, and the Miller effect is reduced considerably.

## 3.5   THE CURRENT MIRROR

In integrated circuits, a stable dc reference current source is designed, then used to generate other dc currents which are multiples or fractions of this source at other points of the circuit for biasing the transistors. The current mirror shown in Figure 3.14 is a universal circuit for producing a current $I_0$ which is proportional to a reference current $I_{ref}$. It comprises two enhancement MOSFET's $Q_1$ and $Q_2$ with the same $V_t$ but possibly different aspect ratios. In the circuit of Figure 3.14, both $Q_1$ and $Q_2$ operate in saturation and due to the parallel connection both transistors have the same $V_{GS}$. Thus, we can write

$$I_{ref} = K_1(V_{GS} - V_t)^2 \tag{3.46}$$

$$I_0 = K_2(V_{GS} - V_t)^2 \tag{3.47}$$

assuming the output resistance of $Q_2$ to be infinite. Hence

$$\begin{aligned} I_0 &= I_{ref}\left(\frac{K_2}{K_1}\right) \\[2mm] &= I_{ref}\frac{(W_2/L_2)}{(W_1/L_1)} \end{aligned} \tag{3.48}$$

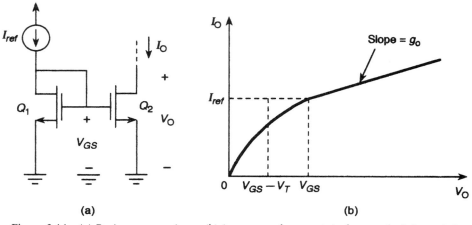

**Figure 3.14**   (a) Basic current mirror, (b) its output characteristic for matched $Q_1$ and $Q_2$.

However, due to the finite output resistance ($= r_0$ of the transistor $Q_2$) the above expression is only approximate. This output resistance of the current mirror can be increased by using the cascode mirror of Figure 3.15(a). For this circuit the incremental resistance of each of the diode-connected $Q_1$ and $Q_4$ is $1/g_m$ which is relatively small. Replacing $Q_2$ by $r_{02}$ and using the equivalent circuit of $Q_3$ we obtain the circuit of Figure 3.13(b) which gives for the output resistance

$$R_0 = v/i = r_{03} + r_{02} + g_{m3}r_{03}r_{02} \tag{3.49}$$

With $r_{02} = r_{03} = r_0$, we have

$$R_0 = r_0(2 + g_m r_0) \tag{3.50}$$

which is larger than that of the simple mirror of Figure 3.14 by a factor $\approx g_m r_0$.

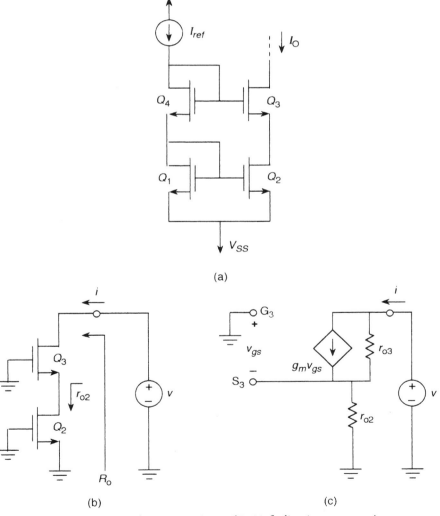

**Figure 3.15**   (a) Cascode current mirror, (b), (c) finding its output resistance.

Another circuit is the Wilson current mirror shown in Figure 3.16 for which

$$r_0 = (g_{m1}r_{01})\left(\frac{g_{m3}}{g_{m2}}\right)r_{03} \tag{3.51}$$

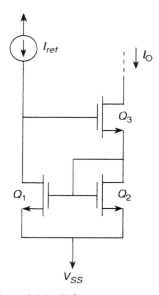

**Figure 3.16**   Wilson current mirror.

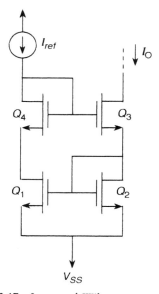

**Figure 3.17**   Improved Wilson current mirror.

with a typical value

$$r_0 \approx (100)(1)(10^5)$$

$$\approx 10\,\text{M}\Omega$$

However, this circuit suffers from the fact that the drain to source voltage drops of $Q_1$ and $Q_2$ are unequal; these can be equalized by adding the diode-connected transistor $Q_4$ as shown in Figure 3.17. The output resistance of this improved Wilson current source is still given by (3.51).

## 3.6   THE CMOS AMPLIFIER

Complementary n-channel and p-channel devices are used together in integrated circuit CMOS technology yielding greater design flexibility and, in addition, they are fabricated in a manner that eliminates the body effect. They are also used with great effect in the design of switches to minimize an undesirable effect called *clock feedthrough* in switched-capacitor filters, as we shall see shortly

The basic CMOS amplifier is shown in Figure 3.18(a), and has the following features:

1. $Q_2$ and $Q_3$ are matched p-channel devices forming a current source with the $v$ against $i$ characteristic shown in Figure 3.18(b). $Q_2$ is forced to operate in saturation by ensuring that its drain voltage is lower than its source voltage $V_{DD}$ by at least $V_{SG} - |V_{tp}|$, where $V_{SG}$ is the dc bias voltage corresponding to a drain current of $I_{ref}$. In saturation, $Q_2$ has the high output resistance

$$r_{02} = \frac{|V_A|}{I_{ref}} \tag{3.52}$$

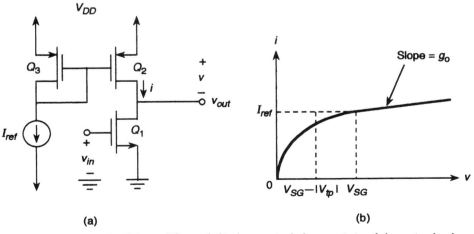

(a)                                                    (b)

**Figure 3.18**   (a) CMOS amplifier and (b) the terminal characteristic of the active load.

2. $Q_2$ is an active load for the amplifying transistor $Q_1$.

3. When $Q_1$ is in saturation, the small signal voltage gain is given by

$$A_v = -(g_{m1})(r_{01}\|r_{02})$$
(3.53)

and since $Q_1$ operates at the dc bias current $I_{ref}$, then using (2.24) we have

$$g_{m1} = \sqrt{2(\mu_n C_{ox})(W/L)I_{ref}}$$
(3.54)

Thus, (3.52) and (3.53) give, with $r_{01} = r_{02}$,

$$A_v = -\frac{\sqrt{K_n}|V_A|}{\sqrt{I_{ref}}}$$
(3.55)

## CONCLUSION

This chapter dealt with elementary integrated circuit components which are necessary for the construction of more elaborate circuits. In particular, the use of MOS transistors as active resistors and loads, and as simple amplifying or gain stages were reviewed. The effect of parasitic capacitances on the high frequency performance of the circuits was discussed. The various types of current mirror were also given and used in the design of simple CMOS amplifiers.

## PROBLEMS

3.1  Design an NMOS common source amplifier with an enhancement-type load device, to have a voltage gain of 10 and an output resistance of $1\,\mathrm{k\Omega}$ taking the effect of the substrate into account. The minimum channel length in the process is $10\,\mu\mathrm{m}$ and for both devices $\chi = 0.2$, $K' = 10\,\mu\mathrm{A/V^2}$, $V_t = 1\mathrm{V}$. Assume the bias value of $V_{GS} = 2\mathrm{V}$.

3.2  Design an NMOS common source amplifier with a depletion-load transistor as a load, to have a gain of 100 taking the effect of the substrate into account. The minimum channel length in the process is $3\,\mu\mathrm{m}$ and $\chi = 0.1$.

3.3  The source follower MOS stage shown in Figure 3.9 uses a transistor with $W/L = 10$, $K' = 50\,\mu\mathrm{A/V^2}$, $\chi = 0.1$, $V_t = 2\mathrm{V}$, $V_A = 100\,\mathrm{V}$, $I = 20\,\mathrm{mA}$. Find the voltage gain and output resistance of the stage.

3.4  Derive equation (3.38) for the gain of a common source amplifier taking into account the parasitic capacitances.

3.5  Design a cascode amplifier to have a gain of 100, taking into account the effect of the substrate for which $\chi = 0.2$ The minimum channel length for the process is $5\,\mu\mathrm{m}$.

3.6  In the current mirror of Figure 3.14, the transistors have $V_t = 1\,\mathrm{V}$, $K = 100\,\mu\mathrm{A/V^2}$ and $V_A = 30\,\mathrm{V}$. If $I_{ref} = 50\,\mu\mathrm{A}$, $V_{SS} = -5\,\mathrm{V}$, $V_o = 5\,\mathrm{V}$, find $I_o$.

3.7 In the cascode current mirror of Figure 3.15, the devices have the same parameter values as given in Problem 3.6. Also $V_{SS}$ and $V_o$ are the same as given in Problem 3.6. Find $I_o$ and compare this value with that obtained for the mirror of Problem 3.6.

3.8 Derive expression (3.51) for the output resistance of the Wilson current mirror of Figure 3.16. Also show that the same expression also applies for the modified circuit of Figure 3.17.

3.9 The Wilson current mirror of Figure 3.16 employs transistors with $K = 100\,\mu A/V^2$, $V_t = 1\,V$, $V_A = 50\,A$. The reference current is $50\,\mu A$ and $V_{SS} = 0$. Find $I_o$ and the output resistance of the mirror.

3.10 Design a CMOS amplifier of the type shown in Figure 3.18 with a voltage gain of 100. Take the transistor parameters to be any reasonable set, based on the experience gained in solving the previous problems and use $I_{ref} = 100\,\mu A$.

# 4

# Two-stage CMOS Operational Amplifiers

## 4.1  INTRODUCTION

A most critical basic building block in the design of switched-capacitor filters is the CMOS Operational Amplifier (Op Amp) [31], [32]. It may be regarded as a composite building block employing the elementary circuits and fundamental concepts given in the previous chapter. This chapter deals with the design of integrated CMOS Op Amps and gives complete design examples. The chapter begins by a summary of Op Amp performance parameters and the fundamentals of feedback amplifier character- istics, then proceeds to give a discussion of the CMOS differential pair which is the first stage in the Op Amp. Next, the very popular two-stage CMOS Op Amp archi- tecture is developed and a detailed account of the design considerations is given together with a summary of the design equations for easy reference and use by the reader. This is followed by a complete design example starting from the design specifications and showing in detail how to arrive at the structure and element values of the final design.

## 4.2  OP AMP PERFORMANCE PARAMETERS

Ideally, an Op Amp is a voltage-controlled voltage source as shown in Figure 4.1 whose output $v_0$ and two inputs $v_1$, $v_2$ are related by

$$v_o = A(v_1 - v_2) \tag{4.1}$$

where $A$ is a constant, independent of frequency and is very large (ideally $= \infty$). In practice, however, $A$ is frequency-dependent and there are a number of non-ideal effects in real MOS Op Amps. The performance (and specifications) of an Op Amp is usually measured by the following parameters:

1. *The open loop dc-gain $A_0$*. This is the value of the voltage gain at zero frequency, i.e. when the input is a constant. Typically $A_0 \approx 10^3 - 2 \times 10^4$.

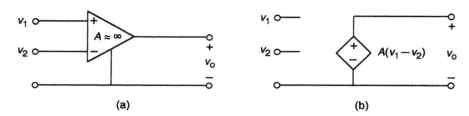

**Figure 4.1**   The ideal Op Amp: (a) symbol, (b) equivalent circuit

2. *Output voltage swing.* This is within $\pm 1$ V of the dc supply voltages; e.g. for $\pm 5$ V supplies, the linear range would be $-4V < v_{out} < 4V$.

3. *Input offset voltage.* Ideally, for $v_1 = v_2$ in (4.1) $v_0 = 0$. However, in practice $v_0 \neq 0$ for zero differential input. The voltage at the input which is required to reduce the output to zero is called the input offset voltage. Typically, this is $\approx 1$–15 mV.

4. *Common-mode rejection ratio (CMRR).* Ideally, $v_0$ depends only on the differential input $(v_1 - v_2)$. In reality, $v_0$ is also affected by the average or common mode voltage

$$v_{CM} = \frac{v_1 + v_2}{2} \qquad (4.2)$$

as well as the differential voltage

$$v_d = (v_1 - v_2) \qquad (4.3)$$

and we can write for the output

$$v_0 = A_d v_d + A_{CM} v_{CM} \qquad (4.4)$$

where $A_d$ is the differential gain and $A_{CM}$ is the common-mode gain. The common-mode rejection ratio is defined as

$$\mathrm{CMRR} = 20 \log \frac{A_d}{A_{CM}} \, \mathrm{dB} \qquad (4.5)$$

which, for an ideal Op Amp should be infinite. Practical values of MOS Op Amps are in the range 60–80 dB.

5. *Common-mode range (CMR).* This is the range of the common-mode voltage over which the CMRR remains acceptably high. Typical values are $\pm(2-3)$ V.

6. *Power supply rejection ratio.* Any additive noise at either terminal of the voltage supply appears at the output $v_0$ with a gain of $A_{ps}$. The power supply rejection ratio is defined by

$$\mathrm{PSRR} = 20 \log \frac{A_d}{A_{ps}} \qquad (4.6)$$

with typical values of 60–90 dB at low frequencies.

7. *The unity-gain bandwidth.* As the frequency increases, the gain of the Op Amp rolls off. The frequency at which the gain reaches unity (or 0 dB) is the unity-gain bandwidth. For simple Op Amps this is typically ≈1–5 MHz, while for sophisticated versions this could be as large as 100 MHz.

8. *Settling time.* This is the time required for the output to reach its final value (with an error of typically 0.1–1%) when the input voltage changes over the linear range of operation. The settling time is related to the unity gain bandwidth, the output impedance and load. Typically, its values are in the range 0.05–5 $\mu$s.

9. *Slew rate.* As the input varies by a jump discontinuity, the output cannot follow the difference instantaneously. The maximum rate of change $dv_0/dt$ is called the slew rate (SR). It is basically a non-linear effect, and for simple Op Amps has values in the range of 1–10 V/$\mu$s.

10. *Output resistance.* Ideally, this should be zero. Actual Op Amps have a finite resistive output impedance. For Op Amps with an output buffer stage, this could be in the range 100–5000 $\Omega$. Unbuffered Op Amps may have output resistances of up to 1 M$\Omega$. The large value affects the charging time of a capacitor connected to the output, thus limiting the highest signal frequency.

11. *Dynamic range.* The real Op Amp has a finite linear range in its transfer characteristic. Therefore, there is a maximum signal amplitude $v_{in\,max}$ which it can handle without excessive non-linear distortion.

    If the power supply is $\pm V_{cc}$, then an optimistic estimate of $v_{in\,max} \cong V_{cc}/A$ where $A$ is the open loop gain.

    Also, due to spurious signals such as noise, there is a minimum signal value $v_{in\,min}$ that is not lost in noise. The dynamic range of the Op Amp is defined as

$$\text{Dynamic range} = 20\log(v_{in\,max}/v_{in\,min}) \tag{4.7}$$

In switched-capacitor filters, dynamic range values of 80–100 dB are possible.

12. *dc power dissipation.* Typical values are in the range 0.25–100 mW.

The general structure of a two-stage Op Amp is shown in Figure 4.2. The input differential amplifier stage is designed to provide a high input impedance, large CMRR, a large PSRR, low noise, low offset voltage and high gain. The next stage can be designed to perform one or more of a number of functions: (a) level shifting to compensate for the dc voltage change in the input stage, ensuring proper dc bias for the following stages; (b) additional gain, and (c) differential-to-single-ended conversion. The output buffer may be needed in some applications. In switched-capacitor filters most Op Amps used are required to drive on-chip capacitances of low value; in these cases the output stage is not needed. However, a small number of Op Amps are required to drive large capacitive or resistive loads. In this case the output buffer stage is needed, which provides the Op Amp with a low output impedance.

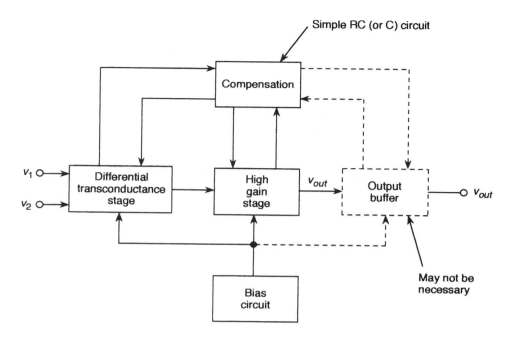

**Figure 4.2**    General structure of a two-stage Op Amp.

## 4.3   FEEDBACK AMPLIFIER FUNDAMENTALS

Negative feedback is employed almost invariably in the design of operational amplifiers to achieve greater control over the frequency and time responses of the amplifier. As expected, this is accomplished at the expense of reduced gain. Figure 4.3 shows the general feedback amplifier topology, where $G(s)$ is the open-loop gain (without feedback) and $\beta(s)$ is the transfer function of the feedback network. The latter could be as simple as a capacitor or an RC network. The transfer function of the feedback amplifier is given by

$$A(s) = \frac{G(s)}{1 + \beta(s)G(s)} \qquad (4.8)$$

and if we define the loop gain $L(s)$ as

$$L(s) = -\beta(s)G(s) \qquad (4.9)$$

then

$$A(s) = \frac{G(s)}{1 - L(s)} \qquad (4.10)$$

If $\omega_0$ is the frequency at which the argument of the loop gain is zero, then stability (no sustained oscillations) requires

$$|L(\omega_o)| < 1 \qquad (4.11)$$

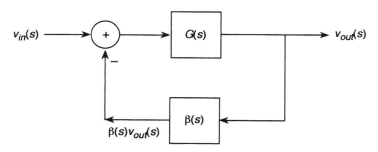

**Figure 4.3** Feedback amplifier topology.

Alternatively, if the 3 dB frequency $\omega_{o3\,dB}$ is defined by

$$|L(j\omega_{o3\,dB})| = 1 \tag{4.12}$$

then, stability requires

$$\arg[L(j\omega_{o3\,dB})] > 0 \tag{4.13}$$

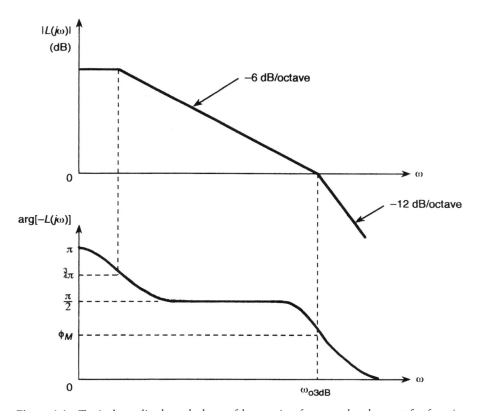

**Figure 4.4** Typical amplitude and phase of loop gain of a second-order transfer function.

These relations are illustrated in Figure 4.4 for a second-order amplifier transfer function which shows the amplitude and phase of the loop gain. Stability requires that the magnitude crosses the 0 dB point before the phase reaches zero. A standard measure of stability is the phase margin $\phi_M$ defined as the value of the phase at the frequency where the magnitude reaches unity. Thus

$$\phi_M = \arg[L(j\omega_{o3\,dB})] \tag{4.14}$$

Since the phase margin is determined by the pole-zero pattern of the transfer function, it also determines the time response of the amplifier. Figure 4.5 shows examples for a second-order function for several values of the phase margin. Since too much ringing is undesirable, a phase margin higher than 45° is usually sought. A value of 60° is generally considered acceptable for most applications.

In determining the frequency response of operational amplifiers, one usually reaches a transfer function of the form

$$A(s) = A(0)\frac{\prod(1 + s/\omega_{zi})}{\prod(1 + s/\omega_{pi})} \tag{4.15}$$

Usually the zeros are at frequencies so high that they do not affect the 3 dB frequency. If, in addition, one of the poles $\omega_{pk}$, say, is at a frequency much lower than all the others, this is called a *dominant pole* and we can write

$$A(s) \cong \frac{\omega_{pk}A(0)}{(s + \omega_{pk})} \tag{4.16}$$

so that

$$\omega_{3\,dB} \cong \omega_{pk} \tag{4.17}$$

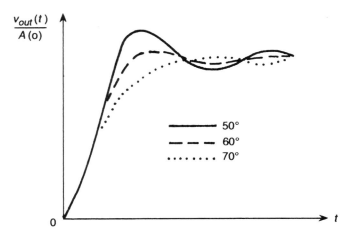

**Figure 4.5**   Step response of a typical second-order feedback amplifier for different values of phase margin.

The unity gain frequency is also given by

$$\omega_t = A(0)\omega_{3\,dB} \qquad (4.18)$$

## 4.4   THE CMOS DIFFERENTIAL PAIR

The basic CMOS differential amplifier is shown in Figure 4.6 as composed of two NMOS transistors $Q_1$ and $Q_2$.

Ideally $Q_1$ and $Q_2$ are perfectly matched. Neglecting the output resistances of $Q_1$ and $Q_2$ we can write in saturation

$$i_{D1} = K(v_{GS1} - V_t)^2 \qquad (4.19)$$

$$i_{D2} = K(v_{GS2} - V_t)^2 \qquad (4.20)$$

where

$$K = \tfrac{1}{2}\mu_n C_{ox}(W/L) \qquad (4.21)$$

Solving for $v_{GS1}$ and $v_{GS2}$ and subtracting we have

$$\sqrt{i_{D1}} - \sqrt{i_{D2}} = \sqrt{K}\,v_{id} \qquad (4.22)$$

with

$$v_{id} = v_{GS1} - v_{GS2} \qquad (4.23)$$

But

$$i_{D1} + i_{D2} = I \qquad (4.24)$$

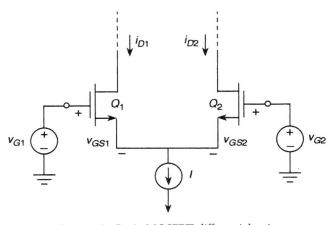

**Figure 4.6**   Basic MOSFET differential pair.

and solving for $i_{D1}$ and $i_{D2}$ we have

$$i_{D1} = \frac{I}{2} + \sqrt{2KI}\left(\frac{v_{id}}{2}\right)\sqrt{1 - \frac{(v_{id}/2)^2}{(I/2K)}} \tag{4.25}$$

$$i_{D2} = \frac{I}{2} - \sqrt{2KI}\left(\frac{v_{id}}{2}\right)\sqrt{1 - \frac{(v_{id}/2)^2}{(I/2K)}} \tag{4.26}$$

But at the bias point $v_{id} = 0$, so that

$$i_{D1} = i_{D2} = I/2 \tag{4.27}$$

and

$$v_{GS1} = v_{GS2} = V_{GS} \tag{4.28}$$

with

$$I/2 = K(V_{GS} - V_t)^2 \tag{4.29}$$

which, when used in (4.25)–(4.26) give

$$i_{D1} = \frac{I}{2} + \left(\frac{I}{V_{GS} - V_t}\right)\left(\frac{v_{id}}{2}\right)\sqrt{1 - \left(\frac{v_{id}/2}{V_{GS} - V_t}\right)^2} \tag{4.30}$$

$$i_{D2} = \frac{I}{2} - \left(\frac{I}{V_{GS} - V_t}\right)\left(\frac{v_{id}}{2}\right)\sqrt{1 - \left(\frac{v_{id}/2}{V_{GS} - V_t}\right)^2} \tag{4.31}$$

For $v_{id} \ll V_{GS} - V_t$ (small signal)

$$i_{D1} \cong \frac{I}{2} + \left(\frac{I}{V_{GS} - V_t}\right)\left(\frac{v_{id}}{2}\right) \tag{4.32}$$

$$i_{D2} \cong \frac{I}{2} - \left(\frac{I}{V_{GS} - V_t}\right)\left(\frac{v_{id}}{2}\right) \tag{4.33}$$

But a MOSFET biased at $I_D$ has $g_m = 2I_D/(V_{GS} - V_t)$. Thus for $Q_1$ or $Q_2$,

$$g_m = \frac{2(I/2)}{V_{GS} - V_t}$$

$$= \frac{I}{V_{GS} - V_t} \tag{4.34}$$

Figure 4.7 shows a graphical representation of equations (4.32) and (4.33). We also note that for differential input signals, each one of $Q_1$ and $Q_2$ has an output resistance of $r_0$.

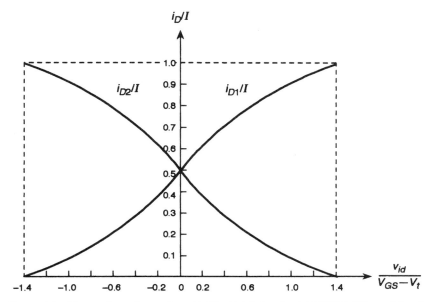

**Figure 4.7** Illustration of equations (4.32)–(4.33) for the MOSFET differential pair.

Next, we consider the use of a current mirror composed of the transistors $Q_3$ and $Q_4$ as a load for the differential pair, as shown in Figure 4.8. The mirror uses PMOS transistors while the differential pair uses NMOS types; therefore the result is a simple yet popular CMOS differential amplifier configuration. Analysis of the circuit gives

$$i = g_m(v_{id}/2) \tag{4.35}$$

with

$$g_m = \frac{I}{V_{GS} - V_t} \tag{4.36}$$

Also

$$v_{out} = 2i(r_{02}||r_{04}) \tag{4.37}$$

with

$$r_{02} = r_{04} = r_0 \tag{4.38}$$

the voltage gain is given by

$$A_V = \frac{v_{out}}{v_{id}} = g_m \frac{r_0}{2}$$

$$= \frac{V_A}{V_{GS} - V_t} \tag{4.39}$$

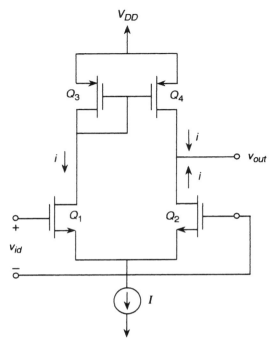

**Figure 4.8**    Active-loaded CMOS differential amplifier.

To calculate the common-mode gain and hence the common-mode rejection ratio we include the bias source conductance $G$ as shown in Figure 4.9(a), from which the equivalent circuit of Figure 4.9(b) is obtained in which the subscript $i$ is used to denote the parameters of the input differential stage devices while the subscript $L$ is used to indicate the parameters of the load devices. Under the assumptions $g_{mi}, g_{mL} \gg G, g_{dsi}$ we obtain:

$$A_d \cong \frac{g_{mi}}{g_{dsL} + g_{dsi}} \tag{4.40}$$

and

$$A_{cm} \cong \frac{-Gg_{dsi}}{2g_{mL}(g_{dL} + g_{dsi})} \tag{4.41}$$

so that

$$\text{CMRR} \cong 2\frac{g_{mi}g_{mL}}{Gg_{dsi}} \tag{4.42}$$

If $g_{dsL} = g_{dsi} = g_o$, then

$$A_d \cong \frac{g_{mi}}{2g_o} \tag{4.43}$$

$$A_{cm} \cong \frac{G}{4g_{mL}} \tag{4.44}$$

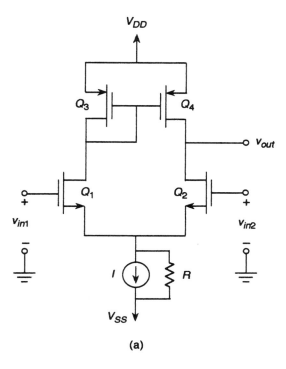

**(a)**

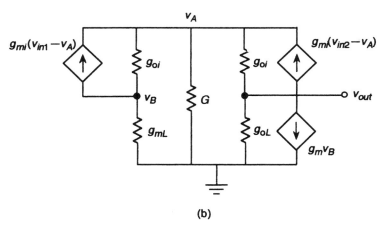

**(b)**

**Figure 4.9** (a) CMOS differential amplifier, (b) equivalent circuit.

so that

$$\text{CMRR} \cong \frac{2g_{mi}g_{mL}}{Gg_o} \qquad (4.45)$$

Also

$$r_{out} \cong \frac{1}{g_{dL} + g_{dsi}} \cong \frac{1}{2g_o} \qquad (4.46)$$

## 4.5  THE TWO-STAGE CMOS OP AMP

In order to increase the gain, a two-stage topology as shown in Figure 4.10 may be used, leading to the two-stage CMOS operational amplifier. It has the following properties:

1. $Q_8$ and $Q_5$ form a current mirror supplying the differential pair $Q_1$, $Q_2$ with its bias current.

2. The aspect ratio of $Q_5$ is chosen to give the required input stage bias.

3. $Q_3$ and $Q_4$ form a current mirror as an active load for the input differential pair.

4. The second stage is a common-source amplifier consisting of $Q_6$ actively-loaded by the current source transistor $Q_7$.

5. The capacitor $C$ is used for frequency compensation.

6. The resistor $R_z$ is usually realized using transistors and is employed to control the location of the zero of the amplifier transfer function.

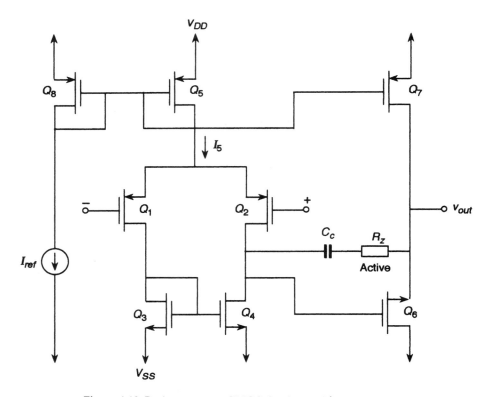

Figure 4.10  Basic two-stage CMOS Op Amp with compensation.

The equivalent circuit of the Op Amp is shown in Figure 4.11 in which $C_1$ is the total capacitance at the interface between the first and second stages, while $C_2$ is the total capacitance at the output node, including the load capacitance $C_L$. Thus $C_1$ and $C_2$ include the various parasitic capacitances discussed in Section 3.4.1. Also, in Figure 4.11

$$G_{mi} = g_{m1} = g_{m2}$$

$$R_1 = r_{02} \| r_{04} = \frac{1}{g_{ds2} + g_{ds4}} = \frac{2}{I_5(\lambda_2 + \lambda_4)}$$

$$G_{m0} = g_{m6}$$

$$R_2 = r_{06} \| r_{07} = \frac{1}{g_{ds6} + g_{ds7}} = \frac{1}{I_6(\lambda_6 + \lambda_7)}$$

(4.47)

where $\lambda_i$ is the channel length modulation parameter for each transistor.

### 4.5.1 The dc Voltage Gain

The dc voltage gain of the input stage is given by (4.35) and (4.37) as

$$A_{v1} = -g_{m1}(r_{02} \| r_{04})$$

(4.48)

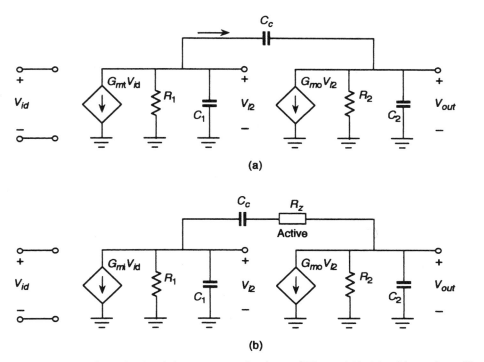

(a)

(b)

**Figure 4.11** Equivalent circuit of the two-stage Op Amp of Figure 4.10: (a) without the nulling resistor; (b) with the nulling resistor.

and the second stage has a gain of

$$A_{v2} = -g_{m6}(r_{06} \| r_{07})$$
(4.49)

In terms of the bias current $I_5$ and the channel length modulation parameters, the differential voltage gain becomes

$$A_{v1} = \frac{g_{md}}{g_{ds2} + g_{ds4}} = \frac{g_{m2}}{g_{ds2} + g_{ds4}} = \frac{2g_{m2}}{I_5(\lambda_2 + \lambda_4)}$$
(4.50)

The voltage gain of the output stage is given by

$$A_{v2} = -\frac{g_{m6}}{g_{ds6} + g_{ds7}} = -\frac{g_{m6}}{I_6(\lambda_6 + \lambda_7)}$$
(4.51)

so that the open loop gain of the operational amplifier is $A_{v1}A_{v2}$ i.e.

$$A_v = \frac{g_{m2}g_{m6}}{(g_{ds2} + g_{ds4})(g_{ds6} + g_{ds7})}$$
(4.52)

or

$$A_v = \frac{2g_{m2}g_{m6}}{I_5 I_6(\lambda_2 + \lambda_4)(\lambda_6 + \lambda_7)}$$
(4.53)

### 4.5.2   The Frequency Response

The frequency response of the operational amplifier in Figure 4.10 can be evaluated using the equivalent circuit of Figure 4.11. Analysis of the circuit is undertaken with the assumptions of $C_1 \ll C_c$, $C_1 \ll C_L$, $C_1 \ll C_2$ and $C_2 \approx C_L$. For the circuit of Figure 4.11(a) This gives:

Poles at

$$s = -\omega_{p1} \cong -\frac{1}{G_{mo}R_2 C_c R_1} \cong -\frac{G_{mi}}{A_v C_c}$$
(4.54)

$$s = -\omega_{p2} \cong -\frac{G_{mo}C_c}{C_1 C_2 + C_c(C_1 + C_2)} \cong -\frac{G_{mo}}{C_L}$$
(4.55)

and a zero at

$$s = z_1 = \frac{G_{mo}}{C_c}$$
(4.56)

To make $\omega_{p1}$ the dominant pole we let it approximate the 3 dB point so that

$$A_v \omega_{p1} \cong A_v \omega_{3dB} = \omega_t$$
(4.57)

or

$$GB = \omega_t = \frac{G_{mi}}{C_c} \qquad (4.58)$$

where $GB = \omega_t$ is the unity gain frequency or the gain-bandwidth product.

Now, since $G_{mi}$ is of the same order as $G_{m0}$, the zero will be close to $\omega_t$, and introduces a phase shift that will decrease the phase margin, thus impairing the amplifier stability. If the zero is placed at least ten times higher than $\omega_t$, then in order to achieve a desirable $60°$ phase margin, the second pole $\omega_2$ must be placed at least 2.2 times higher than $\omega_t$. From these requirements the following relations are obtained:

$$\frac{g_{m6}}{C_c} > 10\left(\frac{g_{m2}}{C_c}\right) \qquad (4.59)$$

i.e.

$$g_{m6} > 10 g_{m2} \qquad (4.60)$$

Also

$$\frac{g_{m6}}{C_L} > 2.2\frac{g_{m2}}{C_c} \qquad (4.61)$$

Combining both relations (4.60) and (4.61) leads to the following condition:

$$C_c > 0.22 C_L \qquad (4.62)$$

### 4.5.3 The Nulling Resistor

Another method for eliminating the effect of the RHP zero is to add the nulling resistor $R_z$ (made up of MOSFET transistors) as indicated in Figure 4.11(b) in series with the compensating capacitor. The new zero location becomes

$$s = z_1 = \frac{1}{C_c\left(\dfrac{1}{G_{m0}} - R_z\right)} \qquad (4.63)$$

which $\to \infty$ as $R_z \to 1/G_{m0}$. A good choice would be $R_z > 1/G_{m0}$, placing the zero on the negative real axis so that it adds to the phase margin.

A consequence of the addition of the nulling resistor is that a third pole results at the location

$$p_3 = -\frac{1}{R_z C_1} \qquad (4.64)$$

Alternatively, the resistor $R_z$ can be chosen such that the RHP zero coincides with the pole $\omega_{p2}$, i.e. a pole-zero cancellation is achieved. Equating (4.55) with (4.63) and solving for $R_z$ leads to:

$$R_z = \frac{1}{g_{m6}} \left( \frac{C_L + C_c}{C_c} \right) \tag{4.65}$$

The resistor $R_z$ is realized as a transistor $Q_8$ operating in the linear (triode) region in which $0 < V_{GS} - V_t < V_{DS}$. Figure 4.12 shows a complete Op Amp circuit including the bias arrangement for $Q_8$ as well as for the entire amplifier in which we have changed the input devices to NMOS type. The value of the nulling resistor is

$$R_z = \frac{1}{g_{ds8}} \tag{4.66}$$

where

$$g_{ds8} = K'_8 \left( \frac{W}{L} \right)_8 (|V_{GS8}| - |V_t|) \tag{4.67}$$

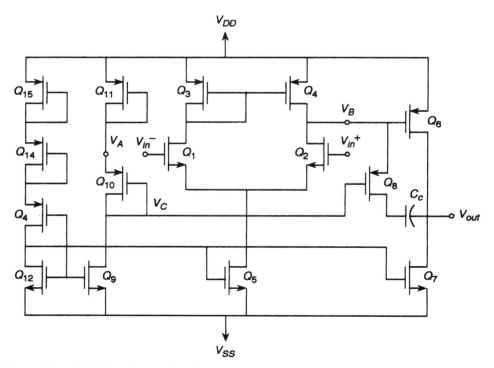

Figure 4.12   CMOS Op Amp with nulling resistor and compensation, including the bias circuits.

The bias circuit for $Q_8$ consists of the devices $Q_9$, $Q_{10}$ and $Q_{11}$ as shown in Figure 4.12. The circuit is designed such that $V_A = V_B$ resulting in

$$|V_{GS10}| - |V_t| = |V_{GS8}| - |V_t| \tag{4.68}$$

Using (2.20) for the transistor in saturation for $Q_{10}$, (4.66) and (4.68) lead to the following design equation:

$$\left(\frac{W}{L}\right)_8 = \left[\left(\frac{W}{L}\right)_6 \left(\frac{W}{L}\right)_{10} \left(\frac{I_6}{I_{10}}\right)\right]^{\frac{1}{2}} \left[\frac{C_C}{C_L + C_C}\right] \tag{4.69}$$

In order to satisfy (4.66), $V_{GS11}$ and $V_{GS6}$ must be equal. This gives the required aspect ratio of $Q_{11}$ as

$$\left(\frac{W}{L}\right)_{11} = \left(\frac{I_{10}}{I_6}\right) \left(\frac{W}{L}\right)_6 \tag{4.70}$$

With the bias circuit for the simulated resistor, the overall power consumption is given as:

$$P_{diss} = (I_5 + I_6 + I_9 + I_{12})(V_{DD} + |V_{SS}|) \tag{4.71}$$

### 4.5.4 The Slew Rate and Settling Time

The slew rate $(S_R)$ of the CMOS Op Amp is limited by the differential stage and is determined by the maximum current that can be sunk or sourced into $C_c$ which is $I_{3(max)} = I_5$. Thus

$$S_R = \frac{I_5}{C_c} \tag{4.72}$$

Therefore, in order to obtain the required slew rate, $I_5$ has to be fixed accordingly, since the value of $C_c$ is determined from the requirement on the phase margin and stability of the amplifier. With

$$\omega_t = \frac{g_{m1}}{C_c}$$

$$= \frac{I_5}{(|V_{GS}| - |V_t|)C_c} \tag{4.73}$$

where $|V_{GS}|$ is the gate-to-source voltage of $Q_1$ and $Q_2$, we have

$$S_R = (|V_{GS}| - |V_t|)\omega_t \tag{4.74}$$

If the slew rate is not specified, but instead the settling time $T_s$ is given, then an equivalent slew rate value is estimated based on the approximation that the amplifier slews half the supply rail voltage at a rate several times faster than the settling time. This leads to the estimate of the slew rate $S'_R$ given by

$$S'_R = \alpha \left( \frac{V_{DD} + |V_{SS}|}{2T_s} \right) \tag{4.75}$$

where $\alpha$ is in the range of 2 to 10. If both $T_s$ and $S_R$ are given then $S'_R$ is calculated and the worst-case value is used to evaluate $I_5$ in (4.72)

### 4.5.5 The Input Common-mode Range and CMRR

These are also determined by the differential stage. For the p-channel input differential stage shown in Figure 4.10, the lowest possible input-voltage (negative CMR) at the gate of $Q_1$ or $Q_2$ is given by $V_{G1(min)} = V_{in(min)}$ as

$$V_{in(min)} = V_{SS} + V_{GS3} + V_{SD1} - V_{SG1} \tag{4.76}$$

In saturation, the minimum value of $V_{SD1}$ is

$$V_{SD1} = V_{SG1} - |V_{t1}| \tag{4.77}$$

so that

$$V_{in(min)} = V_{SS} + V_{GS3} - |V_{t1}| \tag{4.78}$$

Using $V_{GS} = (2I_{DS}/2K)^{\frac{1}{2}} + V_t$ and $2I_1 = I_5$ leads to:

$$V_{in(min)} = V_{SS} + \left( \frac{I_5}{2K_3} \right)^{\frac{1}{2}} + V_{t03} - |V_{t01}| \tag{4.79}$$

where $V_{to}$ is the threshold voltage for $V_{DS} = 0$. In the above expression, the first two terms are determined by the designer. The last ones are fixed by the process and the way the substrate is connected for $Q_1$. Assuming an n-well process with the sources of $Q_1$ and $Q_2$ connected to this well, expression (4.79) becomes

$$V_{in(min)} = V_{SS} + \left( \frac{I_5}{2K_3} \right)^{\frac{1}{2}} + V_{t03(max)} - |V_{t01(min)}| \tag{4.80}$$

in which the worst-case $V_t$ spread as specified by the process, is used by the designer to adjust $I_5$ and $K_3$. In this case, the spread is a high n-channel threshold and a low p-channel threshold.

Similar analysis is used to obtain the highest possible input voltage (positive CMR) as

$$V_{in(max)} = V_{DD} - V_{SD5} - V_{SG1} \tag{4.81}$$

or

$$V_{in(max)} = V_{DD} - V_{SD5} - \left( \frac{I_5}{2K_1} \right)^{\frac{1}{2}} - |V_{t1(max)}| \tag{4.82}$$

The above expressions allow the designer to maximize the common-mode range by making the aspect ratios of $Q_1$ (and $Q_2$) as well as $Q_3$ (and $Q_4$) as large as possible, and minimizing $V_{SD5}$. Also, a small $I_5$ leads to a large CMR.

The corresponding expressions for the n-channel differential input stage (as in Fig. 4.12) are obtained from those above by interchanging $V_{in(min)}$ and $V_{in(max)}$. Thus for the n-channel input differential stage we have

$$V_{in(max)} = V_{DD} - \left(\frac{I_5}{2K_3}\right)^{\frac{1}{2}} - V_{t03(max)} + |V_{t01(min)}| \tag{4.83}$$

$$V_{in(min)} = V_{SS} + V_{SD5} + \left(\frac{I_5}{2K_1}\right)^{\frac{1}{2}} + |V_{t1(max)}| \tag{4.84}$$

The common mode rejection ratio is determined by the input differential stage, and the expression is given by (4.45).

The two stage CMOS Op Amp discussed so far is very popular in the design of switched-capacitor filters. It gives good performance when the loads are capacitive of low value ($<10\,\mathrm{pF}$) which is the case for the Op Amps used in driving on-chip capacitors of the filter itself.

If the two-stage Op Amp is required to drive large loads, such as off-chip capacitors, a buffer stage must be added which provides a low output impedance for the Op Amp. Without this stage, a large capacitive load causes the non-dominant pole to decrease, thus decreasing the phase margin. Also without the buffer stage, a large resistive load will decrease the open-loop gain.

We now give a summary of the design equations of the two-stage Op Amp.

### 4.5.6 Summary of the Two-stage CMOS Op Amp Design Calculations

The process parameters, the supply voltage, and temperature range are fixed conditions for the Op Amp. In addition, the specifications on the performance of the required Op Amp are usually given in terms of specifications of the following parameters:

dc gain $= A_v$

Unity-gain bandwidth $= f_t$

Input common-mode range: $V_{i(max)}, V_{i(min)}$

Load capacitance $C_L$

Slew rate $= S_R$

Settling time $= T_s$

Output voltage swing: $V_{0\,max}, V_{0\,min}$

Power dissipation $= P_{ss}$

The design steps with reference to Figure 4.12 are as follows:

1. Select the smallest channel length that will keep the channel length modulation parameter constant and give good matching for current mirrors.

2. Calculate the minimum compensation capacitor $C_c$ according to the inequality

$$C_c > 0.22 C_L \tag{4.85}$$

the lower limit gives $60°$ phase margin.

Find $I_5$ from the slew-rate and/or settling time requirements as

$$I_5 = \max[(S_R)C_c], \ \alpha \left( \frac{V_{DD} + |V_{SS}|}{2T_s} C_c \right) \tag{4.86}$$

4. For an n-channel differential input stage find $(W/L)_3$ from the specification on the maximum input voltage according to the relation

$$(W/L)_3 = \frac{I_5}{K_3'[V_{DD} - V_{i\,max} - |V_{t03}|_{max} + V_{T_{1\,min}}]^2} \geqslant 1 \tag{4.87}$$

where the indicated lower limit is required to reduce the gate capacitance keeping the effect on the phase margin a minimum.

5. Find $(W/L)_2$ to meet the required value of the unity gain bandwidth $f_t$

$$g_{m2} = \omega_t \cdot C_c \tag{4.88}$$

so that using (2.10) and (2.24)

$$(W/L)_2 = \frac{g_{m2}^2}{K_2' I_5} \tag{4.89}$$

6. For an n-channel input stage, find $(W/L)_5$ from the specification on the minimum input voltage $V_{in(min)}$, in the following two steps

   (a) From (4.80) the saturation voltage of $Q_5$ is

   $$V_{DS5(sat)} = V_{in(min)} - V_{ss} - \left[ \frac{I_5}{2K_1} \right]^{0.5} - V_{t1(max)} \tag{4.90}$$

   (b) Then, the aspect ratio of $Q_5$ is calculated using (2.17) as

   $$(W/L)_5 = \frac{2I_5}{K_5'[V_{DS5}(sat)]^2} \tag{4.91}$$

7. Find $(W/L)_6$ by choosing the second pole $p_2$ to be $= 2.2\omega_t$, i.e. using (4.71) and (4.74) to write

$$g_{m6} = 2.2g_{m2}(C_L/C_C) \tag{4.92}$$

and assuming

$$V_{DS6} = V_{DS6(min)} = V_{DS6}(sat)$$
$$= V_{DD} - V_{out(max)} \tag{4.93}$$

so that

$$(W/L)_6 = \frac{g_{m6}}{K_6' V_{DS6}(sat)} \tag{4.94}$$

8. Find $I_6$ according to the following relation

$$I_6 = \max \left[ \frac{g_{m6}^2}{2K_6'(W/L)_6}, \frac{(W/L)_6}{(W/L)_3} I_1 \right] \tag{4.95}$$

9. Find $(W/L)_7$ to achieve the required current ratios,

$$\frac{(W/L)_7}{(W/L)_5} = I_6/I_5 \tag{4.96}$$

10. Check the gain and power dissipation from

$$A_V = \frac{2g_{m2}g_{m6}}{I_5(\lambda_2 + \lambda_3)I_6(\lambda_6 + \lambda_7)} \tag{4.97}$$

$$P_{diss} = (I_5 + I_6)(V_{DD} + |V_{SS}|) \tag{4.98}$$

11. If the gain specification is not met, then increase $(W/L)_2$ and/or $(W/L)_6$ Alternatively $I_5$ and $I_6$ may be decreased.

12. If the power dissipation is too high, the only solution is to reduce $I_5$ and $I_6$. However, this may require a corresponding increase of some of the $(W/L)$ ratios for satisfying the input and output swings.

### *The bias circuit*

The design of the biasing stage is accomplished as follows. First, $V_{GS5} = V_{GS12}$ is calculated. Secondly, the following condition has to be satisfied:

$$V_{DS15} + V_{DS14} + V_{DS13} + V_{DS12} = V_{DD} + |V_{SS}| \tag{4.99}$$

The aspect ratios of the transistors $Q_{12} - Q_{15}$ and the biasing current $I_{12}$ are chosen such that a suitable configuration is obtained. A reasonable arrangement could be to set the aspect ratios of the p-channel transistors $Q_{13} - Q_{15}$ to one, then calculate the bias current and the aspect ratio of $Q_{12}$.

### The nulling resistor

If compensation for the effect of the RHP zero is desired using the nulling resistor $R_z$, then the procedure is as follows:

1. In order to establish the biasing current (set $V_A$ to $V_B$), $Q_3$ and $Q_{11}$ and their drain currents are matched.

$$\left(\frac{W}{L}\right)_{11} = \left(\frac{W}{L}\right)_3 \qquad (4.100)$$

$$I_{11} = I_{10} = I_3 \qquad (4.101)$$

2. The aspect ratio of $Q_{10}$ is not dependent on the other components, so it will be chosen with minimum value.

3. The aspect ratio of $Q_9$ is determined by the ratios of the two currents $I_5$ and $I_{10}$ as:

$$\left(\frac{W}{L}\right)_9 = \left(\frac{I_{10}}{I_5}\right)\left(\frac{W}{L}\right)_5 \qquad (4.102)$$

4. From (4.68) the aspect ratio of $Q_8$ is determined.

5. Once the compensation circuit is designed, it may be useful to check the location of the RHP zero. First, $V_{GS8}$ is calculated from (2.17) and (4.72) as

$$|V_{GS8}| = |V_{GS10}| = \left[\frac{2I_{10}}{K'_{10}\left(\frac{W}{L}\right)_{10}}\right]^{\frac{1}{2}} + |V_T| \qquad (4.103)$$

Then $R_z$ is calculated using (4.65) and (4.66). This is used to calculate the zero location by (4.63) and if the pole zero cancellation has been successful, this value should be equal to (4.55).

### 4.5.7  A Complete Design Example

Consider the design of a two-stage Op Amp with the following specifications

- $A_v >$ several thousands

- GB=1 MHz

- SR $>3\,\text{V}/\mu\text{s}$

- CMR$= \pm 3$ V

- $C_L = 22.5$ pF

- Supply voltage$= \pm 5$ V

- $V_{out}$ swing$= \pm 4$ V

- $P_{diss} < 10$ mW

1. The process parameters are

$$K'_p = 2.4 \times 10^{-5} \text{ A/V}^2, \qquad K'_n = 5.138 \times 10^{-5} \text{ A/V}^2$$
$$\lambda_p = 0.01 \text{ V}^{-1}, \qquad \lambda_n = 0.02 \text{ V}^{-1}$$
$$V_{tp} = 0.9 \text{ V}, \qquad V_{tn} = 0.865$$

2. A uniform channel length is chosen as $10 \, \mu$m.

3. From (4.85) the minimum value for $C_c$ is calculated as

$$C_c = 0.22 \times C_L = 0.22 \times 22.5 \text{ pF} = 4.95 \text{ pF}$$

and adjusted to 6 pF.

4. From (4.86) the minimum value for the total current $I_5$ is

$$I_5 = S_R \times C_c = 3 \frac{V}{\mu s} \times 6\text{pF} = 18 \, \mu\text{A}$$

5. The aspect ratio of $Q_3$ is calculated using (4.87) as:

$$\left(\frac{W}{L}\right)_3 = \frac{18 \times 10^{-6} \text{ A}}{\left(2.4 \times 10^{-5} \frac{A}{V^2}\right)[5 - 3 - 0.9 + 0.865]^2 \text{ V}^2} = 0.1943$$

which is increased to $(W/L)_3 = 1$.

6. (4.88) is used to obtain the transconductance $g_{m2}$ which is necessary to calculate the aspect ratio of $Q_2$ from (4.89)

$$g_{m2} = GB \times C_c = \left(2 \times 10^6 \pi \frac{1}{s}\right)(6 \times 10^{-12} \text{ F}) = 37.69 \, \mu\text{S}$$

$$\left(\frac{W}{L}\right)_2 = \frac{(37.69 \times 10^{-6} \text{ S})^2}{\left(5.138 \times 10^{-5} \frac{A}{V^2}\right)(18 \times 10^{-6} \text{ A})} = 1.5367$$

7. The saturation voltage of $Q_5$ is given by (4.90) as:

$$V_{DS5}(sat) = \left[ -3 + 5 - \sqrt{\frac{18 \times 10^{-6}\,\text{A}}{\left(5.138 \times 10^{-5}\frac{\text{A}^2}{\text{V}}\right)(1.54)}} - 0.865 \right] \text{V} = 657.53\,\text{mV}$$

The aspect ratio is calculated using (4.91) as

$$\left(\frac{W}{L}\right)_5 = \frac{36 \times 10^{-6}\,\text{A}}{\left(5.138 \times 10^{-5}\,\frac{\text{A}}{\text{V}^2}\right)(657.53 \times 10^{-3}\,\text{V})^2} = 1.6232$$

8. The transconductance $g_{m6}$ and the aspect ratio of $Q_6$ are given by (4.92) and (4.94) as:

$$g_{m6} = (2.2 \times 37.69\,\mu\text{s})\left(\frac{22.5}{6}\right) = 310.94\,\mu\text{s}$$

$$\left(\frac{W}{L}\right)_6 = \frac{311 \times 10^{-6}\,\text{s}}{2.4 \times 10^{-5}\,\dfrac{\text{V}^2}{\text{A}} \times 1} = 12.95$$

9. From (4.95)

$$I_6 = \frac{(311 \times 10^{-6}\,\text{s})^2}{(2)\left(2.4 \times 10^{-5}\,\dfrac{\text{A}^2}{\text{V}}\right)(12.95)} = 155.54$$

or

$$I_6 = \left(\frac{12.95}{1}\right)(9 \times 10^{-6}\,\text{A}) = 116.5\,\text{A}$$

and the larger value is taken.

10. The aspect ratio of $Q_7$ is given by (4.96) as:

$$\left(\frac{W}{L}\right)_7 = \frac{155.5\,\mu\text{A}}{18\,\mu\text{A}} \times 1.62 = 13.97$$

11. The gain obtained is calculated from (4.97) as

$$A_v = \frac{2(37.69 \times 10^{-6}\text{s})(276.32 \times 10^{-6}\text{s})}{(18 \times 10^{-6}\text{A})(0.01 + 0.02)(115.54 \times 10^{-6}\text{s})(0.01 + 0.02)} = 9334$$

The aspect ratios of for the initial design are as follows:

$$\left(\frac{W}{L}\right)_1 = \left(\frac{W}{L}\right)_2 = \frac{15}{10}$$

$$\left(\frac{W}{L}\right)_3 = \left(\frac{W}{L}\right)_4 = \frac{10}{10}$$

$$\left(\frac{W}{L}\right)_5 = \frac{16}{10}$$

$$\left(\frac{W}{L}\right)_6 = \frac{130}{10}$$

$$\left(\frac{W}{L}\right)_7 = \frac{150}{10}$$

The design was optimized by simulation using a program such as CAzM to give the following values:

$$\left(\frac{W}{L}\right)_1 = \left(\frac{W}{L}\right)_2 = \frac{25}{10}$$

$$\left(\frac{W}{L}\right)_3 = \left(\frac{W}{L}\right)_4 = \frac{15}{10}$$

$$\left(\frac{W}{L}\right)_5 = \frac{16}{10}$$

$$\left(\frac{W}{L}\right)_6 = \frac{120}{10}$$

$$\left(\frac{W}{L}\right)_7 = \frac{150}{10}$$

The calculations for the RHP-zero compensation using the nulling resistor are as follows:

1. From (4.100)

$$\left(\frac{W}{L}\right)_{11} = 1.5$$

2. The currents are matched as in (4.101), so that

$$I_{11} = I_{10} = I_3 = 9\,\mu A$$

3. The aspect ratio of $Q_{10}$ is essentially free, so

$$\left(\frac{W}{L}\right)_{10} = 1$$

4. From (4.102)

$$\left(\frac{W}{L}\right)_9 = \frac{9\,\mu A}{18\,\mu A}\,1.6 = 0.8$$

5. The aspect ratio of $Q_8$ is obtained from (4.69) as:

$$\left(\frac{W}{L}\right)_8 = \sqrt{12 \times 1 \times \frac{155\,\mu A}{9\,\mu A}} \times \frac{6\,\text{pF}}{[6 + 22.5]\,\text{pF}} = 3$$

The calculations were modified taking into account the following factors:

● It is important to match the operating points of the differential stage and the output stage. As the expected open-loop gain of the Op Amp is in the range of several thousands, the range of the input signal over which the Op Amp produces a high gain, is very small within a few millivolts. In order to achieve a high overall gain, the output range of the differential stage has to cover the input range of the output stage.

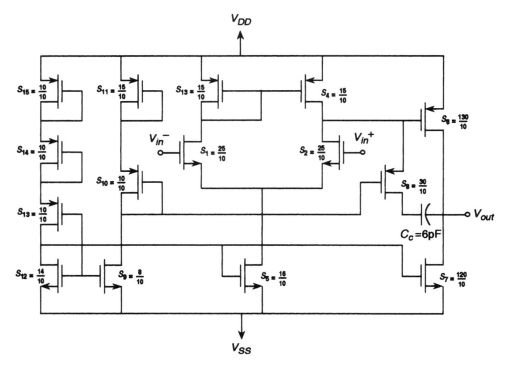

**Figure 4.13**   Op Amp of the example in Section 4.4.7, $S_i = (W/L)_i$.

**Table 4.1**  *Distribution of component areas using MAGIC*

|  | $\mu^2$ | % of total Op Amp area | % of effective Op Amp area |
|---|---|---|---|
| Total Op Amp area | 73 948 | 100.0 | N/A |
| Used Op Amp area | 60 000 | 81.1 | 100.0 |
| Transistors | 10 858 | 14.7 | 18.1 |
| Capacitors | 36 230 | 49.0 | 60.4 |
| Wiring and other | 12 912 | 17.4 | 21.5 |

- Other factors considered are the values of the compensating capacitor and the nulling resistor as they have an impact on slew rate and settling time. The design was modified until satisfactory results concerning these Op Amp characteristics were achieved. This involves several simulations.

The final design is shown in Figure 4.13.

Table 4.1 gives an estimate of the distribution of the component areas having used a layout tool such as MAGIC.

## 4.6  PRACTICAL CONSIDERATIONS AND OTHER NON-IDEAL EFFECTS IN OPERATIONAL AMPLIFIER DESIGN

In addition to the non-ideal effects which have been discussed so far, there are others which the designer should be aware of and attempt to minimize. Some of these effects can be reduced by careful design, but for a considerable improvement, special circuit techniques have to be employed as will be discussed in the next chapter. Here we point out the most important non-ideal effects other than those discussed earlier [31], [32].

### 4.6.1  Power Supply Rejection

The power supply rejection ratio (PSRR) is of great importance in the use of MOS circuits to implement switched-capacitor circuits. First, the clock signals couple to the power supply rails . Second, if digital circuits coexist on the same chip with switched-capacitor circuits, digital noise also couples to the supply lines. If this type of noise couples into the signal paths, they are aliased into the useful frequency bands resulting in degradation of the signal to noise ratio.

In the case of an Op Amp, the PSRR is the ratio of the voltage gain from input to output to the gain from the supply to the output. The basic two-stage Op Amp of Figure 4.10 is particularly prone to high frequency noise from the negative power supply. This is because as the frequency increases, the impedance of the compensation capacitor decreases, effectively connecting the drain of $Q_6$ to its gate. In this case, the gain from the negative supply to the output approaches unity. The same mechanism causes the gain from the positive supply to fall with frequency as the open-loop gain does, therefore the positive PSRR remains relatively constant with frequency. The negative supply PSRR falls to about unity at the unity gain frequency of the amplifier.

### 4.6.2   dc Offset Voltage

This consists of two components, random and systematic. Random offset voltages are a result of mismatches of theoretically identical devices. The systematic type of offset voltage is a result of the circuit design and will always exist even if all devices were perfectly matched. Techniques for reducing dc offset voltages will be discussed in the next chapter.

### 4.6.3   Noise Performance

Due to the relatively high $1/f$ noise of MOS transistors, the noise performance of CMOS amplifiers is an important design consideration. In Figure 4.14(a) the four transistors in the input differential stage contribute to the equivalent input noise as depicted in Figure 4.14(b) Direct calculation of the output noise for each circuit and equating the results, we obtain

$$\langle v_{eq}^2 \rangle = \langle v_{eq1}^2 \rangle + \langle v_{eq2}^2 \rangle + (g_{m3}/g_{m1})(\langle v_{eq3}^2 \rangle + \langle v_{eq4}^2 \rangle) \tag{4.104}$$

in which it is assumed that $g_{m1} = g_{m2}$ and $g_{m3} = g_{m4}$. It follows that the input stage devices contribute directly to the noise, while the contribution of the load devices is reduced by the ratio of their transconductance to that of the input devices.

#### *Input-referred 1/f noise*

The equivalent flicker noise density of a typical MOS transistor is given by (2.53). Then, (4.104) can be used to find for the Op Amp

$$\langle v_{1/f}^2 \rangle = \frac{2K_{fp}}{W_1 L_1 C_{ox}} \left( 1 + \frac{K_{fn}\mu_n L_1^2}{K_{fp}\mu_p L_3^2} \right) \left( \frac{\Delta f}{f} \right) \tag{4.105}$$

where $K_{fp}$ and $K_{fn}$ are the flicker noise coefficients of the p-channel and n-channel, respectively, whose relative values are determined by the process details. The first term is the noise due to the input devices, while the second term is the increase due to the load devices. It is clear that the contribution to the noise by the load devices can be reduced by taking the channel lengths of the load devices longer than that of the input devices. Then the input devices can be made wide enough to achieve the desired performance. Note that, on the other hand. increasing the width of the channel of the load devices does not reduce the $1/f$ noise.

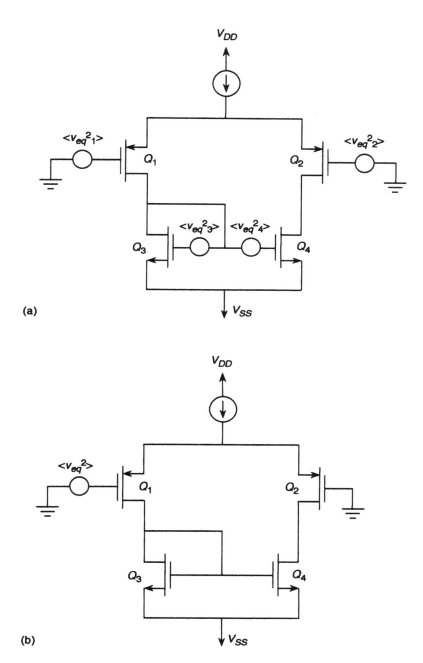

**Figure 4.14**  CMOS Op Amp input stage noise calculations: (a) device contributions; (b) equivalent input noise.

## Thermal noise

The input referred thermal noise of a MOS transistor is given by

$$\langle v_n^2 \rangle = 4kT \left( \frac{2}{3g_m} \right) \Delta f \tag{4.106}$$

Using the same procedure as in the case of flicker noise, we have for the Op Amp:

$$\langle v_{eq}^2 \rangle = 4kT \left( \frac{4/3}{\sqrt{2\mu_p C_{ox}(W/L)_1 I_D}} \right) \left( 1 + \sqrt{\frac{\mu_n(W/L)_3}{\mu_p(W/L)_1}} \right) \tag{4.107}$$

The first term represents the thermal noise from the input devices, and the second term in parentheses represents the increase in noise due to the loads. The latter will be small if the aspect ratios are chosen such that the transconductance of the input devices is much larger than that of the loads. Under this condition, the input noise is determined by the transconductance of the input devices.

## CONCLUSION

The design of two-stage CMOS operational amplifiers was the focus of the discussion in this chapter, together with the inherent non-ideal effects. In switched-capacitor filter design, Op Amp design techniques occupy much time and effort. In particular, the quest for high performance Op Amp designs is a major occupation of integrated circuit design engineers. The relatively simple design of this chapter is satisfactory in many applications, and it also forms the basis for other more elaborate techniques, which will be discussed in the next chapter.

## PROBLEMS

4.1   A basic differential pair of the type shown in Figure 4.6 employs NMOS transistors with $r_{ds} = 100\,\Omega$, $K' = 100\,\mu A/V^2$ and $V_t = 1\,V$. Calculate the differential gain for bias current values of $50\,\mu A$, $100\,\mu A$ and $200\,\mu A$.

4.2   Design the input differential stage of an operational amplifier to operate at $V_{GS} = 1.3\,V$ and provide a transconductance of $0.1\,mA/V$. For the devices, $V_t = 1\,V$, $K' = 10\,\mu A/V^2$. For the design, it is required to find the aspect ratios of the devices as well as the bias current.

4.3   Derive the exact expression for the transfer function of the two-stage Op Amp of Figure 4.10. Show that the pole and zero locations are approximately as given in Section 4.5.2 under the assumptions made.

4.4  Design a two-stage CMOS Op Amp of the type shown in Figure 4.10 with the following specifications:

Low-frequency gain >2000

Settling time = 2 $\mu s$

Unity gain frequency = 1 MHz

Load capacitance = 10 pF

Supply voltage= ±5 V

$CMR = \pm 4$ V

Output swing = ±4 V

Power dissipation < 20 mW

Use the device parameters as:

$$K'_p = 20\,\mu A/V^2, \qquad K'_n = 50\,\mu A/V^2$$
$$\lambda_p = 0.01\,V^{-1}, \qquad \lambda_n = 0.01\,V^{-1}$$

# 5

# High-performance
# Operational Amplifiers

## 5.1  INTRODUCTION

Although the basic two-stage Op Amp discussed in the previous chapter has a performance which is satisfactory in many applications, it suffers from a number of disadvantages which were pointed out in Section 4.5. Improvements in the performance with regards to one or more of these non-ideal effects, such as obtaining higher gain, better PSRR, reduced offset voltage, lower noise, better settling time and slew rate, require the use of special techniques which may entail a modification of the Op Amp structure. These techniques are discussed in this chapter.

## 5.2  CASCODE CMOS OP AMPS

Figure 5.1 shows the first stage of a cascode Op Amp constructed with the main objective of increasing the gain. The two *common-gate* transistors $Q_{1c}$ and $Q_{2c}$ form the *cascode* for the differential pair $Q_1$, $Q_2$. The output resistance at $Q_{2c}$ is

$$R_{02c} \cong g_{m2c} r_{02c} r_{02} \tag{5.1}$$

which is typically $\approx 100$ times greater than the value without the cascode transistors. Naturally, to make full use of the high output resistance, we must also increase the active load resistance; thus a Wilson current mirror is used in Figure 5.1. The output resistance of the Wilson mirror was given by (3.51 )as

$$R_{04c} \cong g_{m4c} r_{04c} r_{03} \tag{5.2}$$

Therefore, the output resistance of the stage is

$$R_0 = R_{02c} \| R_{04c}$$
$$= (g_{m2c} r_{02c} r_{02}) \| (g_{m4c} r_{04c} r_{03}) \tag{5.3}$$

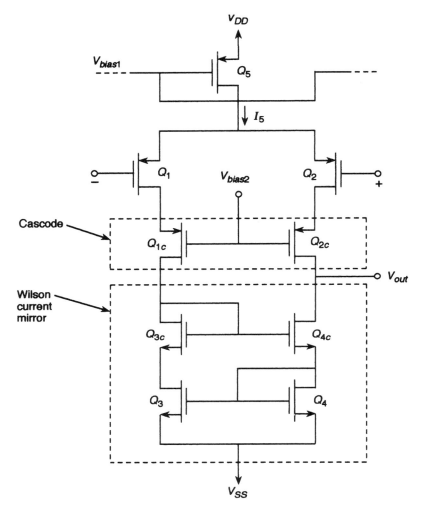

**Figure 5.1**   Input stage of a CMOS Op Amp employing the cascode configuration.

But the voltage gain of the stage is

$$A_1 = -g_m R_0 \tag{5.4}$$

Therefore, the increase in $R_0$ is reflected as an increase in $A_1$ by the same factor. Gains of $\approx 5000 - 10^4$ are possible with this topology. In fact, since the gain from this single stage is quite high, a single stage Op Amp is possible using a modification of the cascode idea. This is discussed below and also leads to a high power supply rejection ratio and better input common-mode range.

## 5.3   THE FOLDED CASCODE OP AMP

Starting from Figure 5.1, each of the six transistors below $Q_1$, $Q_2$ is replaced by its complement and the entire group is disconnected from $V_{SS}$, and folded over to be connected to $V_{DD}$, resulting in the folded cascode of Figure 5.2. The operation of the circuit is similar to the simple cascode except that the input common-mode range is larger since only three transistors are stacked in the input between the two power supplies by comparison with five in the original cascode. $Q_6$ and $Q_7$ are added current sources.

The voltage gain of the folded cascode is given by

$$A = g_{m1} R_0 \tag{5.5}$$

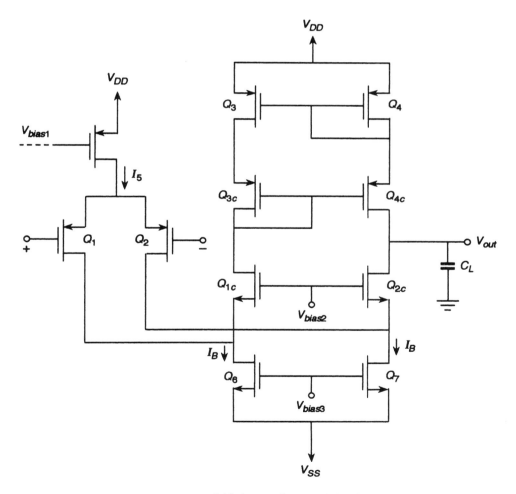

**Figure 5.2**   A folded, cascode CMOS Op Amp.

where $R_0$ is its output resistance given by

$$R_0 = R_{02c} \| R_{04c}$$

$$= [g_{m2c} r_{02c} (r_{07} \| r_{02})] \| [g_{m4c} r_{04c} r_{03}]$$

(5.6)

and due to the high gain value, the circuit can be used as a single stage Op Amp. In fact, if the devices are matched in pairs, with $Q_6$ and $Q_7$ used to establish $I_B = I_5$, the gain of the folded cascode can be put in the explicit form

$$A = \frac{|V_A|^2}{I_5} \frac{\sqrt{\mu_n \mu_p} C_{ox} \sqrt{W_1/L_1}}{3/\sqrt{(W_{2c}/L_{2c})} + \sqrt{\mu_u/\mu_p}/\sqrt{(W_{4c}/L_{4c})}}$$

(5.7)

The dominant pole is determined by the total capacitance at the output node $C_L$ which includes the load capacitance. This is given by

$$\omega_d = 1/R_0 C_L$$

(5.8)

and the unity gain frequency is

$$\omega_t = A\omega_d$$

$$= \frac{g_{m1}}{C_L}$$

(5.9)

The folded cascode has a better power supply rejection ratio than the two-stage Op Amp since the compensation capacitor and load capacitor are the same element in this case. Assuming that the load capacitance or part of it is not connected to the power supply, the circuit does not suffer from the degradation of the high frequency power supply rejection problem inherent in the compensated two-stage Op Amp discussed in Chapter 4. However, due to the fact that cascode transistors are used at the output, the output swing of this circuit is lower than that of the two-stage amplifier; this disadvantage will be remedied in a later section. This Op Amp structure is commonly used in the design of high-frequency switched-capacitor filters. We shall have more to say about this shortly.

## 5.4   LOW-NOISE OPERATIONAL AMPLIFIERS

The signal-to noise ratio (S/N) is of primary importance in communication circuits. This is closely related to the dynamic range which is the ratio of the maximum to the minimum signals which the circuit can process without distortion. The maximum is usually determined by the power supplies and the large signal swing limits. The minimum is determined by the noise or ripple injected by the power supply. For example, suppose we require an 80 dB dynamic range with a $\pm 5$ V power supply and swing of $\pm 1$ V of the supplies. The maximum signal of 4 V is equivalent to 2.83 V r.m.s. Dividing by $10^4$ (equivalent to 80 dB) gives 283 $\mu$V r.m.s. With a 10 kHz flat bandwidth, we obtain the noise value of 2.83 $\mu$V/(Hz)$^{1/2}$.

In the design of Op Amps for good noise performance, two approaches are possible. The first consists in optimizing the device geometries and characteristics to yield the lowest noise possible; the relevant concepts were presented in Section 4.5.3. The second approach uses independent techniques such as *correlated double sampling* and *chopper stabilization* which also reduce the input offset voltage. We first give an example of the first approach, then introduce the new techniques in the following subsections.

### 5.4.1   Low-noise Design by Control of Device Geometries

A low-noise amplifier [33] is shown in Figure 5.3(a) which is the two-stage design of Figures 4.10 and 4.11 with cascode devices $Q_8$ and $Q_9$ to improve the PSRR. The input differential stage is composed of PMOS devices due to their better noise performance. The noise model of this Op Amp is shown in Figure 5.3(b) in which the noise due to the dc current sources is ignored since their gates are usually connected to a low impedance . Also the contribution to the noise by the noise source at the gates of $Q_8$ and $Q_9$ is neglected due to the large impedances seen by the sources. Thus, the total output noise spectral density is

$$\langle v_n^2 \rangle = g_{m6}^2 R_2^2 [\langle v_{n6}^2 \rangle + R_1^2 (g_{m1}^2 \langle v_{n1}^2 \rangle + g_{m3}^2 \langle v_{n3}^2 \rangle + g_{m4}^2 \langle v_{n4}^2 \rangle)] \qquad (5.10)$$

where $R_1$ and $R_2$ are the output resistances of the first and second stages respectively. The equivalent input-referred noise spectral density is obtained from the above expression by dividing by the squared differential gain: $g_{m1} R_1 g_{m6} R_2$ to obtain

$$\langle v_{eq}^2 \rangle = \langle v_{n6}^2 \rangle / g_{m1}^2 R_1^2 + 2 \langle v_{n1}^2 \rangle [1 + (g_{m3}/g_{m1})^2 (v_{n3}/v_{n1})^2] \qquad (5.11)$$

from which the noise contribution by the second stage is divided by the gain of the first, and may therefore be neglected.

Now, in order to minimize the noise in (5.11), we take $g_{m1} > g_{m3}$ so that the input noise is dominated by the input devices which have inherently low noise. The thermal noise contribution can be reduced by increasing the transconductance of the input devices. This can be achieved by increasing the drain current and/or the aspect ratios. The $1/f$ noise can be reduced by reducing both $W$ and $L$. For the circuit of Figure 5.3(a), the dominant pole is at 100 Hz. With a flat noise of 130 nV/(Hz)$^{1/2}$ over 100 Hz, we have a noise voltage of 13 $\mu$V r.m.s. With a peak voltage swing of 4.3 V, we obtain a dynamic range of 107 dB [33].

### 5.4.2   Noise Reduction by Correlated Double Sampling

This is a technique for reducing the $1/f$ noise density at low frequencies, as illustrated in Figure 5.4 [12]. The amplitude spectrum of the noise is multiplied by a function with amplitude $2\sin(\omega T/4)$. This suppresses the noise at zero frequency and at even multiples of the sampling frequency. The viability of this technique depends on the practicability of integrated circuit implementation of the sample-and-hold circuit and the summer without imposing undue demands on the time response of the Op Amp.

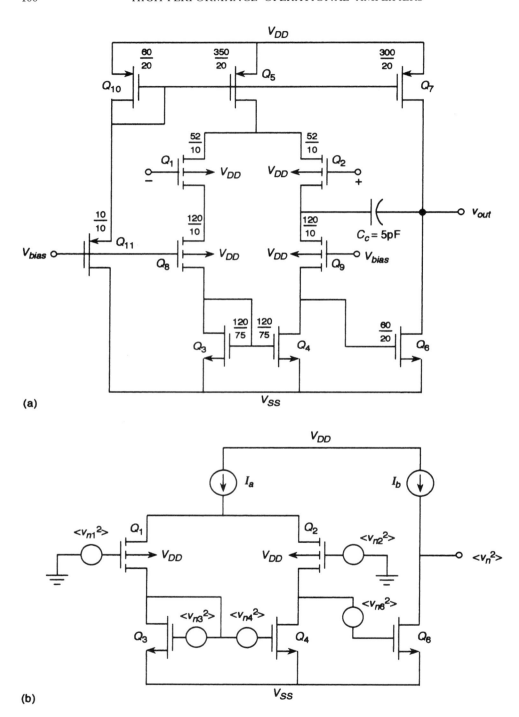

**Figure 5.3**   (a) Example of a low-noise CMOS Op Amp, (b) noise model.

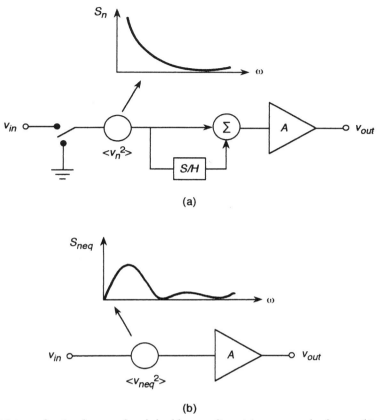

**Figure 5.4** Noise reduction by correlated double sampling: (a) conceptual scheme; (b) equivalent input noise.

### 5.4.3 Chopper-stabilized Operational amplifiers

This technique can be used in conjunction with any Op Amp design to reduce the input offset voltage and $1/f$ noise. The basic concept is illustrated in Figure 5.5 [12], [34] as applied to a two-stage amplifier. $V_{in}$ is the input signal spectrum and $V_n$ is the undesirable noise spectrum. The two multipliers are driven by a chopping square wave of amplitude $\pm 1$V. Figure 5.5 illustrates clearly the results of the chopping operation and if the chopping frequency is taken to be sufficiently higher than the baseband signal $V_{in}$, then the spectrum of the undesirable signal will be shifted to a location well outside the baseband. This unwanted signal is composed of the dc offset and the $1/f$ noise. Thus the effect of these signals on the performance is reduced.

Figure 5.6(a) shows the implementation of chopper stabilization as applied to a CMOS Op Amp. The multipliers are realized by the switches which are controlled by a two-phase clock. With $\phi_1$ on and $\phi_2$ off, we have from Figure 5.6(b)

$$V_{neq}(\phi_1) = V_{n1} + V_{n2}/A_1 \qquad (5.12)$$

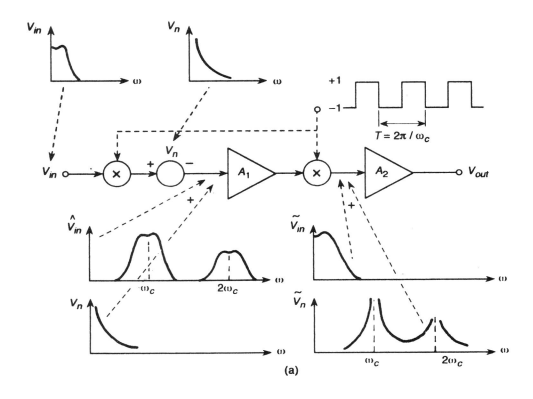

(a)

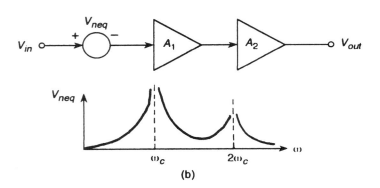

(b)

**Figure 5.5** Noise reduction by chopper stabilization.

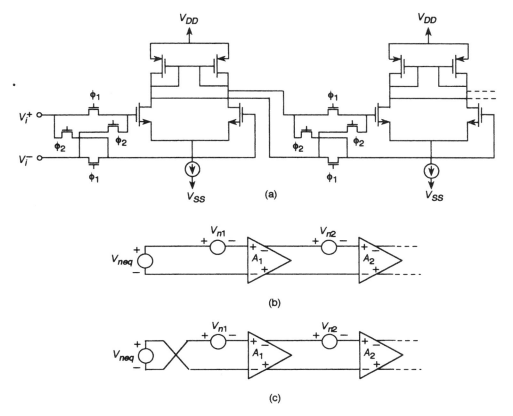

**Figure 5.6**  Application of chopper stabilization to a CMOS Op Amp: (a) circuit; (b) conditions with $\phi_1$ ON and $\phi_2$ OFF; (c) conditions with $\phi_1$ OFF and $\phi_2$ ON.

and when $\phi_1$ is off and $\phi_2$ is on we have from Figure 5.6(c)

$$V_{neq}(\phi_2) = -V_{n1} + V_{n2}/A_1 \tag{5.13}$$

Thus, the average value of the input-referred noise over one period is

$$V_{neq}(av) = 1/2[V_{neq}(\phi_1) + V_{neq}(\phi_2)]$$
$$= V_{n2}/A_1 \tag{5.14}$$

Hence, the equivalent undesirable signal, in particular the $1/f$ noise is cancelled. Also, if $A_1$ is sufficiently high, the second-stage contribution to the noise is reduced.

## 5.5   HIGH-FREQUENCY OPERATIONAL AMPLIFIERS

At high frequencies, which in turn require higher sampling rates, the operational amplifiers have two important requirements: high gain and fast settling time. The former requirement has been discussed, and we now turn to the problem of designing amplifiers with fast settling time [12], [34].

### Settling time considerations

Consider a two-stage Op Amp of the type discussed in Section 4.4 with identical input devices having the same drain resistance and transconductance. We have seen that it has a dominant pole at $-1/r_o g_m C_c r_o$ and a nondominant pole at $-g_m/C_L$.

For a single-stage cascode Op Amp of the type described in Section 5.2, the dominant pole is at

$$s_d = -1/r_o C_L g_m r_o \tag{5.16}$$

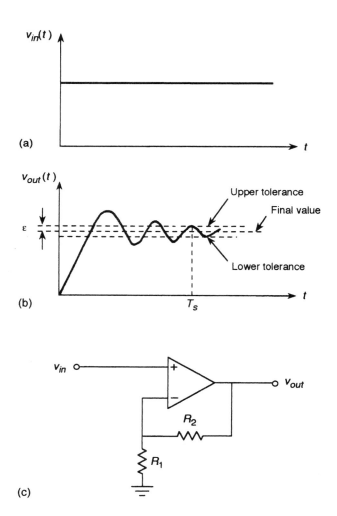

**Figure 5.7**    Settling time: (a) input step; (b) output; (c) circuit for measuring $T_s$.

while the non-dominant pole is at

$$s_n = -g_m/C_p \tag{5.17}$$

where $C_p$ denotes the total capacitance at the cascode node.

Now, if these amplifiers are used in a closed-loop configuration, as they are invariably employed in switched-capacitor filters, then by reference to Figure 5.7, it can be shown [11] that the settling time $T_s$ is determined by the non-dominant pole $s_n$. Specifically, as the loop gain increases, $s_d$ and $s_n$ converge to form a complex pair at

$$s = -\tfrac{1}{2}s_n \tag{5.18}$$

so that,

$$T_s \sim 2/\text{Re}(s_n) \tag{5.19}$$

Thus, comparing the pertinent expressions for the poles of the two-stage and single-stage cascode amplifiers we have

$$\frac{T_s(\text{two-stage})}{T_s(\text{single-stage, cascode})} \sim \frac{C_L}{C_p} \tag{5.20}$$

But $C_p$ is of the order of 0.1–0.2 $C_L$ so that

$$\frac{T_s(\text{two-stage})}{T_s(\text{single-stage, cascode})} \sim 5\text{--}10 \tag{5.21}$$

It follows that the single-stage cascode has faster settling time by a factor of 5 to 10 than the two-stage design.

*From the above considerations, it follows that in order to obtain high gain and high speed, the folded cascode design of Section 5.2 is a good choice.*

## 5.6 FULLY DIFFERENTIAL BALANCED TOPOLOGY

Now, the single-ended operational amplifier architectures discussed so far have disadvantages when the noise from power lines and adjacent digital and switching circuits is to be minimized. This is particularly the case when digital and analog circuits exist on the same chip, but is true in general if power supply rejection is of prime importance. In these situations, the use of a fully differential balanced topology is advantageous. Figure 5.8 shows the three main types of Op Amps: (a) single-ended output; (b) differential output; and (c) balanced fully differential output. The latter is the most useful when the noise considerations discussed above are of importance. Note that the balanced Op Amp is such that the two outputs, relative to ground, are accurately balanced. Thus, such an amplifier needs a fifth terminal to act as a reference for balancing the output. Figures 5.9 and 5.10 show two possible implementations of the balanced Op Amp. Figure 5.9 is basically a single-ended Op Amp together

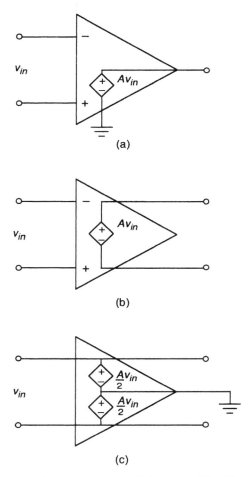

**Figure 5.8** Three types of Op Amps: (a) single-ended output; (b) differential output; (c) balanced fully differential output.

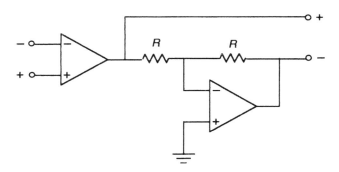

**Figure 5.9** A differential output balanced Op Amp using an Op Amp and an inverter.

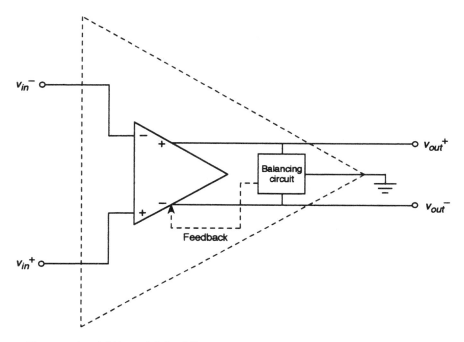

**Figure 5.10**   A balanced fully differential Op Amp using common-mode feedback.

with an inverter. Since both are available as standard cells in many IC design libraries, this is a relatively straightforward realization. However, it has the disadvantage that at high frequencies, the phase shift introduced by the inverter may destroy the balance of the two outputs.

The balanced design of Figure 5.10 requires a differential output Op Amp as well as additional circuits to sense the common-mode (average) output then compares it to ground and applies feedback to correct for its deviation from the desired zero value. Thus, this topology is symmetric and at high frequencies additional phase shift is present with equal amounts at both outputs and proper balancing is maintained.

Before giving a complete design for a balanced fully-differential Op Amp, we summarize the advantages of such a design:

1.  Noise from the power supply lines appears as a common-mode signal, and with proper design, can be reduced.

2.  The effective output swing is doubled, while retaining the same input circuit. This results in an increase of the dynamic range by about 6dB relative to the single-ended case.

3.  When the fully balanced differential-output Op Amp is used in conjunction with switches and capacitors to form switched-capacitor circuits, this results in a reduction of the clock feedthrough noise since it appears as a common-mode signal. This is a distinct advantage at high frequencies because clock feedthrough becomes more pronounced since we have to increase the device sizes to reduce

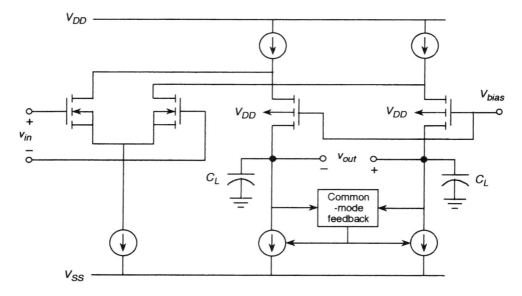

**Figure 5.11**   Details of the concept illustrated in Fig. 5.10.

charging time constants. This, in turn, increases the charge injection into the signal paths. The particulars of this advantage will be discussed at the appropriate point in the book.

4. Systematic offset voltages are reduced.

5. The chopper-stabilization technique can be used in conjunction with the fully differential topology to reduce the $1/f$ noise, thus resulting in a superior performance suitable for use in high-frequency high-precision VLSI communication circuits.

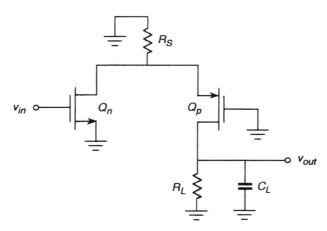

**Figure 5.12**   Small-signal equivalent differential half-circuit of the Op Amp in Figure 4.22.

Having discussed the basic ideas and advantages of the fully differential balanced design, we consider the implementation of the idea as shown schematically in Figure 5.11 [34].

First note that $C_L$ functions as a compensation capacitor and consider the small-signal differential half-circuit shown in Figure 5.12 which gives for the gain

$$A_v = -(g_m r_{on})(g_{mp} r_{op})[1/1 + (r_{on}/R_s) + (r_{on}/R_{on})][1/1 + (r_{on}/R_L)] \qquad (5.22)$$

where

$$R_{on} = (1/g_{mp})[1 + (R_L/r_{op})] \qquad (5.23)$$

is the effective resistance looking into the source terminal of the p-channel cascode device.

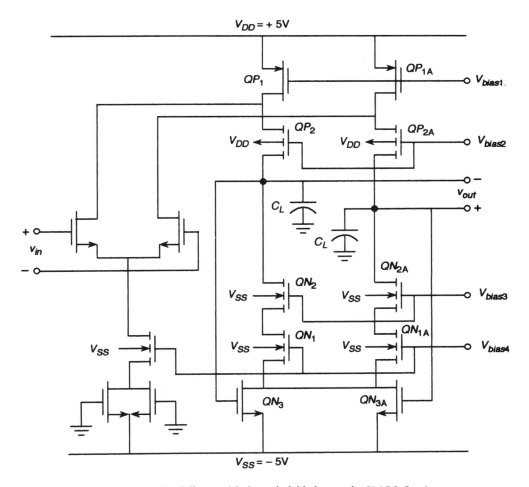

**Figure 5.13**   Fully differential balanced, folded cascode CMOS Op Amp.

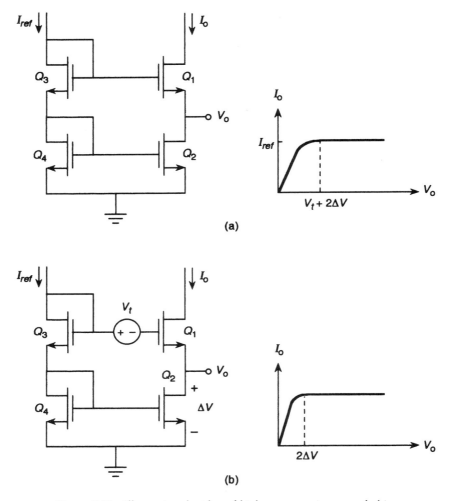

**Figure 5.14**   Illustrating the idea of high output swing cascode bias.

Now, Figure 5.13 shows a fully differential folded cascode Op Amp with common-mode feedback balancing network and the following points are helpful in explaining its operation:

(a) Transistors $QP_1$ and $QP_{1A}$ supply the bias current to the amplifier.

(b) $QP_2$ and $QP_{2A}$ are the cascode elements.

(c) The channel length of $QP_1$ and $QP_{1A}$ is taken longer than that of $QP_2$ and $QP_{2A}$ in order to keep the output resistance relatively high.

(d) Transistors $QN_1$, $QN_{1A}$, $QN_2$ and $QN_{2A}$ realize the high impedance current source loads. The resulting output resistance is of the order of 16 M$\Omega$.

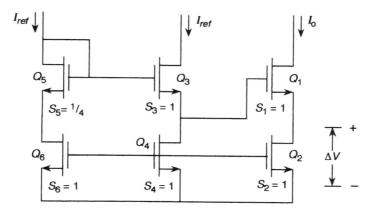

**Figure 5.15**  Implementation of the high-swing cascode bias.

(e) The common-mode feedback network is composed of the transistors $QN_3$ and $QN_{3A}$. These sample the common-mode output signal and feed back a correc- tive common-mode signal into the source terminals of $QN_2$ and $QN_{2A}$. This compensating signal is amplified by the cascode elements ($QN_1$, $QN_{1A}$, $QN_2$, and $QN_{2A}$) to restore the common-mode output voltage to its required original level (ground). Thus the CMFB is essential for the precise definition of the common-mode output.

There is one drawback of this cascode amplifier, namely the reduced output swing. To appreciate this point, consider Figure 5.14(a). If the cascode devices $Q_1$ and $Q_2$ are biased from $Q_3$ and $Q_4$, the voltage across the cascode can swing only within $V_t + 2V_{Dsat}$ from the negative supply line, before $Q_1$ goes into the triode region. The swing can be improved by inserting a level-shifting dc source of strength $V_t$ between the gates of $Q_1$ and $Q_3$ as shown in Figure 5.14(b). This forces $Q_2$ to be biased at the edge of saturation with $V_{DS} = V_{Dsat}$. In this case, the voltage across the cascode can swing to $2V_{Dsat}$ from the negative supply rail before $Q_1$ is pulled out of saturation. Figure 5.15 shows a practical realization of the high-swing cascode circuit. All devices conduct equal biasing currents $I_o$. They all have the same aspect ratio except $Q_5$ for which $(W/L) = 0.25$ of the others. This results in $Q_2$ being biased at the edge of saturation with $V_{DS} = V_{Dsat}$. Under these conditions, the voltage across $Q_1$ and $Q_2$ can swing to within $2\Delta V$ of the negative supply voltage.

## CONCLUSION

This chapter introduced special design techniques of operational amplifiers with the objective of improved performance with regard to one or more criterion such as high gain, low noise, and fast settling time. In particular we note that the folded cascode balanced fully differential design and the use of chopper stabilization result in an excellent Op Amp suitable for high precision applications in communication VLSI circuits.

# PROBLEMS

5.1 A CMOS operational amplifier is to be designed in the folded cascode configuration of the type shown in Figure 5.2 according to the following specifications:

Supply voltages $\pm 5$V

Output voltage swing $\pm 1$V

Input common mode range $\pm 1.5$V

dc gain $\geqslant 70$ dB

For the design, an n-well process is used with minimum channel length of $10\,\mu$m, and the process parameters are: $K_n = 50\,\mu\text{A/V}^2$, $K_p = 20\,\mu\text{A/V}^2$, $V_{tn} = 1$V, $V_{tp} = -1$V, $\lambda = 0.01\,\text{V}^{-1}$ and $\chi = 0.1\,\mu$m/V.

5.2 Calculate the equivalent input noise voltage for the amplifier designed in Problem 5.1 at 100 Hz, 1 kHz and 50 kHz.

5.3 Apply chopper stabilization to the design of Problem 5.1 using two identical stages, with a chopping frequency of 20 MHz. Calculate the equivalent input noise voltage at the same frequencies as Problem 5.2 for the design and compare with that obtained without applying the technique.

5.4 Convert the Op Amp designed in Problem 5.3 into a fully differential design using an additional inverter.

# 6

# Capacitors, Switches and the Occasional Passive Resistor

## 6.1  INTRODUCTION

Switched-capacitor circuits are constructed using operational amplifiers, capacitors and analog switches. We have dealt with CMOS Op Amps in Chapters 4 and 5, and it now remains to examine the design of the other building blocks. This chapter presents the various integrated circuit versions of capacitors and switches and discusses the corresponding non-ideal effects in these components in relation to their use in switched-capacitor circuits. Furthermore, resistors are sometimes also required as on-chip components. These can be realized as active devices, as explained in Chapter 3; alternatively some situations call for the high degree of linearity associated with passive components. The possible passive resistor MOS structures are also reviewed.

## 6.2  MOS CAPACITORS

### 6.2.1  Capacitor Structures

The second building block of integrated switched-capacitor filters is the MOS capacitor. The dielectric used is almost invariably $S_iO_2$ which is a very stable insulator with $\epsilon_{0x} \cong 3.9$ and a high breakdown electric field of about $8 \times 10^6$ V/cm. The choice of electrodes for the capacitor varies according to the available technology that is used for the fabrication of the entire integrated circuit. This leads to the following types:

1. *Metal (or polysilicon) over diffusion structure*. This is shown in Figure 6.1(a) and is formed by growing a thin oxide $S_iO_2$ layer over a heavily-doped region in the substrate. In a metal gate process, the top plate of the capacitor is formed by covering the $S_iO_2$ with metal in the same processing step of providing the metalization for the gate and leads of the entire circuit. In a silicon gate process, heavily-doped polycrystalline silicon (polysilicon) is used as the gate electrode and also to form the top plate of the

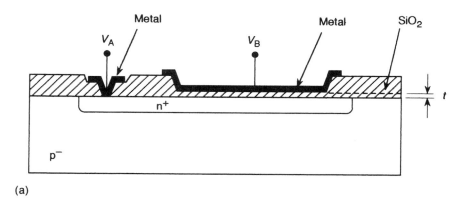

(a)

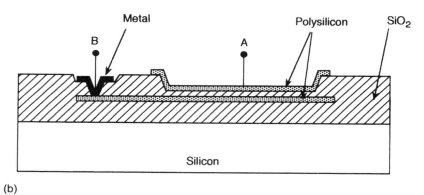

(b)

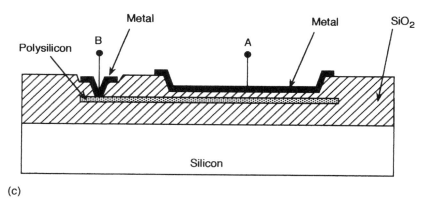

(c)

**Figure 6.1**    MOS capacitors: (a) metal (or polysilicon) over diffusion; (b) polysilicon over polysilicon; (c) metal over polysilicon.

capacitor. Ideally, if the electrodes are perfect conductors, the capacitance per unit area is

$$C_0 = \epsilon_{ox}/t \tag{6.1}$$

However, the actual capacitance is voltage dependent and is of the form

$$C = C_0[1 + b(V_A - V_B)]^{-\frac{1}{2}} \tag{6.2}$$

where $b$ is a constant, inversely proportioned to the doping density. For heavily doped $n^+$ layers, the voltage dependence of the capacitor is slight. Achieved capacitance values using this structure are in the range of 0.35 to 0.5 fF/$\mu$m$^2$.

The tolerance on the value of the capacitance is usually $\approx \pm 15\%$. However, two identical capacitance values can be matched to within 0.1–1%.

2. *Polysilicon-over-polysilicon capacitors.* In a silicon gate 'poly-poly' process, a second layer of high conductivity polysilicon is available for use in forming inter-connect elements. These two layers can be also used as the top and bottom plates of a capacitor, as shown in Figure 6.1(b). A disadvantage of this structure is that, due to the irregularity of the polysilicon surface, the oxide thickness has a large random variation. Typical values achieved are 0.3 to 0.4 fF/$\mu$m$^2$.

3. *Metal-over-polysilicon capacitors.* This is shown in Figure 6.1(c), and its properties are similar to those of Figure 6.1(b).

### 6.2.2  Parasitic Capacitances

Parasitic capacitances are inherent in all the MOS capacitor structures discussed above. Figure 6.2 shows a model of the MOS capacitor including the main parasi-tics. There is an inevitable large parasitic capacitance from the bottom plate to the substrate, and hence to the substrate bias supply. For a metal (or polysilicon) over diffusion capacitor, the bottom plate is embedded into the substrate, and the stray bottom plate parasitic can be $\approx$ 15–30% of the nominal required value. For the

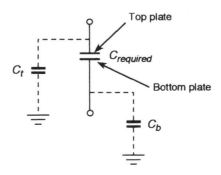

**Figure 6.2**  MOS capacitor model showing the parasitic capacitance.

structure of Figure 6.1(b)–6.1(c), this stray capacitance is of the order of 5–20% of the required capacitance.

Another source of stray capacitance is due to the leads used to connect the capacitor plates in the rest of the circuit. This is the source of the top plate parasitic capacitance. Furthermore, in switched-capacitor filters, one or both plates of a capacitor are connected to the source or drain diffusion of a MOS switch. The pn junction of the diffusion contributes a depletion layer capacitance between the capacitor plate and substrate. All these parasitics depend on the capacitor size as well as the fabrication technology and layout.

### 6.2.3  Capacitor-ratio Errors

The performance of switched-capacitor filters is determined by capacitor ratios, therefore the errors involved in these ratios constitute an important design consideration. The sources of these errors are now discussed briefly.

#### *Random edge variations*

Figure 6.3 shows a top view of a MOS capacitor, in which the edges of the electrodes exhibit random variations. The nominal capacitance value is

$$C = \frac{\epsilon A}{t_{ox}} \tag{6.3}$$

and due to the random variation in the area we have

$$\Delta C = \frac{\epsilon}{t_{ox}}[(W + \Delta W)(L + \Delta L) - WL] \tag{6.4}$$

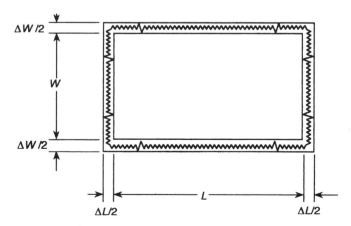

**Figure 6.3**  Schematic of random edge variations in a MOS capacitor.

so that

$$\frac{\Delta C}{C} = \frac{\Delta W}{W} + \frac{\Delta L}{L} \tag{6.5}$$

It can be assumed that $\Delta W$ and $\Delta L$ are independent random variables with equal standard deviation $\sigma_L = \sigma_W$. This leads to the standard deviation of $\Delta C$ being

$$\sigma_c = C\sigma_L \sqrt{W^{-2} + L^{-2}} \tag{6.6}$$

But $C$ (and hence $WL$) is fixed, therefore the relative error $\sigma_c/C$ is a minimum for $W = L$. In this case, the relative capacitance error is

$$\left.\frac{\sigma_c}{C}\right|_{\min} = \sqrt{2}\,\sigma_L/L \quad \text{for } W = L \tag{6.7}$$

which means that the capacitor shape should be square.

The above considerations hold for capacitor ratios also. Suppose the nominal capacitor ratio is

$$\alpha = \frac{C_1}{C_2} \geqslant 1 \tag{6.8}$$

and the dimensions of $C_1$ are $W_1$, $L_1$ while those of $C_2$ are $W_2$, $L_2$. Assuming all dimensions to have the same standard deviation $\sigma$, then

$$\frac{\sigma_\alpha}{\alpha} = \sigma\sqrt{L_1^{-2} + W_1^{-2} + L_2^{-2} + W_2^{-2}} \tag{6.9}$$

which is a minimum if

$$L_1 = W_1 = \sqrt{\alpha}\,L_2 = \sqrt{\alpha}\,W_2 \tag{6.10}$$

with a minimum value of

$$\left.\frac{\sigma_\alpha}{\alpha}\right|_{\min} = \left(\frac{\sqrt{2}\sigma}{L_1}\right)\sqrt{1 + \alpha} \tag{6.11}$$

which implies that for best accuracy, the capacitor ratio should be 1.

### Undercut error

This results from the uncontrollable lateral etching of the plates of the capacitor along its perimeter in the fabrication process as shown in Figure 6.4. This gives rise to a decrease in the value of the capacitance, proportional to the perimeter. Again with the required capacitor ratio

$$\alpha_0 = \frac{C_1}{C_2} = \frac{W_1 L_1}{W_2 L_2} \tag{6.12}$$

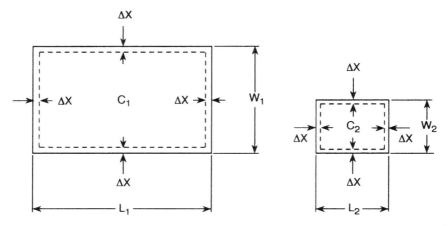

**Figure 6.4**   Undercut error.

the undercut results in the actual ratio being

$$\alpha \cong \frac{W_1 L_1 - 2(W_1 + L_1)\Delta x}{W_2 L_2 - 2(W_2 + L_2)\Delta x} \tag{6.13}$$

where $\Delta x$ is the depth of the undercut, which is assumed to be uniform along the perimeter.

As before, assuming (6.10) to be valid, we can show that with $W_2 = L_2$ and

$$W_1 = L_2(\alpha - \sqrt{\alpha^2 - \alpha})$$

$$L_1 = L_2(\alpha + \sqrt{\alpha^2 - \alpha}) \tag{6.14}$$

the undercut error is zero. For this choice the standard deviation is

$$\frac{\sigma_\alpha}{\alpha} = \left(\frac{\sigma}{L_2}\right)\sqrt{6 - 2\alpha^{-1}} \tag{6.15}$$

However, a very common technique to avoid the undercut error is to connect identical 'unit capacitors' in parallel to construct larger ones. This leads to the ratio of area-to-perimeter to be nearly the same for any two capacitors and the actual ratio is almost the same as the nominal one.

### 6.2.4   Design of Capacitor Ratios

We have seen that, for best accuracy in capacitor ratios, the capacitors should be identical. So, if the ratio is different from one, the two capacitors should be construed as integral multiples of a unit capacitor. If the errors in each unit are the same, the ratio will be free of error.

## Example 6.1

It is required to design two capacitors with ratio of 4.5, with maximum accuracy. To this end we write $4.5 = 4.5\,\text{pF}/1\,\text{pF}$ and determine the greatest common multiple of the numerator and denominator of this fraction. This is $0.5\,\text{pF}$ which is taken as the size of the unit capacitor. Therefore, we use one capacitor made up of 9 units of $0.5\,\text{F}$ each, and another with 2 units. Maximum accuracy is achieved if the errors in the unit capacitors are the same.

Often, the unit capacitor value calculated according to the above procedure is too small for practical implementation. In this case more than one unit capacitor with different values are used provided they differ only slightly.

## Example 6.2

It is required to design two capacitors with a ratio of 4.7. In this case the choice of a unit capacitor of $0.1\,\text{pF}$ is possible but difficult to realize in practice. Thus, we write

$$\alpha = (9 \times 0.5\,\text{pF} + 0.2)/(2 \times 0.5\,\text{pF})$$

or

$$\alpha = (8 \times 0.5\,\text{pF} + 0.7)/(2 \times 0.5\,\text{pF})$$

in which the second is more practical than the first since a $0.7\,\text{pF}$ capacitor is more practical than a $0.2\,\text{pF}$ one.

## 6.3   THE MOS SWITCH

### 6.3.1   A Simple Switch

This is the third component of switched-capacitor filters together with the Op Amp and the capacitor. Figure 6.5(a) shows the simplest MOSFET switch, while Figure 6.5(b) shows its equivalent circuit including the associated parasitic capacitances. Figure 6.5(c) shows the clock signal which drives the switch. The switch is in the ON state when the gate has a sufficiently high voltage (positive for NMOS and negative for PMOS). In this case the voltage $v_{DS}$ will cause a current $i_D$ to flow between the switch terminals A and B. Since the gate voltage $v_\phi$ is usually large ($\approx 5$–$15\,\text{V}$) while the signal voltage across the terminals A and B is $\leqslant 1\,\text{V}$, the MOS-FET can be assumed to be in the triode region, so that equation (2.6) gives for the current in an NMOS switch

$$i_D = K[2(v_{GS} - V_t)v_{DS} - v_{DS}^2] \tag{6.16}$$

Normally

$$|v_{GS} - V_t| \gg |v_{DS}| \tag{6.17}$$

and the switch behaves as a linear resistance

$$R_{on} \cong \frac{1}{2K(v_{GS} - V_t)} \tag{6.18}$$

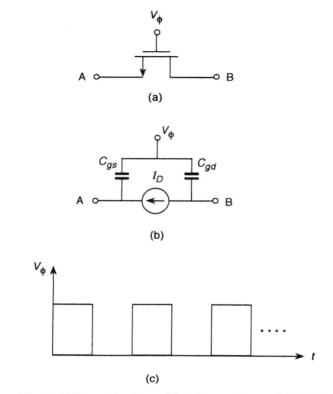

**Figure 6.5**    (a) MOS switch, (b) model with parasitics, and (c) driving clock.

### 6.3.2   Clock Feed-through

Due to the parasitic capacitances shown in Figure 6.5(b), an undesirable phenomenon may occur if the simple switch realization is used. To illustrate this effect, consider Figure 6.6 and suppose that the capacitance loading terminal A is $C_A$ while that loading terminal B is $C_B$. Then the clock signal $v_\phi$ will be transmitted to nodes A and B as

$$v_A = \frac{C_{gs}}{C_{gs} + C_A} v_\phi \tag{6.19}$$

$$v_B = \frac{C_{gd}}{C_{gd} + C_B} v_\phi \tag{6.20}$$

Typically $C_{gs} \approx C_{gd} \approx 0.02\,\mathrm{pF}$ and $C_A \approx C_B \approx 2\,\mathrm{pF}$ so that $v_A \approx v_B \approx 0.01 v_\phi \approx 0.1\,\mathrm{V}$. This means that a signal of this value and a frequency equal to the clock frequency is transmitted to nodes A and B.

    This is called *clock feed-through* and should be minimized. An obvious method would be to connect another transistor with an equal and opposite contribution to the

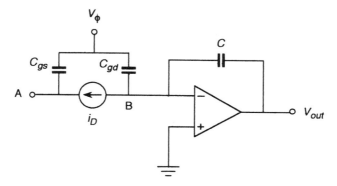

**Figure 6.6**  The clock feed-through effect.

clock feedthrough. One possibility is to add a dummy transistor as shown in Figure 6.7 with its drain and source connected to the signal line and the control voltage opposite in phase to the gate voltage of the main switch. This transistor performs the compensation operation only and has no switching function.

Furthermore, the effect of feedthrough can be reduced by designing the switches as small as possible while the capacitors are designed as large as possible.

A more elaborate compensation scheme employs both NMOS and PMOS transistors to form the CMOS switch or *transmission gate* shown in Figure 6.8. For this circuit the feed-through signals at each node cancel each other, in principle. Another advantage of this switch is that its ON resistance tends to be more linear as a result of the parallel connection of the NMOS and PMOS devices. Moreover, the dynamic signal range in the ON state is increased. This is because a very large input signal, and consequently an equally large output signal, causes one transistor to be OFF, since the gate-to-source voltage does not become sufficiently large. The same signal causes the complementary switch to be fully ON. At a given instant, at least one switch is ON.

For the design of a CMOS switch, complementary clock signals are needed but with the added flexibility of introducing slight delays relative to each other in order to minimize the clock feedthrough. To achieve this, the system clock has to be inverted and delayed such that the gate voltages at the CMOS switch are the complements of each other. Figure 6.9 shows such a scheme.

Despite these precautions, in practice, $C_{gs}$ contains a non-linear component and it is difficult to achieve perfect compensation. Furthermore, the MOS switch introduces

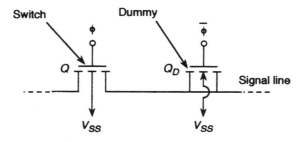

**Figure 6.7**  Reduction of clock feed-through using a dummy MOSFET.

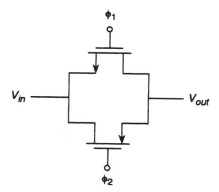

**Figure 6.8**   CMOS switch: transmission gate.

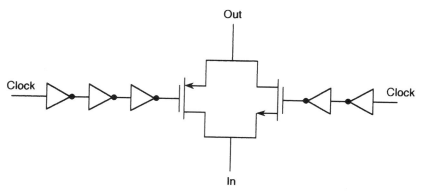

**Figure 6.9**   CMOS switch with inverter-delays.

the two non-linear capacitances $C_{sb}$ and $C_{db}$, which cause harmonic distortion and couple noise from the substrate into the signal path. Also, after the switch is turned OFF, the drain and source currents become the leakage currents associated with reverse-biased pn junctions. Although this current is very small, it may result in a steadily growing drain or source charge, unless there is a dc path to ground connected, at least occasionally, from the drain and source terminals to ground. Finally, as in all MOS circuits, MOS switches are susceptible to thermal noise of the type discussed in Chapter 2. More recent contributions show some effective schemes for the minimization of the effect of the non-ideal behaviour of switches, including the clock feed-through; this will be discussed in detail in a later chapter.

## 6.4   MOS PASSIVE RESISTORS

Although the *passive* resistor is not part of a switched-capacitor circuit, an occasional passive resistor may be needed as an on-chip component for implementing subsidiary filters at the input and output of switched-capacitor filters. Therefore, we give a description of some types of resistors which are available in a CMOS process. These are illustrated in Figure 6.10. The diffused resistor may be constructed using source-drain

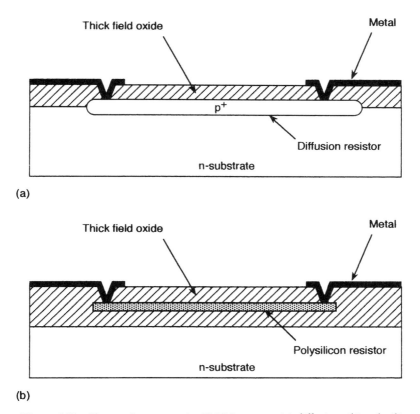

**Figure 6.10**   Two resistor types in CMOS process (a) diffusion, (b) polysilicon.

diffusion, as shown in Figure 6.10(a). Its sheet resistance is typically in the range 10–100 $\Omega/\square$. It also has a voltage dependence of the order of 10–500 ppm/V. The associated parasitic capacitance of the diffusion resistor is also voltage-dependent.

The polysilicon resistor is shown in Figure 6.10(b) and is surrounded by thick oxide. Its typical values are in the range 30–200 $\Omega/\square$. The associated parasitics are quite small and voltage-independent. Such a resistor can be trimmed by blowing links using an electric current or a laser.

By means of an additional masking step, an ion-implanted resistor can be constructed which is quite similar to the diffusion resistor. Sheet resistances in the range 500–2000 $\Omega/\square$ are typical. However, this type of resistor has a pronounced voltage dependence and its parasitics are also voltage dependent.

## CONCLUSION

This chapter concludes the discussion of the integrated circuit components of switched-capacitor filters by studying the design of switches and capacitors. Particular attention was given to the non-ideal effects such as clock feed-through in the switches and capacitor ratio errors. Mothods for reducing these effects were also presented.

# Part III
## Design of
## Switched-capacitor Filters

'Nothing is so dangerous as being too modern. One is apt to grow old-fashioned
quite suddenly.'

**Oscar Wilde**
*An Ideal Husband*

## OUTLINE

General filter design fundamentals are introduced and applied to the design of passive
filters which form reference designs for most other types, particularly switched-capacitor filters. Then the design of amplitude-oriented switched-capacitor filters using a
variety of techniques is discussed in detail. Emphasis is laid on low sensitivity ladder
structures, but cascade designs are also discussed. Next, selective linear-phase design
techniques are given together with the design of filters for data transmission. Finally,
some practical considerations are discussed such as scaling for maximum dynamic
range and minimum capacitance, high frequency filters, balanced fully differential
designs and layout considerations. The design techniques in this part are facilitated
considerably by the use of the computer package **ISICAP**.

# 7

# Continuous-time Filter Models

## 7.1 INTRODUCTION

In this chapter, the principles of filter design are introduced in relation to continuous-time passive filters [1], [36]. The filter design problem is discussed in its most general form dealing with the realizability, approximation and synthesis aspects in some detail. A good understanding of these topics is essential to the study of switched-capacitor filters for two reasons. First, the design of a switched-capacitor filter is very often accomplished by, first, designing a passive prototype upon which the switched-capacitor filter is modelled. Secondly, the concepts and techniques involved in the filter design problem in general have their roots in passive filter synthesis and are used extensively in this book, particularly when a new non-standard class of switched-capacitor filters is sought.

## 7.2 IDEAL FILTERS

Consider a system or network, as shown in Figure 7.1, whose input is $f(t)$ and output is $g(t)$. With the Fourier Transform pairs

$$f(t) \leftrightarrow F(j\omega) \tag{7.1}$$

and

$$g(t) \leftrightarrow G(j\omega) \tag{7.2}$$

The transfer function of the system is

$$H(j\omega) = \frac{G(j\omega)}{F(j\omega)}$$

$$= |H(j\omega)|e^{j\psi(\omega)} \tag{7.3}$$

where $|H(j\omega)|$ is the *amplitude response* and $\psi(\omega)$ is the *phase response*. The system is called an *ideal filter* if its amplitude response is constant (unity for simplicity) within

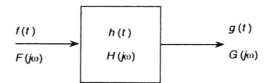

**Figure 7.1**   General continuous-time system or filter.

certain frequency bands and exactly zero outside these bands. In addition, in the
bands where the amplitude is constant, the phase is a linear function of $\omega$. The
amplitude response of the ideal filter is shown in Figure 7.2 for the four main types
of low-pass, high-pass, band-pass, and band-stop filters. The ideal phase response for
the low-pass case is shown in Figure 7.3 and is given by

$$\psi(\omega) = -k\omega \quad |\omega| \leqslant \omega_0 \tag{7.4}$$

where $k$ is a constant.

To appreciate why these are the required ideal characteristics, consider the low-pass
case described by the transfer function

$$H(j\omega) = e^{-jk\omega} \quad 0 \leqslant |\omega| \leqslant \omega_0$$
$$= 0 \qquad |\omega| > \omega_0 \tag{7.5}$$

Then,

$$G(j\omega) = H(j\omega)F(j\omega) \tag{7.6}$$

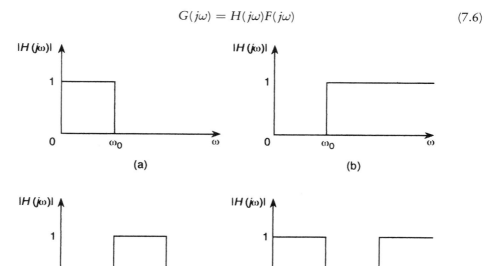

**Figure 7.2**   The ideal amplitude filter characteristics: (a) low-pass; (b) high-pass; (c) band-pass; and
(d) band-stop.

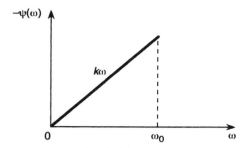

**Figure 7.3**  The ideal low-pass phase response.

so that for $|\omega| > \omega_0$, $G(j\omega) = 0$. In the passband $(0 \leqslant |\omega| \leqslant \omega_0)$

$$G(j\omega) = e^{-jk\omega}F(j\omega) \tag{7.7}$$

Taking the inverse Fourier transform of the above expression we obtain

$$g(t) = f(t - k) \tag{7.8}$$

which means that the output is an exact replica of the input, but delayed by the constant $k$. It follows that any input signal with spectrum lying within the passband of the ideal filter, will be transmitted without attenuation and without any distortion in its phase spectrum, the signal is merely delayed by a constant time value.

Now, the ideal filter characteristics cannot be obtained using realizable (causal) transfer functions and must, therefore, be approximated. This can be easily appreciated if we consider again the ideal low-pass characteristic described by (7.6). Taking the inverse Fourier transform of $H(j\omega)$ we obtain the impulse response of the ideal filter as

$$h(t) = \frac{\sin \omega_0(t - k)}{\pi(t - k)} \tag{7.9}$$

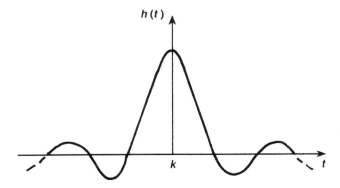

**Figure 7.4**  Impulse response of ideal low-pass filter.

which is shown in Figure 7.4 and clearly exists for negative $t$ so that the required system is non-causal. Similar conclusions can readily be reached for the band-pass case.

Any deviation from the ideal amplitude characteristic is called *amplitude distortion*, while any deviation from the ideal (linear) phase characteristic is called *phase distortion*. In some applications, such as voice communication, filters are designed on an amplitude basis only since it is claimed that the human ear is relatively insensitive to phase distortion. Other applications tolerate some amplitude distortion while requiring a close approximation to the ideal (linear) phase response. However, in modern high capacity communication systems, filters are required to possess good amplitude (highly selective) as well as good phase characteristics.

## 7.3   THE FILTER DESIGN PROBLEM

The central problem in passive filter design is concerned with a lossless two-port operating between a resistive source and a resistive load as shown in Figure 7.5. The most meaningful description of this doubly-terminated lossless two-port, is in terms of the scattering matrix. This is because the transfer function of major interest is the forward transmission coefficient given by

$$S_{21}(s) = 2\sqrt{\frac{R_g}{R_L}} \frac{V_2(s)}{V_g(s)} \tag{7.10}$$

For a physical network, causality is implied and in the sinusoidal steady state $(s \rightarrow j\omega)$

$$|S_{21}(j\omega)|^2 = \frac{|V_2(j\omega)|^2/R_L}{|V_g(j\omega)|^2/4R_g}$$

$$= \frac{P_L(\omega^2)}{P_g(\omega^2)} \tag{7.11}$$

which is the ratio of the power delivered to the load to the maximum power available from the source. Therefore, expression (7.11) determines the amplitude response of

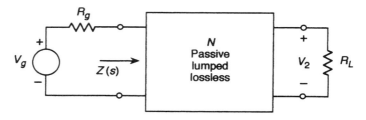

**Figure 7.5**   The doubly-terminated lossless filter.

the network. We may also write

$$S_{21}(j\omega) = |S_{21}(j\omega)|e^{j\psi(\omega)} \qquad (7.12)$$

where $\psi(\omega)$ is the phase function. Clearly any lossless two-port inserted between the source and load will operate as a filter whose characteristics are determined by the particular choice of a suitable $S_{21}(s)$.

### 7.3.1 Statement of the Problem

The filter design problem consists of the following stages:

(a) Given a set of specifications on the response of the network, these are translated into a mathematical expression for the transfer function $S_{21}(s)$. These specifications can be either on the amplitude or the phase response, or they can be on both aspects simultaneously. However, in determining the expression for $S_{21}(s)$, it must possess certain properties which guarantee its realizability, i.e. it must correspond to a physical network made up of the specified types of building blocks, and possibly in a particular structure. In our present discussion, we concentrate on passive lumped networks in which the two-port in Figure 7.5 is also lossless and reciprocal. This stage of the design process is called the *approximation problem*.

(b) Having obtained the transfer function $S_{21}(s)$ which describes the doubly-terminated lossless two-port of Figure 7.5, we proceed with the realization of the network in the preferred structure, i.e. obtaining the structure and element values of the network whose characterizing $S_{21}(s)$ was obtained in stage (a). This is the *synthesis problem*.

Figure 7.6 shows the steps and problems involved in the filter design problem. From this definition of the problem and the preceding discussion, it is clear that before we can proceed with the design of a filter, we must know the answers to the following questions:

1. What are the necessary and sufficient conditions which a function $S_{21}(s)$ must satisfy in order for it to be realizable as the transmission coefficient of a doubly-terminated lossless two-port of the type shown in Figure 7.5?

2. Having obtained a realizable $S_{21}(s)$, how do we obtain the structure and element values of the lossless two-port?

We now give the answers to the above problems, using arguments that are rigorous but rather concise. For a comprehensive discussion of these topics the reader may consult reference [1].

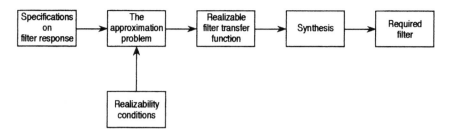

Figure 7.6   Filter design procedure.

### 7.3.2   Realizability Conditions

We begin by, heuristically, developing the realizability conditions for the transfer function $S_{21}(s)$. A rigorous treatment is found in [1]. First, $S_{21}(s)$ is a real rational function of $s$, since it describes a network composed of real elements. Second, like all functions describing *stable* systems, it must be analytic in the right half-plane. Finally, for a lossless two port, $|S_{21}(j\omega)| \leqslant 1$. Putting all these properties together we have the following result.

*Theorem 7.1*

The transfer function $S_{21}(s)$ of a passive lumped lossless two-port, referred to real terminations, satisfies the following conditions:

(a)
$$S_{21}(s) = \frac{P_m(s)}{Q_n(s)}$$
(7.13)

is a real rational function of $s$, i.e. both $P_m(s)$ and $Q_n(s)$ are polynomials with real coefficients.

(b) $S_{21}(s)$ is analytic in the closed right half of the complex s-plane, thus $Q_n(s)$ is a strictly Hurwitz polynomial. Note that a strictly Hurwitz polynomial is one which has all its zeros in the *strict left half plane*, i.e. excluding the $j\omega$-axis.

(c)
$$|S_{21}(j\omega)| \leqslant 1 \quad -\infty \leqslant \omega \leqslant \infty$$
(7.14)

A function satisfying the conditions of Theorem 3.1 is called a *bounded real* function. In fact the *definition* of a bounded real function is as follows:

*Definition 7.1*

A function $S(s)$ is said to be bounded real if it satisfies

(1) $S(s)$ is real for $s$ real

(2) $|S(s)| < 1$ in $\mathrm{Re}\, s > 0$

Using the theory of functions of a complex variable, it can be shown that (2) above is equivalent to (b) and (c) of Theorem 7.1 for a rational function.

So, the transmission coefficient of a lossless two-port must be a bounded real function. However, Theorem 7.1 allows the lossless two-port to be non-reciprocal. For a reciprocal realization we must add the condition [1]:

$$P_m(s) \text{ is an even (or odd) polynomial}$$

i.e. the numerator of $S_{21}(s)$ contains only even (or odd) powers of $s$.

In fact, the converse of Theorem 7.1 is also true and we have the following important result.

### Theorem 7.2

An arbitrary rational bounded real function $S_{21}(s)$, with an even or odd numerator, is realizable as the transmission coefficient of a lossless reciprocal two-port.

*Example 7.1*   Test the transfer function

$$S_{21}(s) = \frac{1}{(s+1)(s^2+s+1)}$$

for bounded real character.

*Solution*   We verify that $S_{21}(s)$ satisfies the conditions of Theorem 7.1. First, $S_{21}$ is the ratio of two polynomials with real coefficients; thus condition (a) is satisfied. Secondly, the poles of the function occur at

$$s = -1$$

$$s = \frac{-1 \pm j\sqrt{3}}{2}$$

which are in the open left half-plane. Therefore condition (b) is satisfied.
   Finally, we evaluate

$$|S_{21}(j\omega)|^2 = S_{21}(j\omega) \cdot S_{21}(-j\omega)$$

$$= \frac{1}{(j\omega+1)(-\omega^2+j\omega+1)} \frac{1}{(-j\omega+1)(-\omega^2-j\omega+1)}$$

$$= \frac{1}{(\omega^2+1)(\omega^4-\omega^2+1)}$$

or

$$|S_{21}(j\omega)|^2 = \frac{1}{1+\omega^6} \leqslant 1 \quad \text{for all } \omega$$

It follows that $S_{21}(s)$ satisfies condition (c) of Theorem 7.1. Thus $S_{21}(s)$ is bounded real. In addition, the numerator of $S_{21}(s)$ is unity, a trivial even polynomial. Thus, $S_{21}(s)$ is realizable as a lossless *reciprocal* two-port, according to Theorem 7.2.

Now, consider the transfer function $S_{21}(s)$ written as the ratio of two polynomials, as given by (7.13) where $m$ is the degree of $P_m(s)$ and $n$ is the degree of $Q_n(s)$. We know that $S_{21}(s)$ cannot have poles in the right half-plane or on the $j\omega$-axis, *including the point at infinity*. Therefore the degree of $P_m(s)$ must be at most equal to the degree of $Q_n(s)$ otherwise $S_{21}(s)$ would have a pole at infinity. Thus,

$$m \leqslant n \tag{7.15}$$

From (7.10), at the zeros of $S_{21}(s)$ there is no signal at the output for any input. Therefore, the zeros of $S_{21}(s)$, are called the *zeros of transmission*. In the general theory of network synthesis [1], it is shown that these zeros determine the structure that realizes the transfer function $S_{21}(s)$.

Consider Figure 7.5 in which the input impedance of the resistor- terminated lossless two-port is $Z(s)$. The *input reflection coefficient*, relative to the source resistance is given by

$$S_{11}(s) = \frac{Z(s) - R_g}{Z(s) + R_g} \tag{7.16}$$

so that for $R_g = 1 \Omega$

$$S_{11}(s) = \frac{Z(s) - 1}{Z(s) + 1} \tag{7.17}$$

On the $j\omega$-axis, $|S_{11}(j\omega)|^2$ is a measure of the reflected power at the input port due to the mismatch between $Z(s)$ and its maximum-power-transfer value of $R_g(=1)$. Clearly at a frequency where $Z(s) = R_g$, $S_{11} = 0$. The familiar relation between the power transmission and reflection for a lossless two-port is given by

$$|S_{21}(j\omega)|^2 + |S_{11}(j\omega)|^2 = 1 \tag{7.18}$$

Another useful description of the lossless two-port is in terms of its transmission (ABCD) parameters defined by

$$V_1(s) = A(s)V_2 - B(s)I_2(s) \tag{7.19}$$

$$I_1(s) = C(s)V_2 - D(s)I_2(s) \tag{7.20}$$

and the transmission matrix is

$$[T(s)] = \begin{bmatrix} A(s) & B(s) \\ C(s) & D(s) \end{bmatrix} \tag{7.21}$$

It can be shown [1] that all the transmission parameters have the same denominator which allows us to write

$$[T(s)] = \frac{1}{f(s)}[t(s)] \qquad (7.22)$$

where $[t(s)]$ is the *polynomial* transmission matrix given by

$$[t(s)] = \begin{bmatrix} a(s) & b(s) \\ c(s) & d(s) \end{bmatrix} \qquad (7.22)$$

The polynomial transmission parameters of a lossless two-port have the following properties:

(1) $a(s)$ and $d(s)$ are both even or both odd.

(2) $b(s)$ and $c(s)$ are both odd if $a(s)$ is even and both even if $a(s)$ is odd.

(3) $a(s) + b(s) + c(s) + d(s)$ is strictly Hurwitz.

Finally, dividing (7.19) by (7.20) we see that if a lossless two-port is terminated in an impedance $Z_L(s)$, then the input impedance of the terminated network is given by

$$Z(s) = \frac{A(s)Z_L + B(s)}{C(s)Z_L + D(s)} \qquad (7.24)$$

### 7..3.3   Synthesis Procedure

In general, the procedure for realizing the transfer function $S_{21}(s)$ as the transmission coefficient of a doubly-terminated lossless two-port, as shown in Figure 7.5, begins by evaluating the input impedance of the resistor-terminated two-port. The starting function can be one of the following

(a) $$|S_{21}(j\omega)|^2 = \frac{|P(j\omega)|^2}{|Q(j\omega)|^2} \qquad (7.25)$$

or

(b) $$S_{21}(s) = \frac{P(s)}{Q(s)} \qquad (7.26)$$

where $P(s)$ is assumed even (or odd) for a reciprocal realization.

If the transducer power gain in (7.25) is the starting function, then we evaluate $S_{21}(s)\,S_{21}(-s)$ by letting $\omega \to s/j$ in (7.25), i.e.

**Step (1)**

$$\frac{P(s)P(-s)}{Q(s)Q(-s)} = \left\{\frac{|P(j\omega)|^2}{|Q(j\omega)|^2}\right\} \omega \rightarrow s/j \tag{7.27}$$

then $Q(s) \, Q(-s)$ is factored and the open left half-plane zeros are assigned to $Q(s)$ since it must be strictly Hurwitz.

If, on the other hand, $S_{21}(s)$ in (7.26) is the starting function, then we proceed directly to the next step.

**Step 2**
Using (7.18) we have

$$|S_{11}(j\omega)|^2 = 1 - |S_{21}(j\omega)|^2 \tag{7.28}$$

so that

$$S_{11}(s)S_{11}(-s) = 1 - S_{21}(s)S_{21}(-s)$$

$$= 1 - \frac{P(s)P(-s)}{Q(s)Q(-s)} \tag{7.29}$$

$$= \frac{Q(s)Q(-s) - P(s)P(-s)}{Q(s)Q(-s)}$$

Since $S_{11}(s)$ is the input reflection coefficient of a passive network, its poles are restricted to lie in the open left half-plane. Therefore, these are the zeros of $Q(s)$ which are the *same* poles of $S_{21}(s)$, so that

$$S_{11}(s) = \frac{U(s)}{Q(s)} \tag{7.30}$$

where $U(s)$ is obtained by factorization of the numerator of (7.29) and assigning half the zeros to $U(s)$. In doing so, we are free to take any combination of half the zeros of the numerator of (7.29), provided the corresponding factors combine to form a real polynomial. Therefore, this factorization is not unique, resulting in different networks having the same response.

It is noteworthy that $S_{11}(s)$ as obtained above is *also* a *bounded real* function. This is easily seen by noting that it has the same denominator $Q(s)$ as $S_{21}(s)$, hence $S_{11}(s)$ is also analytic in the closed right half-plane. Also by (7.18) $|S_{11}(j\omega)|^2 \leqslant 1$ for all $\omega$ and the degree of $U(s)$ is at most equal to that of $Q(s)$. Thus, $S_{11}(s)$ is also a bounded real function satisfying,

$$|S_{11}(j\omega)|^2 \leqslant 1 \quad \text{for all } \omega \tag{7.31}$$

or

$$|Q(j\omega)|^2 - |U(j\omega)|^2 \geqslant 0 \quad \text{for all } \omega \tag{7.32}$$

**Step (3)**
From $S_{11}(s)$ as obtained in *Step 2* we evaluate the input impedance $Z(s)$, of the resistor-terminated two-port using (7.16) [for $R_g = 1$] for convenience. Hence

$$Z(s) = \frac{1 + S_{11}(s)}{1 - S_{11}(s)}$$

$$= \frac{Q(s) + P(s)}{Q(s) - P(s)} \tag{7.33}$$

$$= \frac{N(s)}{D(s)}$$

This function is the starting point in the synthesis of the network.

Let us now investigate the analytic properties of $Z(s)$ obtained according to the above procedure. First, we note that $Z(s)$ is a real rational function, since both $P(s)$ and $Q(s)$ are real polynomials. Secondly, if we add the numerator and denominator of $Z(s)$ we obtain $2Q(s)$ which is a strictly Hurwitz polynomial. Finally, let us evaluate the expression

$$2\,\mathrm{Re}\,Z(j\omega) = Z(j\omega) + Z(-j\omega)$$

$$= \frac{N(j\omega)}{D(j\omega)} + \frac{N(-j\omega)}{D(-j\omega)} \tag{7.34}$$

which, upon use of (7.30) becomes

$$\mathrm{Re}\,Z(j\omega) = \frac{Q(j\omega)Q(-j\omega) - U(j\omega)U(-j\omega)}{|Q(j\omega) - U(j\omega)|^2} \tag{7.35}$$

Using (7.32) we have

$$\mathrm{Re}\,Z(j\omega) \geqslant 0 \quad \text{for all } \omega \tag{7.36}$$

or

$$\frac{N(j\omega)D(-j\omega) + N(-j\omega)D(j\omega)}{D(j\omega)D(-j\omega)} \geqslant 0 \quad \text{for all } \omega \tag{7.37}$$

Putting together the above properties we obtain the following result.

**Theorem 7.3**
The driving-point impedance of a resistor-terminated lumped passive lossless two-port satisfies the following conditions:

(a) $Z(s) = N(s)/D(s)$ is real and rational

(b) $N(s) + D(s)$ is strictly Hurwitz

(c) the real even polynomial

$$u(\omega^2) = \{N(j\omega)D(-j\omega) + N(-j\omega)D(j\omega)\} \qquad (7.38)$$

is non-negative along the entire $j\omega$-axis.

A function satisfying the above conditions is said to be a *positive real function* (abbreviated p.r.f.). In fact, the *definition* of a positive real function $Z(s)$ is as follows

*Definition 7.2*  A function $Z(s)$ is said to be positive real if it satisfies

(1) $Z(s)$ is real for $s$ real

(2) $\operatorname{Re} Z(s) > 0$ for $\operatorname{Re} s > 0$.

i.e. its real part is positive for all values of $s$ in the open right half-plane.

Using the theory of functions of a complex variable it can be shown that (2) above is equivalent to (b) and (c) of Theorem 7.3 for a rational function.

*Example 7.2*  Test the following function for positive real character.

$$Z(s) = \frac{s^2 + 2s + 2}{s^2 + 2s + 1}$$

*Solution*  We verify that $Z(s)$ satisfies the conditions of Theorem 7.3. Condition (a) is satisfied since $Z(s)$ is real for $s$ real. The sum of the numerator and denominator is $2s^2 + 45 + 3$, which is easily found to be strictly Hurwitz. Thus condition (b) is also satisfied. Next, evaluate the polynomial defined in condition (c) as

$$N(j\omega)D(-j\omega) + N(-j\omega)D(j\omega)$$

$$= (-\omega^2 + 2j\omega + 2)(-\omega^2 - 2j\omega + 1) + (-\omega^2 - 2j\omega + 2)(-\omega^2 + 2j\omega)$$

$$= 2\omega^4 + 2\omega^2 + 4$$

which is non-negative for all $\omega$. Hence the function $Z(s)$ also satisfies condition (c) of Theorem 7.3 and is, therefore, positive real.

Now, the converse of Theorem 7.3 is also true, so that we also have the following important result.

## Theorem 7.4: Darlington's theory

An arbitrary rational positive real function $Z(s)$ is realizable as the driving-point impedance of a passive lumped lossless reciprocal two-port, terminated in a resistor [1].

The most general type of lossless two-port contains coupled-coils and/or ideal transformers and the synthesis techniques of such networks may be found in reference [1]. Here we limit our discussion to some important special cases which are particularly important in filter design. First, some useful properties of positive real functions are stated; the proofs are left to the reader.

(a) If $Z(s)$ is p.r., then $Y(s) = 1/Z(s)$ is also p.r.

(b) The sum of two p.r. functions is also p.r., but the difference may not be p.r.

(c) A p.r.f. $Z(s)$ is analytic in the open right half-plane. It may have poles on the $j\omega$-axis, but these must be simple with positive residues.

### 7.3.4  Extraction of $j\omega$-axis Poles from a p.r.f.

Let us give a network interpretation of the $j\omega$-axis pole of an impedance correspond, to in the actual network described by $Z(s)$? Let $j\omega_i$, $i = 1, 2, \ldots m$ be simple poles on the imaginary axis, of a p.r.f. $Z(s)$. The function may, in addition, have a simple pole at $s = 0$ and a simple pole at $s = \infty$. Therefore, $Z(s)$ may be written as

$$Z(s) = \frac{k_0}{s} + \sum_{i=1}^{m} \frac{2k_i s}{s^2 + \omega_i^2} + k_\infty s + Z_1(s) \tag{7.39}$$

where $Z_1(s)$ is a function devoid of $j\omega$-axis poles, including 0 and $\infty$. The residues at the poles can be obtained from the expressions

$$k_0 = \{sZ(s)\}_{s=0} \tag{7.40}$$

$$k_\infty = \lim_{s \to \infty} \left\{ \frac{Z(s)}{s} \right\} \tag{7.41}$$

$$2k_i = \left\{ \frac{s^2 + \omega_i^2}{s} Z(s) \right\}_{s=j\omega_i} \tag{7.42}$$

Now, in (7.39) we recognize that $k_0/s$ is the impedance of a capacitor, $k_\infty s$ is the impedance of an inductor, and a typical term in the summation $sk_i s/(s^2 + \omega_i^2)$ is the

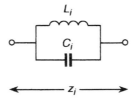

**Figure 7.7**  Circuit for realizing a finite $j\omega$-axis pole in the impedance.

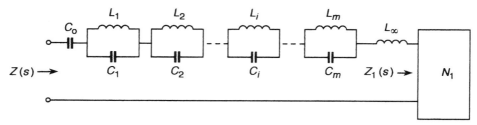

**Figure 7.8**   Network interpretation of (7.39).

impedance of a parallel $L_iC_i$ as shown in Figure 7.7

$$z_i = \frac{(1/C_i)s}{s^2 + 1/L_iC_i} \tag{7.43}$$

Since these terms are added in an impedance expression, we obtain the network interpretation of expression (7.39) shown in Figure 7.8. The element values are obtained by comparison of the various terms in (7.39) with the impedance of a capacitor, an inductor, and that of a parallel $L_iC_i$. Thus

$$\begin{aligned}
C_0 &= 1/k_0 & L_\infty &= k_\infty \\
C_i &= 1/2k_i & L_i &= 2k_i/\omega_i^2
\end{aligned} \tag{7.44}$$

It can be easily shown that $Z_1(s)$ is also a positive real function. In Figure 7.8 we refer to the process of realizing a term in the expansion of $Z(s)$ in (7.39) as the *extraction* of a pole from $Z(s)$.

If $Z(s)$ has *all* its poles on the $j\omega$-axis, then after the extraction of these poles, the remainder $Z_1(s)$ is either a constant or zero, resulting in $N_1$ being either a resistor or a short-circuit respectively. In the latter case, $Z(s)$ is realized completely as a purely lossless one-port, and $Z(s)$ is called a *reactance function*.

In a similar manner, let the admittance $Y(s) = 1/Z(s)$ have poles at $s = j\omega_i$, $i = 1.2.\ldots r$ as well as possible poles at $s = 0$ and $s = \infty$. Then we can write

$$Y(s) = \frac{h_0}{s} + \sum_{i=1}^{r} \frac{2h_i s}{(s^2 + \omega_i^2)} + h_\infty s + Y_1(s) \tag{7.45}$$

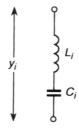

**Figure 7.9**   Circuit for realizing a finite $j\omega$-axis pole in the admittance.

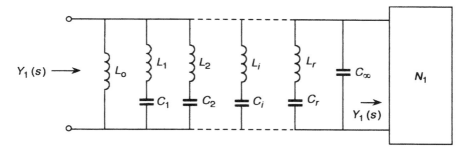

**Figure 7.10** Network interpretation of (7.45).

where $h_0/s$ is the admittance of an inductor, $h_\infty s$ that of a capacitor, and $2h_i s/(s^2 + \omega_i^2)$ is the admittance of a series $L_i C_i$ as shown in Figure 7.9 which is given by

$$y_i = \frac{(1/L_i)s}{s^2 + 1/L_i C_i} \tag{7.46}$$

Hence, the network interpretation of expression (7.45) is shown in Figure 7.10 with the element values

$$\begin{aligned} L_0 &= 1/h_0 & c_\infty &= h_\infty \\ L_i &= 1/2h_i & C_i &= 2h_i/\omega_i^2 \end{aligned} \tag{7.47}$$

Again, if all the poles of $Y(s)$ are on the $j\omega$-axis, then $N_1(s)$ is either a resistor or an open circuit depending on whether $Y_1(s)$ is a constant or zero, respectively. In the latter case, the entire realization is a purely lossless one-port.

## 7.4 DOUBLY-TERMINATED LOSSLESS LADDERS

Let us return to the basic filter configuration shown in Figure 7.5, of a lossless two-port driven by a source of resistive internal impedance (normalized to $1\,\Omega$) and terminated in a resistive load $R_L$. For an arbitrary bounded real $S_{21}(s)$ describing the filter, we derive the input impedance $Z(s)$ of the resistor-terminated lossless two-port which is guaranteed positive real. The synthesis begins from $Z(s)$ and for an arbitrary p.r. function requires the use of coupled coils and/or ideal transformers. On the other hand most filters are realized in ladder structures of the type shown in Figure 7.11, in which all $Z_i$ are lossless impedances, i.e. they contain only self inductors and capacitors. In addition to the desirable running ground, this lossless ladder has the very important property that all its zeros of transmission are included in the open-circuit frequencies of the series arms. They are either poles of $Z_1, Z_3 \ldots Z_{n-1}$ or zeros of $Z_0, Z_2, Z_4, \ldots Z_n$. This is because at such frequencies one (or more) of the series branches becomes an open circuit or one (or more) of the shunt branches becomes a short-circuit. In either case the signal is blocked from reaching the

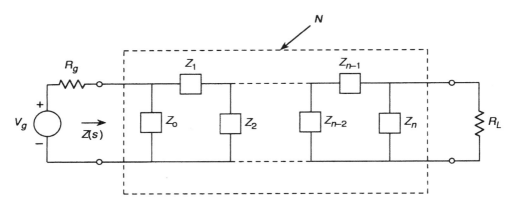

**Figure 7.11**   Doubly-terminated lossless ladder.

output, resulting in a transmission zero. Two basic prototype ladder structures are of particular importance.

### 7.4.1   Simple Low-pass Ladder

Suppose the transfer function $S_{21}(s)$ is a bounded real function of the *specific form*

$$S_{21}(s) = \frac{1}{Q_n(s)} \qquad (7.48)$$

or the transducer power gain has the form

$$|S_{21}(j\omega)|^2 = \frac{1}{|Q_n(j\omega)|^2} \qquad (7.49)$$

Clearly $S_{21}(s)$ has no finite zeros, and all the $n$ zeros of transmission occur at $s = \infty$. Such a transfer function is termed an *all-pole* function. The question we address here is how to realize such a function. First note that since all the zeros of transmission are at $s = \infty$, we expect the series arms of the ladder in Figure 7.11 to be simple inductors, and all the shunt arms to be simple capacitors, as shown in Figure 7.12. This should be clear since each series inductor provides a zero of transmission at $s = \infty$ by degenerating into an open circuit at this point. Similarly each shunt capacitor degenerates into a short-circuit at $s = \infty$. To obtain the element values of the ladder in Figure 7.12 we start from $|S_{21}(j\omega)|^2$ or $S_{21}(s)$ and follow steps (1) to (3) of Section 7.3.3 given earlier for determining the input impedance $Z(s)$. Thus,

$$|S_{11}(j\omega)|^2 = 1 - |S_{21}(j\omega)|^2$$

$$= \frac{Q_n(j\omega)Q_n(-j\omega) - 1}{Q_n(j\omega)Q_n(-j\omega)} \qquad (7.50)$$

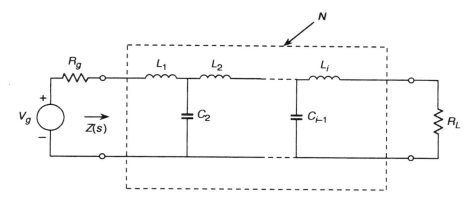

**Figure 7.12** Simple low-pass resistively-terminated lossless ladder.

Letting $j\omega \rightarrow s$ in the above expressions we obtain

$$S_{11}(s)S_{11}(-s) = \frac{Q_n(s)Q_n(-s) - 1}{Q_n(s)Q_n(-s)} \tag{7.51}$$

Performing the factorization of the numerator, we assign half the zeros to $S_{11}(s)$ provided the corresponding factors combine to form an $n$th degree *real polynomial*. Thus we arrive at an expression for $S_{11}(s)$ of the form

$$S_{11}(s) = \frac{U_n(s)}{Q_n(s)} \tag{7.52}$$

where $Q_n(s)$ is strictly Hurwitz. Next, we determine the input impedance of the resistor-terminated lossless ladder from (7.33) as

$$Z(s) = \frac{1 + S_{11}(s)}{1 - S_{11}(s)}$$

$$= \frac{Q(s) + Q(s)}{Q(s) - Q(s)} \tag{7.53}$$

in which we have assumed that the source resistor is $1\,\Omega$ for convenience.

Now, the specific all-pole function in (7.48) is such that

$$\{S_{21}(s)\}_{s \rightarrow \infty} = 0 \tag{7.54}$$

so that from (7.50),

$$\{S_{11}(s)\}_{s \rightarrow \infty} = \pm 1 \tag{7.55}$$

and it follows from (7.53) that

$$\{Z(s)\}_{\rightarrow \infty} = 0 \text{ or } \infty \tag{7.56}$$

Consequently, either $Z(s)$ or $Y(s)$ has a pole at $s = \infty$. Writing

$$Z(s) = \frac{a_n s^n + a_{n-1} s^{n-1} + \cdots + a_0}{b_m s^m + b_{m-1} s^{m-1} + \cdots + b_0} \qquad (7.57)$$

then $m$ and $n$ must differ by one, to give rise to a pole at $s = \infty$ in $Z(s)$ or $Y(s)$. If $n = m + 1$ then the first element in the ladder of Figure 7.12 is a series inductor. If $n = m - 1$, then $Y(s)$ has a pole at $s = \infty$ and the first element is a shunt capacitor. Since all the zeros of transmission occur at $s = \infty$, we expect the ladder in Figure 7.12 to be capable of completely realizing the impedance $Z(s)$, giving $n$ inductors and capacitors where $n$ is the degree of $Z(s)$, which is also the number of transmission zeros. To produce the network we start from $Z(s)$ or $Y(s)$, depending on which function has a pole at $s = \infty$. Assuming $Z(s)$ satisfies this requirement, then we divide the denominator into the numerator to obtain

$$Z(s) = \left(\frac{a_n}{b_m}\right) s + Z_1(s)$$
$$= \alpha_1 s + Z_1(s) \qquad (7.58)$$

where $Z_1(s)$ is a remainder impedance, which is identified as the driving-point impedance of the network *minus* the first series inductor. The value of the inductor $L_1$ is $\alpha_1 = (a_n/b_m)$. The admittance $Y_1(s) = 1/Z_1(s)$ now has a pole at $s = \infty$. Thus we can write (7.58) as

$$Z(s) = \alpha_1 s + \frac{1}{Y_1(s)} \qquad (7.59)$$

and we repeat the same process of dividing the numerator of $Y_1(s)$ by its denominator to obtain

$$Z(s) = a_1 s + \frac{1}{a_2 s + Y_2(s)} \qquad (7.60)$$

where $\alpha_2$ is the value of the shunt capacitor $C_2$ in Figure 7.12. If the process of division and inversion is iterated, we obtain the *continued fraction expansion* of $Z(s)$ as

$$Z(s) = \alpha_1 s + \cfrac{1}{a_2 s + \cfrac{1}{a_3 s + \cfrac{}{\ddots}}} \qquad (7.61)$$

$$a_n s + R_L (\text{or } 1/R_L)$$

where $\alpha_1$ gives the element values. It is also to be noted that if $Z(s)$ did not possess the pole at $s = \infty$, then $Y(s)$ has such a pole and the first step in the expansion is an inversion, so that the first element in the ladder is the shunt capacitor.

*Example 7.3*  Consider the realization of the all-pole transfer function of Example 7.1, i.e.

$$S_{21}(s) = \frac{1}{(s+1)(s^2+s+1)} \tag{7.62}$$

or if the transducer power gain is the given function we start from

$$|S_{21}(j\omega)|^2 = \frac{1}{1+\omega^6} \tag{7.63}$$

We begin by forming

$$S_{11}(s)S_{11}(-s) = 1 - S_{21}(s)S_{21}(-s)$$

$$= 1 - \left\{|S_{21}(j\omega)|^2\right\}_{\omega \to s/j}$$

$$= \left. \frac{\omega^6}{1+\omega^6} \right|_{\omega \to s/j} \tag{7.64}$$

$$= \frac{-s^6}{1-s^6}$$

so that choosing all the left half-plane poles, which are the same poles of $S_{21}(s)$ we have

$$S_{11}(s) = \frac{\pm s^3}{(s+1)(s^2+s+1)} \tag{7.65}$$

If we choose the plus sign, then

$$Z(s) = \frac{1+S_{11}(s)}{1-S_{11}(s)}$$

$$= \frac{2s^3+2s^2+2s+1}{2s^2+2s+1} \tag{7.66}$$

whose continued fraction expansion is

$$
\begin{array}{r|l}
 & s \\
\hline
2s^2+2s+1 & 2s^3+2s+2s+1 \\
 & 2s^3+2s^2+s \\
\hline
\end{array}
$$

$$
\begin{array}{r|l}
 & 2s \\
\hline
s+1 & 2s^2+2s+1 \\
 & 2s^2+2s \\
\hline
\end{array}
$$

$$
\begin{array}{r|l}
 & s \\
\hline
1 & s+1 \\
 & s \\
\hline
\end{array}
$$

$$
\begin{array}{r|l}
 & 1 \\
\hline
1 & 1 \\
\hline
 & 0 \\
\end{array}
$$

which gives the network of Figure 7.13(a).

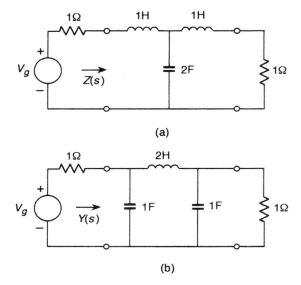

**Figure 7.13**  Networks of Example 7.3.

On the other hand, if we choose the minus sign in (7.65) we have

$$Z(s) = \frac{1 + S_{11}(s)}{1 - S_{11}(s)}$$

$$= \frac{2s^2 + 2s + 1}{2s^3 + 2s^2 + 2s + 1}$$

(7.67)

which has a zero at $s = \infty$, so that $Y(s) = 1/Z(s)$ has a pole at this point. Thus, we start the process of finding the continued fraction expansion from Y. This looks exactly like that in (7.66) above, but the first element is an admittance of a capacitor, etc. This gives the dual network of Figure 7.13(b).

### 7.4.2   Mid-shunt and Mid-series Ladders

Consider a bounded real transfer function with finite zeros on the $j\omega$-axis of the form

$$S_{21}(s) = \frac{\displaystyle\prod_{i=1}^{\ell} (s^2 + \omega_i^2)}{Q_n(s)}$$

(7.68)

where

$$\ell = (n - 1)/2 \quad \text{for } n \text{ odd}$$

$$= n/2 \qquad \quad \text{for } n \text{ even}$$

In general, the realization of these transfer functions requires coupled coils [1]. However, *under certain conditions*, in addition to the bounded real constraint, these functions can be realized in the ladder structure shown in Figure 7.14. The importance of these transfer functions lies in the fact that they are required to realize the optimum amplitude response of filters by introducing finite zeros of transmission on the $j\omega$-axis.

The conditions for the realization of the functions of (7.68) in either of the forms of Figure 7.14, are given in reference [1]. Here we shall only give the synthesis procedure. If $S_{21}(s)$ is given, we form the input impedance $Z(s)$ using the standard procedure discussed in Section 7.3.3. Assuming that $Z(s)$ is free of $j\omega$-axis poles, we produce the mid-shunt ladder of Figure 7.14(a) starting from a $Z(s)$ which has a zero at $s = \infty$. Thus $Y(s)$ has a pole at $s = \infty$. Suppose we extract a shunt capacitor $C$ from the admittance leaving a remainder

$$Y_1 = Y(s) = Cs \qquad (7.69)$$

where $C$ must be chosen such that $Z_1 = Y^{-1}$ has a pole at $j\omega_i$, where $\omega_i$ is a finite zero of transmission, i.e. a zero of the numerator of $S_{21}(s)$ in (7.68). The rule for this choice is

$$C = \min\left\{ \frac{Z^{-1}(j\omega_1)}{j\omega_1}, \frac{Z^{-1}(j\omega_2)}{j\omega_2}, \dots \frac{Z^{-1}(j\omega_1)}{j\omega_1}, \lim_{s\to\infty} \frac{Z^{-1}(s)}{s} \right\} \qquad (7.70)$$

where the minimization is taken only over those quantities which are well-defined and non-negative. This rule for choosing $C$ uncovers a pole of $Z_1(s)$ at $j\omega_1$, a zero of transmission. We then extract this pole as a parallel $LC$ in series, and repeat the procedure until we reach the terminating resistor $R$.

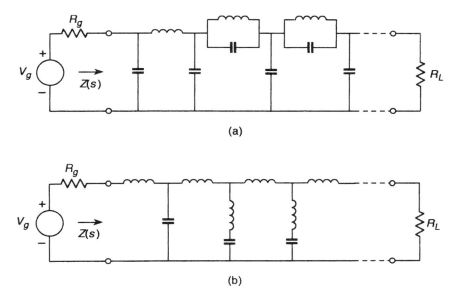

**Figure 7.14**   (a) Mid-shunt and (b) mid-series lossless ladders.

For the mid-series ladder of Figure 7.14(b), we note that in every cycle in the synthesis it is necessary to choose a series inductor such that its extraction uncovers a pole of $Y(s)$ which is also a zero of transmission. The rule for choosing the value of $L$ is given by

$$L = \min\left\{\frac{Z(j\omega_1)}{j\omega_1}, \frac{Z(j\omega)}{j\omega_2}, \ldots \frac{Z(j\omega_\ell)}{j\omega_\ell}, \lim_{s\to\infty}\frac{Z(s)}{s}\right\} \qquad (7.71)$$

*Example 7.4*   Consider the transfer function

$$S_{21}(s) = \frac{(s^2 + 8)}{3s^3 + 5s^2 + 10s + 8}$$

Assuming that the function is realizable as the transmission coefficient of a mid-shunt ladder of the type shown in Figure 7.14(a), find the network.

*Solution*   We begin by forming the input impedance $Z(s)$ according to the steps of Section 7.3.3. This gives

$$Z(s) = \frac{5s^2 + 4s + 8}{6s^3 + 5s^2 + 16s + 8}$$

which is devoid of poles on the $j\omega$-axis, infinity and the origin included. $Y(s) = Z^{-1}(s)$ has a pole at $s = \infty$. The zeros of transmission are at $s = \pm j\sqrt{8}$. The pole at $s = \infty$ is extracted partially according to (7.69) in the form of a shunt capacitor,

$$C = \min\left\{\frac{Z^{-1}(j\sqrt{8})}{j\sqrt{8}}, \lim_{s\to\infty}\frac{Z^{-1}(s)}{s}\right\}$$

$$= \min\{1, 6/5\} = 1\,\mathrm{F}$$

The remainder admittance is

$$Y_1(s) = Y(s) - s = \frac{s^3 + s^2 + 8s + 8}{5s^2 + 4s + 8}$$

i.e.

$$Z_1(s) = \frac{5s^2 + 4s + 8}{(s^2 + 8)(s + 1)} = \frac{4s}{(s^2 + 8)} + \frac{1}{(s + 1)} \qquad (7.72)$$

possessing a pole-pair at $s = \pm j\sqrt{8}$, which are the zeros of transmission. $Z_1(s)$ is realized from the above partial fraction expansion and the complete network is shown in Figure 7.15.

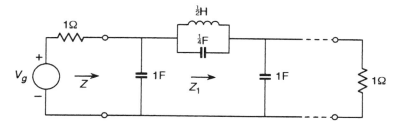

**Figure 7.15** Network of Example 7.4.

## 7.5 AMPLITUDE-ORIENTED DESIGN

In this section, we discuss the amplitude approximation problem for doubly-terminated lossless ladder filters. This consists in finding a realizable transducer power gain $|S_{21}(j\omega)|^2$ which is capable of meeting arbitrary specifications on the amplitude response of the filter. Once this function has been found, the synthesis techniques of Section 7.4 can be employed for the realization. Some commonly used filter characteristics lead to explicit formulae for the element values. These will be also given where appropriate in order to facilitate the design procedure.

### 7.5.1  General Considerations

The attenuation (or loss) function of a filter described by $S_{21}(j\omega)$ is defined as

$$\alpha(\omega) = 10 \log \frac{1}{|S_{21}(j\omega)|^2} \text{ dB} \tag{7.73}$$

It is convenient to begin our discussion by considering the design of low-pass filters, then proceed with the methods of obtaining other types. Thus, if we put

$$|S_{21}(j\omega)|^2 \equiv G(\omega^2) \tag{7.74}$$

then the low-pass approximation problem consists in determining $G(\omega^2)$ such that it meets the typical specifications shown in Figure 7.16. This is a tolerance scheme with the following features

(a) In the passband

$$G_p \leqslant G(\omega^2) \leqslant 1 \qquad 0 \leqslant \omega \leqslant \omega_0 \tag{7.75}$$

or, in terms of the attenuation

$$\alpha(\omega) \leqslant \alpha_p \qquad 0 \leqslant \omega \leqslant \omega_0 \tag{7.76}$$

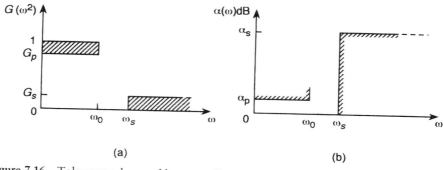

**(a)**                                                        **(b)**

**Figure 7.16**   Tolerance scheme of low-pass filter specifications: (a) amplitude; (b) attenuation.

where

$$\alpha_p = 10 \log G_p^{-1} \, \text{dB} \qquad (7.77)$$

(b) In the stopband

$$0 \leqslant G(\omega^2) \leqslant G_s \qquad \omega \geqslant \omega_s \qquad (7.78)$$

or, in terms of the attenuation

$$\alpha(\omega) \geqslant \alpha_s \qquad \omega \geqslant \omega_s \qquad (7.79)$$

where

$$\alpha_s = 10 \log G_s^{-1} \, \text{dB} \qquad (7.80)$$

(c) In the transition band, the power gain is assumed to decrease nonotonically from $G_p$ at $\omega_0$ to $G_s$ at $\omega_s$. Thus, the attenuation increases monotonically from $\alpha_p$ at $\omega_0$ to $\alpha_s$ at $\omega_s$.

(d) In the passband and stopband, the response lies within the shaded areas in Figure 7.16 (a) and the corresponding areas in Figure 7.16(b).

Now, the transducer power gain of the filter may be written as

$$|S_{21}(j\omega)|^2 = G(\omega)$$

$$= \frac{\displaystyle\sum_{r=0}^{m} a_r \omega^{2r}}{\displaystyle\sum_{r=0}^{n} b_r \omega^{2r}} \qquad (7.81)$$

and the problem may be posed as one of determining the coefficients $a_r$ and $b_r$, such that $|S_{21}(j\omega)|^2$ is capable of meeting an arbitrary set of specifications. Expression

(7.81) may be put in the alternative form

$$|S_{21}(j\omega)|^2 = G(\omega^2)$$

$$= \frac{1}{1 + \chi_n(\omega^2)} \qquad (7.82)$$

where $\chi_n(\omega^2)$ is a real rational function of $\omega^2$ which satisfies

$$\chi_n(\omega^2) \geqslant 0 \qquad -\infty \leqslant \omega \leqslant \infty. \qquad (7.83)$$

The most commonly used solutions to the amplitude approximation problem are now discussed. The transfer functions employed for the approximation are restricted to be of the *minimum-phase* type. A function $S_{21}(s)$ written as

$$S_{21}(p) = \frac{P_m(s)}{Q_n(s)} \qquad (7.84)$$

is said to be a minimum-phase function, if all its zeros lie in the closed left-half of the s-plane, i.e.

$$P_m(s) \neq 0 \quad \text{for Re } s > 0 \qquad (7.85)$$

The designation *minimum-phase* stems from the fact [2] that under the constraint (7.85), the phase shift $\psi(\omega)$ of $S_{21}(j\omega)$ is a minimum over the range $-\infty \leqslant \omega \leqslant \infty$ for a given transducer power gain $|S_{21}(j\omega)|^2$. In amplitude approximation, the constraint in (7.85) does not present any limitation since the formulation of the problem with the most favourable optimality criteria leads automatically to a minimum-phase function.

### 7.5.2   Maximally-flat Response in Both Passband and Stopband

This type of approximation leads to the so-called Butterworth response, the general appearance of which is shown in Figure 7.17. This is obtained by forcing the maximum possible number of derivatives of $|S_{21}(j\omega)|^2$, with respect to $\omega$, to vanish at $\omega = 0$ and $\omega = \infty$. To begin with, the conditions

$$S_{21}(0) = 1 \quad \text{and} \quad S_{21}(\infty) = 0 \qquad (7.86)$$

lead to the following form

$$|S_{21}(j\omega)|^2 = \frac{1 + \sum_{r=1}^{n-1} a_r \omega^{2r}}{1 + \sum_{r=1}^{n} b_r \omega^{2r}} \qquad (7.87)$$

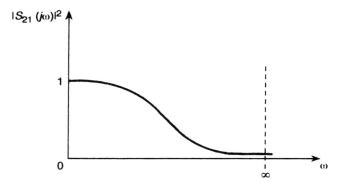

**Figure 7.17**   General appearance of maximally-flat filter response.

Next, to obtain a maximally flat response around $\omega = 0$, we force the first $(2n - 1)$ derivatives of $|S_{21}|^2$ to be zero. To this end, (7.87) is used to write

$$1 - |S_{21}(j\omega)|^2 = \frac{\displaystyle\sum_{r=1}^{n-1}(b_r - a_r)\omega^{2r} + b_n\omega^{2n}}{1 + \displaystyle\sum_{r=1}^{n}b_r\omega^{2r}} \tag{7.88}$$

and force the above function together with its first $(2n - 1)$ derivatives to be zero at $\omega = 0$. For this to occur, the series expansion of (7.88) around $\omega = 0$ must take the form:

$$1 - |S_{21}(j\omega)|^2 = c_n\omega^{2n} + c_{n+1}\omega^{2n+2} + \cdots \tag{7.89}$$

which when used in (7.88) leads to

$$a_r = b_r \quad r = 1, 2, \ldots (n - 1) \tag{7.90}$$

Thus, for a maximally flat response around the origin, the first $(n - 1)$ coefficients in the numerator of $|S_{21}|^2$ must be equal to the corresponding ones in the denominator. Next, it is required to force the maximally-flat conditions at $\omega = \infty$. To this end $|S_{21}(j\omega)|^2$ is re- written in the form

$$|S_{21}(j\omega)|^2 = \frac{\displaystyle\sum_{r=1}^{n-1}a_r\omega^{2r-2n} + \omega^{-2n}}{\displaystyle\sum_{r=1}^{n}b_r\omega^{2r-2n} + \omega^{-2n}} \tag{7.91}$$

whose series expansion around $\omega = \infty$ is forced to be of the form

$$|S_{21}(j\omega)|^2 = d_n\omega^{-2n} + d_{n+1}\omega^{-(2n+2)} \tag{7.92}$$

which when used in (7.91) yields

$$a_r = 0 \quad r = 1, 2, \ldots (n-1) \tag{7.93}$$

Thus, the maximally-flat condition around $\omega = \infty$ requires that the $(n-1)$ coefficients with highest powers in the numerator of $|S_{21}(j\omega)|^2$ be zero. Combining (7.90) with (7.93) and substituting in (7.87), we obtain

$$|S_{21}(j\omega)|^2 = \frac{1}{1 + b_n \omega^{2n}} \tag{7.94}$$

in which the coefficient $b_n$ may be set to unity without affecting the response. Thus

$$|S_{21}(j\omega)|^2 = \frac{1}{1 + \omega^{2n}} \tag{7.95}$$

with the 3 dB-point occurring at $\omega = 1$ for all $n$. This can be arbitrarily scaled, later in the synthesis, to any desired value. Typical responses of filters defined by (7.95) are shown in Figure 7.18, for varying degree $n$. These show that all aspects of the response improve with increasing the degree of the filter.

Later in the book, we shall need the expression for $S_{21}(s)$, which can now be determined from $|S_{21}(j\omega)|^2$. The poles of expression (7.95) occur at

$$\omega^{2n} = -1 = \exp\{j(2r-1)\pi\}, \quad r = 1, 2, \ldots 2n \tag{7.96}$$

i.e. at

$$\omega = \exp\left\{\frac{j(2r-1)\pi}{2n}\right\} \tag{7.97}$$

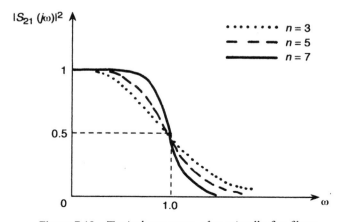

**Figure 7.18**  Typical responses of maximally-flat filters.

or

$$s = j \exp(j\theta_r)$$
$$= -\sin\theta_r + j\cos\theta_r \qquad (7.98)$$

where

$$\theta_r = \frac{(2r-1)}{2n}\pi \qquad (7.99)$$

For a bounded real $S_{21}(s)$ we select the left half-plane poles for a strictly Hurwitz denominator. Hence, the resulting transfer function is obtained as

$$S_{21}(s) = \frac{1}{\displaystyle\prod_{r=1}^{n}\{s - j\exp(j\theta_r)\}} \qquad (7.100)$$

Evidently, $S_{21}(s)$ is an *all-pole* transfer function having all its zeros at $s = \infty$. Hence, the function is realizable as a simple $LC$ ladder using the techniques of Section 7.4.1. Therefore, we first obtain

$$|S_{11}(j\omega)|^2 = 1 - |S_{21}(j\omega)|^2$$
$$= \frac{\omega^{2n}}{1 + \omega^{2n}} \qquad (7.101)$$

and let $\omega \to s/j$ to find

$$S_{11}(s)\, S_{11}(-s) = \frac{(-1)^n s^{2n}}{1 + (-s)^{2n}} \qquad (7.102)$$

Observing that the poles of $S_{11}(s)$ are the same as those of $S_{21}(s)$ we obtain

$$S_{11}(s) = \frac{\pm s^n}{\displaystyle\prod_{r-1}^{n}\{s - j\exp(j\theta_r)\}} \qquad (7.103)$$

Next the input impedance of the resistor-terminated ladder is obtained from (7.33), as

$$Z(s) = \frac{1 + S_{11}(s)}{1 - S_{11}(s)} \qquad (7.104)$$

which is realized as explained in 7.4.1. If the plus sign in (7.103) is chosen, then $Z(s)$ will have a pole at $s = \infty$ and the network of Figure 7.19(a) results. Alternatively, the minus sign in (7.103) may be chosen but in this case $Y(s) = 1/Z(s)$ will have a pole at

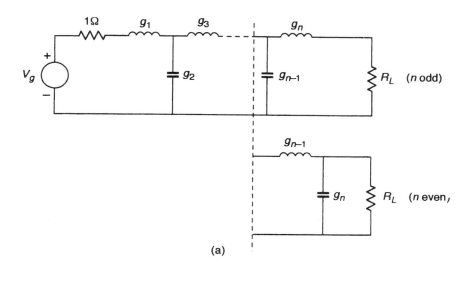

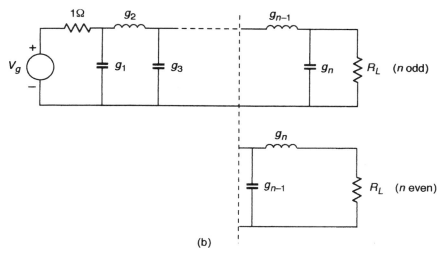

**Figure 7.19** Forms of realization of maximally-flat filters and Chebyshev low-pass prototypes.

$s = \infty$ and the network of Figure 7.19(b) results. In fact, it is possible to derive explicit formulae for the element values of both networks [1]. These are given by

$$g_r = 2 \sin \theta_r \quad r = 1, 2, \ldots n \tag{7.105}$$

where

$$g_r = L_r \text{ or } C_r \tag{7.106}$$

and

$$R_L = 1 \,\Omega \tag{7.107}$$

The above formulae also reveal that the resulting ladder is either *symmetric* for $n$ odd, or *antimetric* for $n$ even. Moreover the formulae give the element values of the equally-terminated low-pass prototype with $R_g = R_L = 1\,\Omega$, with the 3 dB point at $\omega = 1$.

It now remains to determine the degree $n$ of the required filter from a given set of specifications. These may be given in either of the following two forms

(a) 3-dB point at $\omega = 1$

Stopband edge at $\omega = \omega_s$ with $\alpha(\omega) \geqslant \alpha_s$ for $\omega \geqslant \omega_s$.

To obtain the required degree in this case, (7.95) is used to write at the stopband edge

$$10 \log(1 + \omega_s^{2n}) \geqslant \alpha_s \qquad (7.108)$$

which gives

$$n \geqslant \frac{\log(10^{0.1\alpha_s} - 1)}{2 \log \omega_s} \qquad (7.109)$$

in which $\omega_s$ is the actual frequency normalized with respect to the 3 dB point. Usually $\alpha_s$ is sufficiently large for unity to be neglected by comparison with $10^{0.1\alpha}$. Thus

$$n \geqslant \frac{\alpha_s}{20 \log \omega_s} \qquad (7.110)$$

(b) An alternative format for the specifications may be as follows

Maximum passband attenuation $= \alpha_p \quad \omega \leqslant \omega_p$

Minimum stopband attenuation $\alpha_s \quad \omega \geqslant \omega_s$

Ratio of stopband to passband edges $\omega_s/\omega_0 = \gamma$.

In the above format, the passband edge is defined at a frequency which is not necessarily the 3 dB point. Taking the above frequency values to be normalized with respect to the 3 dB point, then $\omega_p$ is a fraction while $\omega_s$ is greater than unity. In this case we require at the passband edge

$$10 \log(1 + \omega_p^{2n}) \leqslant \alpha_p \qquad (7.111)$$

so that

$$2n \log \omega_p \leqslant (10^{0.1\alpha_p} - 1) \qquad (7.112)$$

where the left side is negative. Also (7.109) is still valid so that

$$2n \log \omega_s \geqslant (10^{0.1\alpha_s} - 1) \qquad (7.113)$$

Combining (7.112) with (7.113) we have for the filter degree

$$n \geqslant \frac{\log\left[\dfrac{10^{0.1\alpha_s} - 1}{10^{0.1\alpha_p} - 1}\right]}{2\log(\gamma)} \tag{7.114}$$

### 7.5.3 The Chebyshev Response

A considerable improvement, in the rate of cutoff, over the maximally flat passband response results if we require $|S_{21}(j\omega)|^2$ to be equiripple in the passband while retaining the maximally flat response in the stopband. Typical responses for this type of approximation are shown in Figure 7.20. For a low-pass prototype with passband edge normalized to $\omega = 1$, we still use (7.93) to force the function to have $(2n - 1)$ zero derivatives at $\omega = \infty$. The resulting function takes the form

$$|S_{21}(j\omega)|^2 = \frac{1}{1 + \epsilon^2 T_n^2(\omega)} \tag{7.115}$$

where $T_n(\omega)$ is chosen to be an odd or even polynomial which oscillates between $-1$ and $+1$ the maximum number of times in the passband $|\omega| \leqslant 1$ and is monotonically increasing outside this interval. The desired behaviour of $T_n(\omega)$ is shown in Figure 7.21. This leads to a filter response in which $|S_{21}(j\omega)|^2$ oscillates between the values 1 and $1/(1 + \epsilon^2)$ in the passband $|\omega| \leqslant 1$. The size of the oscillations or *ripple* can be controlled by a suitable choice of the parameter $\epsilon$.

In order to derive the polynomial $T_n(\omega))$ from the above description, it is observed that all points in the passband where $|T_n(\omega)| = 1$ must be maxima or minima. This leads to

$$\left.\frac{\mathrm{d}T_n(\omega)}{\mathrm{d}\omega}\right|_{T_n(\omega)|=1} = 0 \quad \text{except at } |\omega| = 1 \tag{7.116}$$

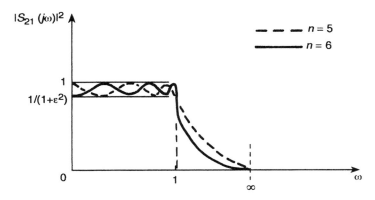

**Figure 7.20**  Typical Chebyshev low-pass filter responses.

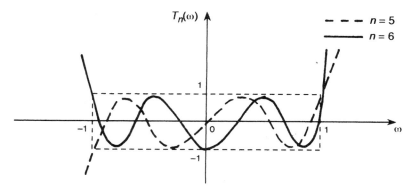

**Figure 7.21**   General appearance of the Chebyshev polynomials.

Consequently, the required polynomial satisfies the differential equation

$$\frac{dT_n(\omega)}{d\omega} = K\frac{\sqrt{1 - T_n^2(\omega)}}{\sqrt{1 - \omega^2}} \tag{7.117}$$

i.e.

$$\frac{dT_n(\omega)}{\sqrt{1 - T_n^2(\omega)}} = K\frac{d\omega}{\sqrt{1 - \omega^2}} \tag{7.118}$$

where $K$ is a constant to be determined such that $T_n(\omega)$ is a polynomial of exact degree $n$. Integrating (7.118) we have

$$\cos^{-1}T_n(\omega) = K\cos^{-1}\omega$$

i.e

$$T_n(\omega) = \cos(K\cos^{-1}\omega)$$

and if we put $K = n$, then $T_n(\omega)$ is an $n$th degree polynomial given by

$$\begin{aligned} T_n(\omega) &= \cos(n\cos^{-1}\omega) \qquad 0 \leqslant \omega \leqslant 1 \\ &= \cosh(n\cosh^{-1}\omega) \quad \omega > 1 \end{aligned} \tag{7.119}$$

This is the *Chebyshev polynomial of the first kind*, which can be conveniently obtained from the recurrence formula

$$T_{n+1}(\omega) = 2\omega T_n(\omega) - T_{n-1}(\omega) \tag{7.120}$$

with

$$T_0(\omega) = 1, \qquad T_1(\omega) = \omega$$

The above formula gives

$$T_2(\omega) = 2\omega^2 - 1$$

$$T_3(\omega) = 4\omega^3 - 3\omega$$

$$T_4(\omega) = 8\omega^4 - 8\omega^2 + 1$$

$$T_5(\omega) = 16\omega^5 - 20\omega^3 + 5\omega \tag{7.121}$$

$$T_6(\omega) = 32\omega^6 - 48\omega^4 + 18\omega^2 - 1$$

$$T_7(\omega) = 64\omega^7 - 112\omega^5 + 56\omega^3 - 7\omega$$

It is observed that

$$T_n(0) = 0 \quad \text{for } n \text{ odd}$$

$$T_n(0) = 1 \quad \text{for } n \text{ even} \tag{7.122}$$

which means that for $n$ odd $|S_{21}(0)|^2 = 1$ whereas for $n$ even $|S_{21}(0)|^2 = 1/(1 + \epsilon^2)$. Furthermore, $T_n(\omega)$ has the property

$$T_n(\omega) \approx 2^{n-1}\omega^n \quad \text{for } \omega \gg 1 \tag{7.123}$$

Therefore, the parameter $\epsilon$ in (7.115) can be chosen freely to determine the passband ripple, while increasing $n$ improves the stopband attenuation since $T_n(\omega)$ increases with increasing $n$ for $\omega > 1$. We shall see, shortly, that a combination of $\epsilon$ and $n$ can always be found to meet arbitrary specification. Now, the Chebyshev approximation is known to be the optimum solution to the problem of determining an $|S_{21}(j\omega)|^2$ which is constrained to lie in a band for $0 \leqslant \omega \leqslant \infty$ for a given degree $n$.

The synthesis of the transfer function defined by (7.114) can be performed using the technique of Section 7.4.1. The result is evidently a ladder of the form shown in Figure 7.19 since $S_{21}(s)$ would be an all-pole function. As before, we form

$$|S_{11}(j\omega)|^2 = 1 - S_{21}(j\omega)|^2$$

$$= \frac{\epsilon^2 T_n^2(\omega)}{1 + \epsilon^2 T_n^2(\omega)} \tag{7.124}$$

and

$$S_{11}(s)\,S_{11}(-s) = \frac{\epsilon^2 T_n^2(s/j)}{1 + \epsilon^2 T_n^2(s/j)} \tag{7.125}$$

which must be factored to form $S_{11}(s)$. The zeros of $S_{11}(s)S_{11}(-s)$ occur at

$$T_n^2(s/j) = \cos^2(n\cos^{-1}s/j) = 0 \tag{7.126}$$

or

$$s = j \cos \theta_r \qquad (7.127)$$

where

$$\theta_r = \frac{(2r - 1)}{2n} \pi \qquad (7.128)$$

The poles of (7.125) occur at

$$\epsilon^2 T_n^2(\omega) = -1 \qquad (7.129)$$

Let an auxiliary parameter $\eta$ be defined as

$$\eta = \sinh\left(\frac{1}{n} \sinh^{-1} \frac{1}{\epsilon}\right) \qquad (7.130)$$

Then, from (7.119), the pole locations satisfy

$$\cos^2(n \cos^{-1} \omega) = -\sin^2(n \sinh^{-1} \eta) \qquad (7.131)$$

or

$$n \cos^{-1} \omega = n \sin^{-1} j\eta + (2r - 1)\pi/2 \qquad (7.132)$$

i.e. the poles occur at

$$s = -j \cos[\sin^{-1} j\eta + \theta_r] \qquad (7.133)$$

Again, for a bounded real $S_{11}(s)$, the open left half-plane poles are selected. Thus, the denominator of $S_{11}(s)$ is given by

$$Q_n(s) = \prod_{r=1}^{n} [s + j \cos(\sin^{-1} j\eta + \theta_r)] \qquad (7.134)$$

In selecting the $n$ zeros of $S_{11}(s)$ from those given by (7.127), we are free to choose any combination provided the corresponding factors give rise to an $n$th degree real polynomial. We may, therefore, choose all the left half-plane zeros resulting in a minimum-phase reflection coefficient $S_{11}(s)$. Alternatively we may take those alternating between the two half-planes. Each choice leads to a network with different element values, but both have the same transducer power gain $|S_{21}(j\omega)|^2$.

Let us obtain a minimum-phase $S_{11}(s)$ by choosing the left half-plane zeros of (7.126). Combining the result with (7.134) we obtain

$$S_{11}(s) = \pm \prod_{r=1}^{n} \left\{ \frac{s + j \cos \theta_r}{s + (\eta \sin \theta_r + j\sqrt{1 + \eta^2} \cos \theta_r)} \right\} \qquad (7.135)$$

It is also observed that $S_{21}(s)$ has the same denominator as $S_{11}(s)$. Thus

$$S_{21}(s) = \frac{\prod_{r=1}^{n}\left\{\eta^2 + \sin^2\left(\frac{r\pi}{n}\right)\right\}^{\frac{1}{2}}}{\prod_{r=1}^{n}\left\{s + \left(\eta\sin\theta_r + j\sqrt{1+\eta^2}\cos\theta_r\right)\right\}} \qquad (7.136)$$

Performing the synthesis by forming the input impedance from (7.104) and (7.135), it is possible to obtain expressions for the element values in the networks of Figure 7.19, for the Chebyshev case, as follows [1]

$$g_1 = \frac{2}{\eta}\sin\left(\frac{\pi}{2n}\right) \qquad (7.137)$$

$$g_r g_{r+1} = \frac{4\sin\theta_r\sin\left(\frac{2r+1}{2n}\right)\pi}{\eta^2 + \sin^2\left(\frac{r\pi}{n}\right)} \qquad (7.138)$$

with

$$g_r = L_r \text{ or } C_r \qquad (7.139)$$

Again, the resulting ladder is symmetric for $n$ odd, or antimetric for $n$ even. For $n$ odd, $S_{21}(0) = 1$ and the load resistor $R_L = R_g = 1\,\Omega$. On the other hand, for $n$ even $|S_{21}(0)|^2 = 1/(1+\epsilon^2)$ and $R_L \neq 1$. To obtain the value of the load resistor, we note that at $\omega = 0$ all inductors become short circuits and all capacitors become open circuits. Therefore, by reference to the networks of Figure 7.19 for $n$ even,

$$|S_{21}(0)|^2 = \frac{4R_L}{(R_L + 1)^2} \qquad (7.140)$$

so that use of (7.115) gives the value of $R_L$ as

(a) $R_L = (\epsilon + \sqrt{1+\epsilon^2})^2$ for $S_{11}(0) > 0$

or

(b) $R_L = (\epsilon + \sqrt{1+\epsilon^2})^{-2}$ for $S_{11}(0) < 0$

Now, it remains to determine an expression for the required degree of the filter, to meet a typical set of specifications. Let these be expressed as follows

Passband attenuation $\alpha(\omega) \leqslant \alpha_p \quad 0 \leqslant \omega \leqslant 1$

Stopband attenuation $\alpha(\omega) \geqslant \alpha_s \quad \omega \geqslant \omega_s$

Therefore, from (7.115), we require in the passband

$$10\log(1 + \epsilon^2) \leqslant \alpha_p$$

or

$$\epsilon^2 \leqslant 10^{0.1\alpha_p} - 1$$

At the stopband edge we require

$$10\log\{1 + [\epsilon\cosh(n\cosh^{-1})w_s]^2\} \geqslant \alpha_s$$

The above conditions are solved for $n$ to give for the required degree

$$n \geqslant \frac{\cosh^{-1}[(10^{0.1\alpha_s} - 1)/(10^{0.1\alpha_p} - 1)]^{0.5}}{\cosh^{-1}w_s} \qquad (7.141)$$

in which $w_s$ is the actual frequency normalized to the passband edge, since the latter is assumed to be at $\omega = 1$.

### 7.5.4   Elliptic Function Response

This is the optimum amplitude response in the sense of minimizing the maximum deviation from specified values in each band of a rational function, for the two-band approximation of Figure 7.16. It gives rise to equiripple responses in both the passband and stopband. Figure 7.22 shows a typical low-pass response of this optimum equiripple case. To obtain such a response, we abandon the restriction on $S_{21}(s)$ to be an all-pole function and allow finite zeros of transmission on the $j\omega$-axis. It is this more general approach that results in the vast improvement in the amplitude response by comparison with the maximally-flat and Chebyshev cases. By reference to the general expression (7.82) for the transducer power gain we now allow $\chi(\omega^2)$ to be the square of a *rational function*, instead of just the square of a *polynomial*. Thus, we write

$$|S_{21}(j\omega)|^2 = \frac{1}{1 + \epsilon^2 F_n^2(\omega)} \qquad (7.142)$$

where $F_n(\omega)$ is required to have the behaviour shown in Figure 7.23, which possesses the following properties

(1)  $F_n(\omega)$ has all its $n$ zeros in the passband $|\omega| < 1$ and all its poles outside this interval.

(2)  $F_n(\omega)$ oscillates between the values $-1$ and $+1$ in the passband $|\omega| < 1$

(3)  $F_n(1) = 1$

(4)  $1/F_n(\omega)$ oscillates between $-1/M$ and $+1/M$ in the stopband $|\omega| > w_s$ where

$$M^2 = \frac{G_s^{-1} - 1}{G_p^{-1} - 1} = \frac{10^{0.1\alpha_s} - 1}{10^{0.1\alpha_p} - 1} \qquad (7.143)$$

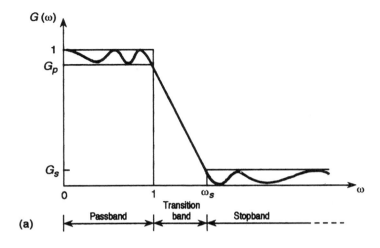

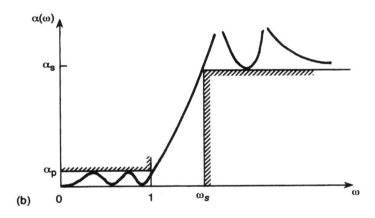

**Figure 7.22** General appearance of elliptic function response: (a) amplitude; (b) attenuation.

and

$$\epsilon^2 = 10^{0.1\alpha_p} - 1 \qquad (7.144)$$

The above properties lead to the following conditions

$$\left.\frac{dF_n(\omega)}{d\omega}\right|_{F_n(\omega)|=1} = 0 \text{ except at } |\omega| = 1 \qquad (7.145)$$

$$\left.\frac{dF_n(\omega)}{d\omega}\right|_{F_n(\omega)=M^{-1}} \text{ except at } |\omega| = \omega_s \qquad (7.146)$$

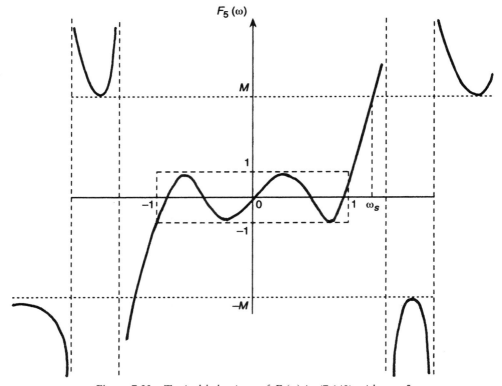

**Figure 7.23**   Typical behaviour of $F_n(\omega)$ in (7.142) with $n = 5$.

It follows that $F_n(\omega)$ satisfies the differential equation

$$\frac{\mathrm{d}F_n(\omega)}{\sqrt{[1 - F_n^2(\omega)][M^2 - F_n^2(\omega)]}} = \frac{A\,\mathrm{d}\omega}{\sqrt{(1 - \omega^2)(\omega_s^2 - \omega^2)}} \tag{7.147}$$

where $A$ is a constant. The solution of the above equation requires the use of elliptic integrals and Jacobian elliptic functions. These are now briefly introduced. The *elliptic integral of the first kind* is defined as

$$u(\phi \mid k) = \int_0^\phi \frac{\mathrm{d}x}{\sqrt{1 - k^2\sin^2x}} \tag{7.148}$$

where $k \leqslant 1$ and is called the *modulus* of the elliptic integral. If $\phi = \pi/2$, the integral (7.148) becomes the *complete elliptic integral of the first kind* denoted by

$$K(k) = u\left(\frac{\pi}{2} \middle| k\right)$$

$$= \int_0^{\pi/2} \frac{\mathrm{d}x}{\sqrt{1 - k^2\sin^2x}} \tag{7.149}$$

The complementary elliptic integral of the first kind is defined by

$$K'(k) = u\left(\frac{\pi}{2}\bigg|k'\right)$$

$$= \int_0^{\pi/2} \frac{dx}{\sqrt{1 - k'^2 \sin^2 x}}$$

(7.150)

where

$$k' = \sqrt{1 - k^2}$$

(7.151)

The elliptic integrals for given moduli can be obtained from extensive sets of tables [37], or by writing a simple computer program [38].

The Jacobian elliptic functions are defined by reference to (7.148)–(7.151) as

$$\text{elliptic sine} \qquad \text{sn}(u \mid k) = \sin \phi$$

$$\text{elliptic cosine} \qquad \text{cn}(u \mid k) = \cos \phi$$

(7.152)

with similar definitions of the elliptic tangent, etc. Again the values of these functions can be obtained from tables [37] or by writing a computer program [38].

With the above definitions of elliptic integrals and functions, the solution to the differential equation (7.147) can now be written.

(1) for $n$ even:

$$F_n(\omega) = \prod_{r=1}^{n/2}\left\{\frac{\omega^2 - \omega_r^2}{\omega^2 - (\omega_s^2/\omega_r^2)}\right\}$$

(7.153)

where

$$\omega_r = \text{sn}\left(\frac{(2r - 1)K(\omega_s^{-1})}{n}\bigg|\omega_s^{-1}\right)$$

(7.154)

and

$$A = \prod_{r=1}^{n/2}\left\{\frac{1 - (\omega_s^2/\omega_r^2)}{1 - \omega_r^2}\right\}$$

(7.155)

(2) for $n$ odd:

$$F_n(\omega) = B\omega \prod_{r=1}^{(n-1)/2}\left\{\frac{\omega^2 - \omega_r^2}{\omega^2 - (\omega_s^2/\omega_r^2)}\right\}$$

(7.156)

where

$$\omega_r = \text{sn}\left(\frac{2rK(\omega_s^{-1})}{n}\bigg|\omega_s^{-1}\right) \tag{7.157}$$

and

$$B = \prod_{r=1}^{(n-1)/2}\left\{\frac{1-(\omega_s^2/\omega_r^2)}{1-\omega_r^2}\right\} \tag{7.158}$$

In the above expressions, the constants $A$ and $B$ are chosen such that $F(1) = 1$.

Now, given the set of specifications $\alpha_p$, $\alpha_s$, and $\omega_s$, the degree of the required elliptic filter is obtained from

$$n \geqslant \frac{K(\omega_s^{-1})K'(M^{-1})}{K'(\omega_s^{-1})K(M^{-1})} \tag{7.159}$$

in which $M$ is calculated from $\alpha_p$ and $\alpha_s$ using (7.143). It is noted that the passband edge is at $\omega = 1$, so that $\omega_s$ is the ratio of the *actual* stopband edge to the actual passband edge.

The vast majority of elliptic filters can be realized using the partial pole extraction techniques in the mid-shunt or mid-series ladder forms discussed in Section 7.4.2. Due to the fact that these filters provide the optimum amplitude response for a given degree, the normalized design has been made available in an extensive set of tables [39]. These may be used directly to obtain the element values from the specifications.

### 7.5.5  Frequency Transformations and Impedance Scaling

Our discussion has, so far, concentrated on low-pass prototype filters in which the passband edge (cut-off frequency) is normalized to $\omega = 1$ and the source resistor is normalized to $1\,\Omega$. We now consider the process of denormalization of these values to arbitrary ones, as well as the design of high-pass, band-pass and band-stop filters, relying on the low-pass prototype. Table 7.1 summarizes the appropriate transformations which are now derived.

#### *Low-pass to low-pass transformation*

In the prototype transfer function, denormalization to an arbitrary cut-off $\omega_0$ can be achieved by means of the transformation

$$\omega \rightarrow \omega/\omega_0 \tag{7.160}$$

or

$$s \rightarrow s/\omega_0 \tag{7.161}$$

In terms of the element values, the above transformation amounts to scaling all the frequency-dependent elements by the factor $1/\omega_0$. Thus, in the prototype network (e.g. Figure 7.19) we let

$$L_r \rightarrow L_r/\omega_0$$

and                                                                                   (7.162)

$$C_r \rightarrow C_r/\omega_0$$

The resulting filter response is given by the following functions

(1) *Maximally Flat*

$$|S_{21}(j\omega)|^2 = \frac{1}{1 + \left(\dfrac{\omega}{\omega_0}\right)^{2n}}$$                                  (7.163)

(2) *Chebyshev*

$$|S_{21}(j\omega)|^2 = \frac{1}{1 + \epsilon^2 T_n^2\left(\dfrac{\omega}{\omega_0}\right)}$$                                  (7.164)

(3) *Elliptic*

$$|S_{21}(j\omega)|^2 = \frac{1}{1 + \epsilon^2 F_n^2\left(\dfrac{\omega}{\omega_0}\right)}$$                                  (7.165)

### Low-pass to high-pass transformation

A high-pass response with passband edge at $\omega_0$ can be obtained from the low-pass prototype transfer function by letting

$$\omega \rightarrow \omega_0/\omega$$                                                    (7.166

or

$$s \rightarrow \omega_0/s$$                                                              (7.167)

This transformation is illustrated in Figure 7.24. It results in the transformation of the prototype element values according to

$$L_r' \rightarrow C_r' = 1/\omega_0 L_r$$                                              (7.168)

and

$$C_r \rightarrow L_r' = 1/\omega_0 C_r$$                                              (7.169)

i.e. inductors become capacitors and vice versa.

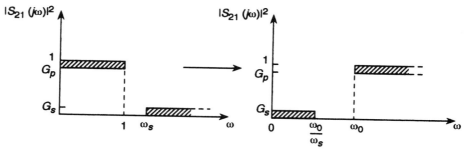

**Figure 7.24**   Low-pass to high-pass transformation.

The resulting filter responses of the three basic types are given by

(1)  *Maximally flat*

$$|S_{21}(j\omega)|^2 = \frac{1}{1 + \left(\frac{\omega_0}{\omega}\right)^{2n}} \tag{7.170}$$

(2)  *Chebyshev*

$$|S_{21}(j\omega)|^2 = \frac{1}{1 + \epsilon^2 T_n^2\left(\frac{\omega_0}{\omega}\right)} \tag{7.171}$$

(3)  *Elliptic*

$$|S_{21}(j\omega)|^2 = \frac{1}{1 + \epsilon^2 F_n^2\left(\frac{\omega_0}{\omega}\right)} \tag{7.172}$$

## Low-pass to band-pass transformation

Starting from the low-pass prototype specifications shown in Figure 7.25(a) in which the negative frequency side is included, we seek a transformation to a band-pass response with passband extending from $\omega_1$ to $\omega_2$ as shown in Figure 7.25(b). In the prototype transfer function let

$$\omega \rightarrow \beta\left(\frac{\omega}{\omega_0} - \frac{\omega_0}{\omega}\right) \tag{7.173}$$

or

$$s \rightarrow \beta\left(\frac{s}{\omega_0} + \frac{\omega_0}{s}\right) \tag{7.174}$$

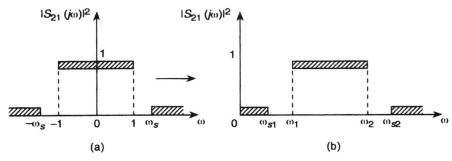

**Figure 7.25**   Low-pass to band-pass transformation.

where $\beta$ and $\omega_0$ are to be determined from $\omega_1$ and $\omega_2$. By reference to Figure 7.25 we use (7.173) to impose the conditions

$$-1 = \beta\left(\frac{\omega_1}{\omega_0} - \frac{\omega_0}{\omega_1}\right) \tag{7.175}$$

$$1 = \beta\left(\frac{\omega_2}{\omega_0} - \frac{\omega_0}{\omega_2}\right) \tag{7.176}$$

which, when solved simultaneously, give

$$\omega_0 = \sqrt{\omega_1\omega_2}$$

$$\beta = \frac{\omega_0}{\omega_2 - \omega_1} \tag{7.177}$$

For example the low-pass maximally-flat response function given by (7.95) is transformed to the band-pass one given by

$$|S_{21}(j\omega)|^2 = \frac{1}{1 + \left\{\beta\left(\dfrac{\omega}{\omega_0} - \dfrac{\omega_0}{\omega}\right)\right\}^{2n}} \tag{7.178}$$

while the Chebyshev low-pass function given by (7.115) becomes

$$|S_{21}(j\omega)|^2 = \frac{1}{1 + \epsilon^2 T_n^2\left\{\beta\left(\dfrac{\omega}{\omega_0} - \dfrac{\omega_0}{\omega}\right)\right\}} \tag{7.179}$$

The elliptic response case is obtained by applying the same transformation (7.173) to (7.142).

Now, the low-pass to band-pass transformation in (7.174) implies that, in the low-pass prototype network, every inductor is transformed into a series resonant circuit, and every capacitor into a parallel resonant circuit, as illustrated in Table 7.1.

CONTINUOUS-TIME FILTER MODELS

Table 7.1

| Normalized low-pass prototype with cut-off at $\omega = 1$ | Low-pass with cut-off at $\omega = \omega_0$ | High-pass with cut-off at $\omega = \omega_0$ | Band-pass with passband edges at $\omega_1$ and $\omega_2$ | Band-stop with stopband edges at $\omega_1$ and $\omega_2$ ($\omega_s$ is the stopband edge in the low-pass prototype) |
|---|---|---|---|---|
| $L_r$ | $L_r/\omega_0$ | $1/L_r\omega_0$ | $\dfrac{\beta L_r}{\omega_0}$    $\dfrac{1}{\beta L_r\omega_0}$ | $\dfrac{L_r\omega_s}{\beta\omega_0}$    $\dfrac{\beta}{L_r\omega_0\omega_s}$ |
| $C_r$ | $C_r/\omega_0$ | $1/C_r\omega_0$ | $\dfrac{1}{\beta C_r\omega_0}$    $\dfrac{\beta C_r}{\omega_0}$ | $\dfrac{\beta}{C_r\omega_0\omega_s}$    $\dfrac{C_r\omega_s}{\beta\omega_0}$ |
| For denormalization to arbitrary source resistor $R_g$, $L \to R_gL$, $C \to C/R_g$, $R_L \to R_gR_L$ |||||

Examination of the transformation in (7.173) reveals that the resulting band-pass response has geometric symmetry about $\omega_0$ which is referred to as the *band centre*. This means that the amplitude has the same value at every pair of frequencies $\overline{\omega}$ and $\overline{\overline{\omega}}$ related by $\overline{\omega}\overline{\overline{\omega}} = \omega_0^2$, i.e.

$$\alpha(\overline{\omega}) = \alpha(\overline{\overline{\omega}}) \tag{7.180}$$

for

$$\overline{\omega}\overline{\overline{\omega}} = \omega_0^2 \tag{7.181}$$

### Low-pass to band-stop transformation

This is illustrated in Figure 7.26, and can be achieved by starting from the low-pass prototype and obtaining a *high-pass* prototype with passband edge at $\omega = 1$ by the transformation $s \to 1/s$, then applying the same band-pass transformation in (7.174) to the high-pass prototype. This gives

$$\omega \to 1 \left/ \left\{ \beta\left(\frac{\omega}{\omega_0} - \frac{\omega_0}{\omega}\right)\right\}\right. \tag{7.182}$$

or

$$s \to 1 \left/ \left\{ \beta\left(\frac{s}{\omega_0} + \frac{\omega_0}{s}\right)\right\}\right. \tag{7.183}$$

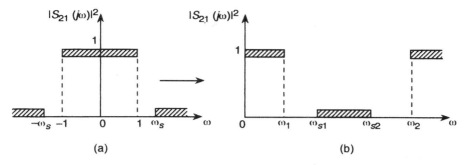

**Figure 7.26** Low-pass to band-stop transformation.

with

$$\omega_0 = \sqrt{\omega_1 \omega_2}, \quad \beta = \frac{\omega_0}{\omega_1 - \omega_2} \tag{7.184}$$

where $\omega_1$ is the lower passband edge and $\omega_2$ is the upper passband edge. The appropriate transformations of the low-pass prototype elements are shown in Table 7.1.

### Impedance scaling

For an arbitrary source resistor $R_g$, all impedance values are scaled by the same value $R_g$. Thus $L \to R_g L$, $C \to C/R_g$ and $R_l \to R_g R_L$. Naturally, if the prototype had equal termination, then $R_L(=1\Omega) \to R_g$.

### 7.5.6   Design Examples

*Example 7.5*   Design a maximally-flat low-pass filter with the following specifications:

   Passband 0–10 kHz, attenuation $\leqslant 3$ dB
   Stopband edge: 40 kHz, attenuation $\geqslant 30$ dB
   Equal terminating resistors of 600 $\Omega$

*Solution*   The normalized value of stopband edge is $\omega_s = 40/10 = 4$, relative to the 3 dB point. Thus the required degree is obtained from (7.110) as

$$n \geqslant \frac{30}{20 \log 4} \geqslant 2.94, \quad \text{or} \quad n = 3$$

The element values of the prototype are obtained from (7.105) and this is shown in Figure 7.27(a). Next this network is impedance scaled by 600 $\Omega$ and the transformation in Table 7.1 is used to locate the cutoff at $\omega_o = 2\pi \times 10^4$. This gives the network of Figure 7.27(b).

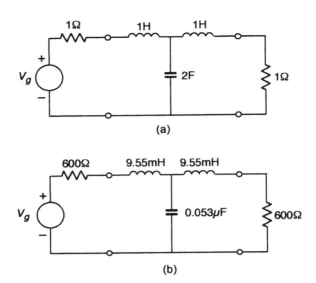

**Figure 7.27**   Filter of Example 7.5: (a) Normalized prototype; (b) required filter.

*Example 7.6*   Design a Chebyshev low-pass filter with the following specifications:

Passband 0–10 kHz, 0.0988 dB ripple
Stopband edge: 40 kHz, attenuation $\geqslant 30$ dB
Equal terminating resistors of 50 Ω

*Solution*   Normalizing the frequencies to the given cutoff of 10 k Hz we have $\omega_s =$ 20/10. Also from (7.115)

$$\epsilon^2 = 10^{0.0098} - 1 = 0.023\epsilon = 0.152$$

Substituting in (7.141) for $\alpha_s = 30$, $\alpha_p = 0.0988$ and $\omega_s = 2$ we obtain n=5. The auxiliary parameter in (7.130) is

$$\eta = \sinh\left[\frac{1}{5}\sinh^{-1}\frac{1}{0.153}\right] = 0.543$$

Therefore, using (7.137) and (7.138) we obtain the element values of the prototype with reference to Figure 7.19(b), n=5. These are

$$C_1 = C_5 = 1.144, \quad L_2 = L4 = 1.372, \quad C_3 = 1.972$$

Finally the above values are scaled in impedance by 50 Ω and the transformation in Table 7.1 is used to give the required cutoff at 10 kHz. This gives the required filter shown in Figure 7.28.

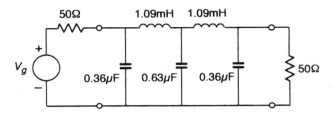

**Figure 7.28** Filter of Example 7.6.

*Example 7.7*  Design a band-pass Chebyshev filter with the specifications illustrated by the tolerance scheme of Figure 7.29 with 50 $\Omega$ equal terminations.

*Solution*  The tranformation of (7.173) and Figure 7.25 produces a response with geometric symmetry around band-centre $\omega_o$. $\alpha(\omega)$ has the same value at every pair of frequencies $f_{s1}f_{s2} = f_o^2$. The specfications of Figure 7.29 do not possess such symmetry: $1.0 \times 1.5 \neq 0.75 \times 2.2$. Therefore, with $f_o^2 = 1.0 \times 1.5 = 1.5$, the filter has to be designed according to the more severe of the two requirements: 20 dB at 0.75 mHz or 20 dB at 2.2 MHz, and the other one will be over-satisfied. If we require $\alpha \geqslant 20$ for $f \geqslant 2.2$, we also obtain $\alpha \geqslant 20$ for $f \leqslant (1.5/2.2) \leqslant 0.68$ and we have failed to satisfy the lower stopband requirement at 0.75. On the other hand, requiring $\alpha \geqslant 20$ for $f \leqslant 0.75$ also gives $\alpha \geqslant 20$ for $f \geqslant (1.5/0.75) \geqslant 2$. Thus the upper stopband requirement at 2.2 is over-satisfied, therefore we use the 20 dB requirement at 0.75 to determine the prototype. From (7.184)

$$\omega_o = \sqrt{\omega_1\omega_2} = 2\pi \times 1.225 \times 10^6, \quad (\omega_2 - \omega_1) = \pi \times 10^6, \quad \beta = \frac{\omega_o}{\omega_2 - \omega_1} = 2.45$$

The frequency 0.75 MHz corresponds to $-\omega_s$ in Figure 7.25(a) of the low-pass prototype and is obtained using (7.182) as

$$-\omega_s = 2.45 \left( \frac{0.75}{1.225} - \frac{1.225}{0.75} \right) = -2.5$$

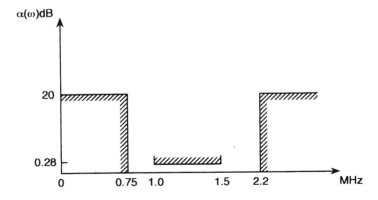

**Figure 7.29**  Specifications on the Filter of Example 7.7.

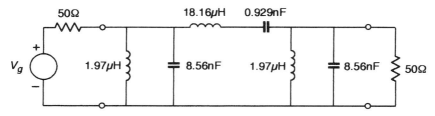

**Figure 7.30**   Filter of Example 7.7.

The determination of the prototype degree is then the same as in the previous example with the above value of $\omega_s$. This gives $n = 3$. The element values with reference to Figure 7.19(b) with $n = 3$ are $C_1 = 1.345$, $L_2 = 1.141$, $C_2 = 1.345$. The transformation in Table 7.1 is used with scaling by $50\omega$. to obtain the required sixth-order band-pass filter of Figure 7.30.

## 7.6   PHASE AND DELAY FUNCTIONS

The ideal (no distortion) phase characteristic is a linear function of $\omega$, as shown in Figure 7.3 for the low-pass case. In the treatment of phase approximation to the ideal characteristic, the problem may be stated directly in terms of the required phase. Thus, if we write

$$S_{21}(j\omega) = |S_{21}(j\omega)|\, e^{j\psi(\omega)} \tag{7.185}$$

$$\psi(\omega) \approx -k\omega \quad |\omega| \leqslant \omega_0 \tag{7.186}$$

Alternatively, the problem may be formulated in terms of either the group delay $T_g(\omega)$ or the phase delay $T_{ph}(\omega)$ defined by

$$T_g(\omega) = -\frac{d\psi(\omega)}{d\omega} \tag{7.187}$$

$$T_{ph}(\omega) = -\frac{\psi(\omega)}{\omega} \tag{7.188}$$

Evidently, for an approximation to the ideal phase characteristic in the passband $|\omega| < \omega_0$, the group-delay in (7.187) is required to approximate a constant within the passband. Taking logarithm of both sides of (7.185) we obtain

$$\ln S_{21}(j\omega) = \ln|S_{21}(j\omega)| + j\psi(\omega)$$
$$= \tfrac{1}{2} \ln\{S_{21}(j\omega)\, S_{21}(-j\omega)\} + j\psi(\omega) \tag{7.189}$$

or

$$\psi(\omega) = j\frac{1}{2} \ln \frac{S_{21}(j\omega)}{S_{21}(-j\omega)} \tag{7.190}$$

so that

$$\frac{d\psi(\omega)}{d\omega} = \frac{1}{2}\left[\frac{d}{d(j\omega)}\ln S_{21}(j\omega) + \frac{d}{d(-j\omega)}\ln S_{21}(-j\omega)\right]$$

$$= -\text{Re}\left[\frac{d}{d(j\omega)}\ln S_{21}(j\omega)\right] \tag{7.191}$$

and we can write

$$T_g(\omega) = -Ev\left[\frac{d}{ds}\ln S_{21}(s)\right]\Bigg|_{s=j\omega} \tag{7.192}$$

Furthermore, if we write $S_{21}(j\omega)$ in the form

$$S_{21}(j\omega) = \frac{E_1(\omega) + jO_1(\omega)}{E_2(\omega) + jO_2(\omega)} \tag{7.193}$$

where $E_{1,2}(\omega)$ are even and $O_{1,2}(\omega)$ are odd polynomials, then the phase $\psi(\omega)$ is an odd function given by

$$\psi(\omega) = \tan^{-1}\frac{O_1(\omega)}{E_1(\omega)} - \tan^{-1}\frac{O_2(\omega)}{E_2(\omega)} \tag{7.194}$$

Defining the generalized phase function as

$$\psi(s) = \frac{1}{2}\ln\left[\frac{S_{21}(s)}{S_{21}(-s)}\right] \tag{7.195}$$

so that

$$\psi(\omega) = -j\psi(s)|_{s=j\omega} \tag{7.196}$$

and the generalized group delay can be defined as

$$T_g(s) = -\frac{d\psi(s)}{ds}$$

$$= \frac{1}{2}\frac{d}{ds}\left[\ln\frac{S_{21}(s)}{S_{21}(-s)}\right] \tag{7.197}$$

$$= \frac{1}{2}\left[\frac{d}{ds}\ln S_{21}(s) + \frac{d}{d(-s)}\ln S_{21}(-s)\right]$$

i.e.

$$T_g(s) = Ev\left[\frac{d}{ds}\ln S_{21}(s)\right] \tag{7.198}$$

so that

$$T_g(\omega) = T_g(s)|_{s=j\omega} \tag{7.199}$$

If $S_{21}(s)$ is written as

$$S_{21}(s) = \frac{P(s)}{Q(s)} \tag{7.200}$$

then use of (7.198) gives

$$T_g(s) = \frac{1}{2}\left\{\frac{P'(s)}{P(s)} + \frac{P'(-s)}{P(-s)} - \frac{Q'(s)}{Q(s)} - \frac{Q'(-s)}{Q(-s)}\right\} \tag{7.201}$$

Now, for reasons which will be discussed later, continuous-time transfer functions approximating the ideal phase characteristics are of no use in the design of sampled-data filters. Therefore, the latter have to be redesigned independently, and we shall not discuss the phase-oriented design of continuous time filters.

## 7.7   DISTRIBUTED NETWORK MODELS

We shall see in a later chapter that there exists a formal analogy between switched capacitor filters and certain classes of distributed networks. These use as building blocks, sections of lossless transmission line each of length $\ell$ and characteristic impedance $Z_o$, which is referred to as a *unit element* [1]. This is shown symbolically in Figure 7.31. A popular structure for filter design is a resistor-terminated cascade of $n$ unit elements having the same length but different characteristic impedances as shown in Figure 7.32. Direct analysis of this circuit gives

$$S_{21}(\lambda) = \frac{(1-\lambda^2)^{n/2}}{D_n(\lambda)} \tag{7.202}$$

where

$$\lambda = \tanh(\tau s) \tag{7.203}$$

and $\tau$ is the one-way propagation delay of the unit element. Clearly $\tau = \ell/c$ where c is the speed of propagation of electromagnetic waves in the medium. In (7.202) $D_n(\lambda)$ is

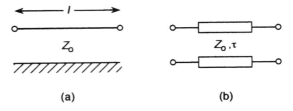

**Figure 7.31**   (a) The unit element (UE) and (b) its symbol.

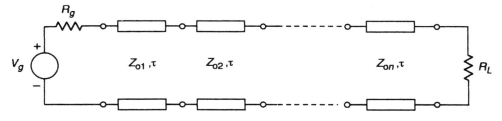

**Figure 7.32**  Resistor-terminated cascade of UE's.

a strict Hurwitz polynomial in $\lambda$. On the $j\omega$ axis, $\lambda \to j\Omega$ where

$$\Omega = \tan(\tau\omega) \tag{7.204}$$

so that

$$|S_{21}(j\Omega)|^2 = \frac{(1+\Omega^2)^n}{|D_n(j\Omega)|^2} \tag{7.205}$$

Noting that

$$\sin^2(\tau\omega) = \Omega^2/(1+\Omega^2) \tag{7.206}$$

and

$$\cos^2(\tau\omega) = 1/(1+\Omega^2) \tag{7.207}$$

expression (7.205) may be put in the form

$$|S_{21}|^2 = \frac{1}{P_n(\sin^2(\tau\omega))} \tag{7.208}$$

For a maximally flat response around $\tau\omega = 0$ and one zero derivative at $\tau\omega = \pi/2$ (i.e. $\Omega = \infty$) we have

$$|S_{21}|^2 = \frac{1}{1+\left(\dfrac{\sin\tau\omega}{\sin\tau\omega_o}\right)^{2n}} \tag{7.209}$$

and the 3 dB point occurs at $\omega_o$.

For an optimum equiripple response in the passband up to $\omega_o$ and one zero derivative at $\tau\omega = \pi/2$ ($\Omega = \infty$) we have

$$|S_{21}|^2 = \frac{1}{1+\epsilon^2 T_n^2\left(\dfrac{\sin\tau\omega}{\sin\tau\omega_o}\right)} \tag{7.210}$$

where $T_n$ is the Chebyshev polynomial of the first kind in $(\sin\tau\omega/\sin\tau\omega_o)$.

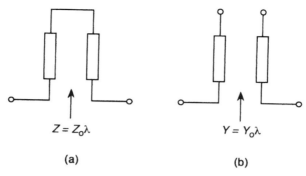

**Figure 7.33**   (a) Short-circuited stub and (b) open-circuited stub.

We shall not examine the realization of these transfer functions, since we are only interested in its formal appearance in relation to switched capacitor filter transfer functions as we shall see in a later chapter.

Now, if a unit element is terminated in a short circuit, it becomes a short-circuited stub: a two-terminal element whose impedance is $Z_o\lambda$. This has the same formal properties with respect to the variable $\lambda$ as those of the inductor with respect to the usual complex frequency variable $s$. Similarly, a unit element terminated in an open circuit is an open-circuited stub with admittance $Y_o\lambda$, and it has the same formal properties with respect to $\lambda$ as those of the capacitor with respect to $s$. Both elements are depicted in Figure 7.33. These properties are highlighted by the one-to-one corre-spondence between the s-plane and the $\lambda$-plane. This will be pursued further in the next chapter. Meanwhile, it suffices to indicate that a distributed network can be obtained from a lumped prototype as shown in Figure 7.34, by transforming every inductor of value $L$ into a short-circuited stub of characteristic impedance $L$, and every capacitor into an open-circuited stub of characteristic admittance $C$. The result-ing network will have the same formal properties with respect to $\lambda$ as the lumped one had with respect to $s$. Thus, all $\lambda$-domain networks can be synthesized using the same techniques of this chapter. In particular the relations pertaining to the Darlington synthesis of resistively terminated two-ports can be used with $s$ replaced by $\lambda$ and $\omega$

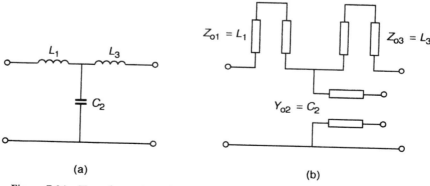

**Figure 7.34**   Transformation of a lumped filter into a distributed all-stub ladder.

replaced by $\Omega$. For example the following relations can be used:

$$|S_{11}(j\Omega)|^2 + |S_{21}(j\Omega)|^2 = 1 \tag{7.211}$$

$$S_{11}(\lambda)S_{11}(-\lambda) + S_{21}(\lambda)S_{21}(-\lambda) = 1 \tag{7.212}$$

$$Z_{in}(\lambda) = R_g \frac{1 + S_{11}(\lambda)}{1 - S_{11}(\lambda)} \tag{7.213}$$

More will be said about this generalization later in the book and we finally note that the factor $\sqrt{1 - \lambda^2}$ in (7.202) will not affect this generalization, since the impedance function itself will be rational.

## CONCLUSION

This chapter dealt with the realizability conditions and synthesis techniques of, mainly, passive lumped filters. The approximation problem in filter design was treated in some detail. The filters discussed in this chapter are useful in themselves, but our main interest is in their use as prototype models from which switched-capacitor filters can be derived. Particular distributed circuit models were also pointed out due to their formal appearance in relation to transfer functions of switched-capacitor filters as will be seen later.

## PROBLEMS

7.1   A lossless two-port has the transmission coefficient

$$S_{21}(S) = \frac{4}{s^3 + 4s^2 + 6s + 4}$$

referred to $1\,\Omega$ terminations. Calculate the input impedance of the $1\Omega$-terminated two-port, and hence find a realization.

7.2   Realize the following transducer power gain as a doubly-terminated lossless two-port:

$$|S_{21}(j\omega)|^2 = \frac{1}{1 + \omega^2(4\omega^2 - 3)^2}$$

7.3   Design a low-pass lumped-element maximally-flat filter with the following specifications:

Passband edge at 1 MHz, with 0.5 dB maximum attenuation
Stopband edge at 2.5 MHz with 30 dB minimum attenuation
Equal terminating resistors of $50\,\Omega$.

7.4   Design a low-pass lumped-element Chebyshev filter with the following specifications:

Passband edge at 1 MHz, with 0.5 dB ripple
Stopband edge at 2.5 MHz with 30 dB minimum attenuation
Equal terminating resistors of $50\Omega$.

7.5  Design a high-pass lumped-element maximally-flat filter with the following specifications:

Passband edge at 4 MHz, with 3 dB maximum attenuation
Stopband edge at 2 MHz with 20 dB minimum attenuation
Equal terminating resistors of 50 Ω.

7.6  Design a high-pass lumped-element Chebyshev filter with the following specifications:

Passband edge at 5 MHz, with 0.05 dB maximum attenuation
Stopband edge at 2.5 kHz with 50 dB minimum attenuation
Equal terminating resistors of 50 Ω.

7.7  Design a band-pass lumped-element maximally-flat filter with the following specifications:

Passband edges at 1.5 MHz, and 4.0 MHz with 3 dB maximum attenuation
Stopband edges at 0.5 MHz and 8 MHz with 20 dB minimum attenuation in both stopbands
Equal terminating resistors of 75 Ω.

7.8  Design a band-pass lumped-element Chebyshev filter with the following specifications:

Passband edges at 4 MHz, and 6 MHz with 0.1 dB maximum attenuation
Stopband edges at 2 MHz and 10 MHz with 40 dB minimum attenuation
Equal terminating resistors of 50 Ω.

7.9  Design a band-stop lumped-element maximally-flat filter with the following specifications:

Passband edges at 1 MHz, and 12 MHz with 3 dB maximum attenuation
Stopband edges at 5 MHz and 9 MHz with 30 dB minimum attenuation
Equal terminating resistors of 50 Ω.

7.10 Design a band-stop lumped-element Chebyshev filter with the following specifications:

Passband edges at 2 MHz, and 12 MHz with 0.1 dB maximum attenuation
Stopband edges at 4 MHz and 10 MHz with 40 dB minimum attenuation
Equal terminating resistors of 50 Ω

# 8

# Switched-capacitor Filter Transfer Functions

## 8.1   INTRODUCTION

A switched-capacitor filter is a *sampled-data* system [36]. That is, the operation of filtering is performed on sampled versions of the continuous-time signals. Such a system is also called a discrete-time, or simply a discrete, system. It is to be noted, however, that both the input and output of the sampled-data filter are still *analog* in nature. The design of a switched-capcitor filter can be achieved using one of two fundamental approaches:

(a) As in the case of continuous-time filters studied in Chapter 7, the specifications on the filter characteristics are used to obtain a description of the filter in terms of a *stable* transfer function. This is the *approximation problem*. Once the transfer function of the filter has been determined, it can be realized using operational amplifiers, capacitors and switches in a prescribed topology. This is the *synthesis* or *realization problem*.

(b) An alternative, and very powerful technique, begins by first designing a passive prototype filter which meets the required specifications. Next, the actual circuit (structure and element values) are transformed into a switched-capacitor filter meeting the same specifications. The advantages of this method are: first, the low-sensitivity properties of passive filter prototypes are mapped into the corresponding switched-capacitor filters. Second, the wealth of material available in the area of passive filter design can be exploited and directly incorporated.

In this chapter, we develop methods for the description of sampled-data filters in general [36], and switched-capacitor filters in particular. It is then explained in detail how switched-capacitor filter transfer functions may be obtained from the continuous-time prototypes discussed in Chapter 7.

## 8.2  SAMPLING

In a switched-capacitor filter, the processing is performed on samples of the signals throughout the filter. This is achieved by means of the switches which are driven by periodic clock pulses. The sampling process is now explained in general.

### 8.2.1  Ideal Impulse Sampling

Although impulses cannot be produced physically, their use in explaining the sampling process is very instructive. Let $f(t)$ be a continuous-time signal and consider its multiplication by the periodic train of impulses (see Figure 8.1),

$$\delta_\infty(t) = \sum_{n=-\infty}^{\infty} \delta(t - nT) \tag{8.1}$$

to obtain the signal

$$f_s(t) = f(t) \sum_{n=-\infty}^{\infty} \delta(t - nT) \tag{8.2}$$

Using a property of the unit impulse, the above expression becomes

$$f_s(t) = \sum_{n=-\infty}^{\infty} f(nT)\delta(t - nT) \tag{8.3}$$

which is another periodic train of impulses each of strength $f(nT)$ which is the value of the function $f(t)$ at the $n$th instant. Figure 8.1 shows a model of this *impulse modulation*

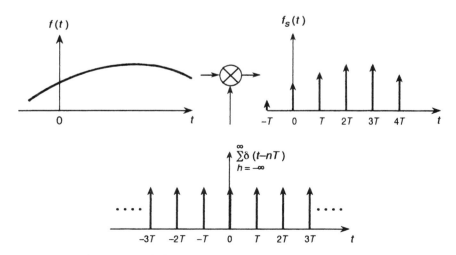

**Figure 8.1**   Ideal impulse sampling of a continuous-time signal.

process together with the resulting impulse train. This impulse train is regarded as a sampled signal. This is because the strength of the $n$th impulse is the sample value of $f(nT)$ and if each impulse is replaced by its strength (area) then we obtain a set of numbers $\{f(nT)\}$ defining the discrete-time signal and the sampling process has been achieved.

Now, consider the effect of the sampling process on the Fourier spectrum of the original continuous-time signal. Let

$$f(t) \leftrightarrow F(\omega) \tag{8.4}$$

and suppose $F(\omega)$ is band-limited to $\omega_m$, i.e.

$$|\mathcal{F}F(\omega)| = 0 \quad |\omega| \geqslant \omega_m$$

Using the frequency convolution relation applied to $f(t)$ and $\delta_\infty(t)$ we have the Fourier transform of the sampled signal

$$\mathcal{F}[f_s(t)] = \mathcal{F}[f(t)\delta_\infty(t)]$$

$$= \frac{1}{2\pi}\mathcal{F}[f(t)] * \mathcal{F}\left(\sum_{r=-\infty}^{\infty} \delta(t - nT)\right) \tag{8.5}$$

where the star denotes the convolution operation. However, the impulse train satisfies the relation

$$\mathcal{F}\left(\sum_{n=-\infty}^{\infty} \delta(t - rT)\right) = \omega_0 \sum_{n=-\infty}^{\infty} \delta(\omega - n\omega_0)$$

$$= \frac{2\pi}{T} \sum_{n=-\infty}^{\infty} \delta\left(\omega - \frac{2n\pi}{T}\right) \tag{8.6}$$

so that (8.5) becomes

$$\mathcal{F}[f_s(t)] = \frac{1}{T}F(\omega) * \sum_{n=-\infty}^{\infty} \delta\left(\omega - \frac{2n\pi}{T}\right) \tag{8.7}$$

But the unit impulse is the identity element in the process of convolution. Hence, using (8.7) we have (with $T = 2\pi/\omega_0$)

$$\mathcal{F}[f_s(t)] = F_s(\omega)$$

$$= \frac{1}{T} \sum_{n=-\infty}^{\infty} F(\omega - n\omega_0) \tag{8.8}$$

Thus, the spectrum of the sampled signal consists of the periodic extension of the spectrum $F(\omega)$ of the original continuous-time signal. This is illustrated in Figure 8.2

which also illustrates very important consequences of the sampling process. In Figure 8.2(b) the periodic spectrum of the sampled signal is shown in the case where $\omega_0 > 2\omega_m$. This means that the sampling frequency

$$\omega_0 = 2\pi/T \tag{8.9}$$

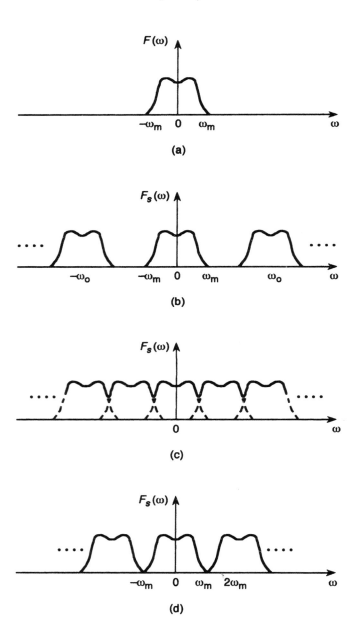

**Figure 8.2** Effect of sampling on the spectrum of a signal: (a) signal spectrum; (b) oversampling; (c) undersampling; (d) critical sampling.

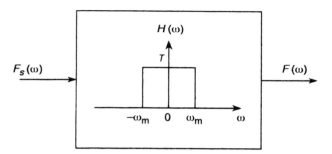

**Figure 8.3**   Signal recovery by an ideal filter.

exceeds twice the highest frequency component of the spectrum $F(\omega)$ of the original signal. In this case, it is clear that $F(\omega)$ can be recovered by passing the spectrum $F_s(\omega)$ through a low-pass filter, which eliminates

$$\sum_{r=-\infty}^{\infty} F(\omega - r\omega_0) \quad \text{for } r = 1, 2 \ldots \tag{8.10}$$

In Figure 8.2(c) the periodic spectrum $F_s(\omega)$ is shown for the case $\omega_0 < 2\omega_m$. This means that the sampling frequency is less than twice the highest frequency component of the band-limited spectrum $F(\omega)$ of the original signal. In this case the periodic parts of the spectrum overlap resulting in the effect called *aliasing*. This makes it impossible to recover the spectrum $F(\omega)$ by filtering.

Figure 8.2(d) shows the case of *critical* sampling with $\omega_0 = 2\omega_m$, i.e. the sampling frequency *just* exceeding twice the highest frequency component of $F(\omega)$. In this case, it is possible in principle to recover $F(\omega)$ by passing $F_s(\omega)$ through an *ideal* low-pass filter, with cut-off at $\omega_m$, as shown in Figure 8.3.

The *minimum* sampling rate required to prevent aliasing must *just* exceed *twice* the highest frequency component in the spectrum $F(\omega)$ of $f(t)$ before sampling. This minimum sampling rate is called the (radian) Nyquist frequency $\omega_N$. These considerations lead to the following fundamental result.

## 8.2.2   The Sampling Theorem

A signal $f(t)$ whose spectrum is band-limited to below a frequency $\omega_m$, can be completely recovered from its samples $\{f(nT)\}$ taken at a rate

$$f_N = \frac{\omega_N}{2\pi}(= 1/T) \quad \text{where } \omega_N = 2\omega_m \tag{8.11}$$

The signal $f(t)$ is determined from its sample values $\{f(nT)\}$ by

$$f(t) = \sum_{n=-\infty}^{\infty} f(nT) \frac{\sin \omega_m(t - nT)}{\omega_m(t - nT)} \tag{8.12}$$

where

$$T = \frac{\pi}{\omega_m} = \frac{2\pi}{\omega_N} = \frac{1}{f_N} \tag{8.13}$$

To prove (8.12) we note that $F(\omega)$ can be recovered from $F_s(\omega)$ by passing it through an ideal low-pass filter of amplitude $T$ and cut-off at $\omega_m$, as shown in Figure 8.3. Thus, assuming critical sampling at twice the highest frequency in $F(\omega)$, the impulse response of the required filter is obtained from

$$h(t) = \mathcal{F}^{-1}[H(j\omega)] \tag{8.14}$$

where

$$H(j\omega) = T \qquad |\omega| \leqslant \omega_m$$
$$= 0 \qquad |\omega| > \omega_m \tag{8.15}$$

Thus,

$$h(t) = T \frac{\sin \omega_m t}{\pi t}$$
$$= \frac{\sin \omega_m t}{\omega_m t} \tag{8.16}$$

and the output of the filter is

$$f(t) = f_s(t) * h(t)$$

$$= \int_{-\infty}^{\infty} \left( \sum_{-\infty}^{\infty} f(nT)\delta(\tau - nT) \right) \frac{\sin \omega_m(t - \tau)}{\omega_m(t - \tau)} d\tau \tag{8.17}$$

$$= \sum_{n=-\infty}^{\infty} \left( \int_{-\infty}^{\infty} f(nT)\delta(\tau - nT)) \frac{\sin \omega_m(t - \tau)}{\omega_m(t - \tau)} d\tau \right)$$

which leads to

$$f(t) = \sum_{n=-\infty}^{\infty} f(nT) \frac{\sin \omega_m(t - nT)}{\omega_m(t - nT)} \tag{8.18}$$

as stated in (8.12). This expression is essentially a formula for the interpolation of the signal values by its values at the sampling points. $F(\omega)$ is, however, recoverable from $F_s(\omega)$ when $F(\omega)$ does not contain any frequency components higher than half the sampling frequency. It follows that $f(t)$ is recoverable from its sample values, at least in principle, using (8.18) if and only if the sampling theorem is satisfied.

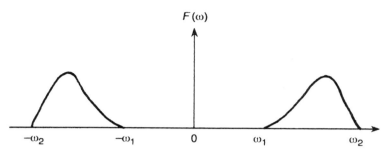

**Figure 8.4**  Spectrum of a band-pass signal.

From the above discussion of the idealized sampling process, we can draw the following important conclusions.

1. The choice of a sampling frequency for a signal is determined by the highest frequency component of the Fourier spectrum $F(\omega)$ of the signal $f(t)$. In practice, the signal is usually band-limited to a frequency $\omega_N/2$, prior to sampling at a frequency of $\omega_N$. This is done by *prefiltering*.

2. Critical sampling with $\omega_N = 2\omega_m$ requires an ideal filter for the reconstruction of a signal having frequency components up to $\omega_m$. Such a filter is non-causal and hence is physically unrealizable. Therefore, in practice the sampling frequency is chosen to be higher than the Nyquist rate in order that the reconstruction filter may have a realizable response. For example, speech signals are bandlimited to 3.4 kHz and sampled at a rate of 8 kHz instead of the critical rate of 6.8 kHz.

3. So far, we have assumed a signal $f(t)$ whose spectrum extends from $\omega = 0$ to $\omega = \omega_m$ that is a *low-pass* signal. In this case the signal is completely determined from its set of values at regularly spaced intervals of period $T = 1/2f_m = \pi/\omega_m$. Now consider a *band-pass* signal whose spectrum exists only in the range

$$\omega_1 < |\omega| < \omega_2 \tag{8.19}$$

as shown in Figure 8.4. It is easy to show that the minimum (radian) sampling frequency in this case is given by

$$\omega_N = 2(\omega_2 - \omega_1) \tag{8.19}$$

The reconstruction of the signal, in this case, must be done by a band-pass filter.

### 8.2.3  Sampled-and-held Signals

In the previous section we considered *instantaneous* sampling of signals, by means of impulses. Though very instructive, leading to valid conclusions about the sampling process, impulse sampling is not feasible in practice.

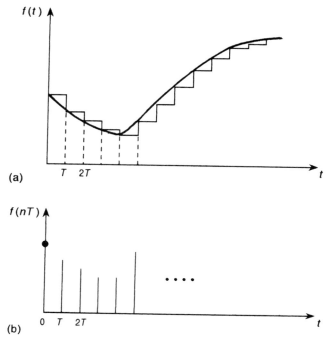

**Figure 8.5** (a) Sampled-and-held signal, (b) the sample values defined at the beginning of each interval.

Therefore, we now give the more practical sampling method employed in switched-capacitor filters, and show that it leads essentially to the same conclusions.

Consider the signal $f(t)$, band-limited to $\omega_m$, and let it be sampled as shown in Figure 8.5(a), such that: at the instant $(nT)$, the signal is sampled and held at its value $f(nT)$ until the next sample is taken at $(n+1)T$. The sample values are defined at the beginning of each interval. Then, the resulting signal is given by

$$f_s(t) = \sum_{-\infty}^{\infty} f(nT)\{u(t-nT) - u[t-(n+1)T]\} \tag{8.20}$$

where $u(t)$ is the unit step function. The function between the curled brackets is a unit pulse starting at $(nT)$ of width $T$, which has a Fourier transform

$$\mathcal{F}[u(t-nT) - u[t-(n+1)T]] = T\left(\frac{\sin \omega T/2}{\omega T/2}\right) e^{j\omega T/2} e^{-jn\omega T} \tag{8.22}$$

Therefore the Fourier transform of $f_s(t)$ is given by

$$\hat{F}_s(\omega) = \left\{T\left(\frac{\sin \omega T/2}{\omega T/2}\right) e^{-j\omega T/2}\right\} \sum_{n=-\infty}^{\infty} f(nT) e^{-jn\omega T} \tag{8.23}$$

But, using (8.2)–(8.8)

$$\sum_{n=-\infty}^{\infty} f(nT)\, e^{-jn\omega T} = \mathcal{F}\left\{ \sum_{n=-\infty}^{\infty} f(nT)\delta(t - nT) \right\}$$

$$= \frac{1}{T} \sum_{n=-\infty}^{\infty} F\left(\omega - \frac{2\pi n}{T}\right) \qquad (8.24)$$

$$= \frac{1}{T} \sum_{n=-\infty}^{\infty} F(\omega - n\omega_0)$$

i.e.

$$\hat{F}_s(\omega) = \left[ \left(\frac{\sin \omega T/2}{\omega T/2}\right) e^{-j\omega T/2} \right] \left[ \sum_{n=-\infty}^{\infty} F(\omega - n\omega_0) \right] \qquad (8.25)$$

$$= H_{SH}(\omega) F_s(\omega)$$

where the function

$$F_s(\omega) = \sum_{n=-\infty}^{\infty} F(\omega - n\omega_0) \qquad (8.26)$$

is the spectrum of the signal when *impulse-sampled* and is the same as that shown in Figure 8.2. On the other hand the spectrum

$$H_{SH}(\omega) = \frac{\sin \omega T/2}{\omega T/2} e^{-j\omega T/2} \qquad (8.27)$$

is produced by the holding effect of the pulses. Therefore, this $(\sin x/x)$ function, shown in Figure 8.6 multiplies the spectrum $F_s(\omega)$. Thus, the spectrum $F_s(\omega)$ will be distorted in amplitude by the factor $(\sin \omega T/2)/\omega T/2$ while the factor $e^{-j\omega T/2}$ will introduce a linear

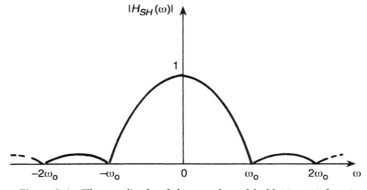

**Figure 8.6**   The amplitude of the sample and hold $(\sin x/x)$ function.

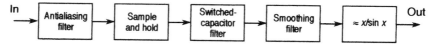

**Figure 8.7**   The switched-capacitor filter in continuous-time environment.

phase shift, or a constant delay. This can be thought of as a system performing the sample-and-hold function whose transfer function is given by (8.27).

From the above discussion it is noted that the sampling theorem is still valid in this case. However, the recovery of the sampled-and-held signal is not possible exactly due to the distortion introduced by the factor in (8.27). Therefore, this effect must somehow be compensated for if perfect reconstruction of the signal is to be achieved.

We end this section by reference to Figure 8.7 which shows a general switched-capacitor filter including the interfaces with the continuous-time environment. The functions of the various blocks are as follows:

1. The anti-aliasing filter ensures that the input signal spectrum is band-limited to half the sampling frequency.

2. The sample-and-hold stage ensures that the input to the sampled-data (switched-capacitor) filter is now of the sampled type (discrete).

3. The smoothing filter provides a continuous-time signal from the sampled-and-held one.

4. The amplitude equalizer is needed so that the reduction in the amplitude spectrum of the output due to the $\sin x/x$ junction in (8.27) is undone. Naturally, this should be a network which approximates the inverse of this effect, i.e. approximates $(x/\sin x)$.

Next, we consider the description of number sequences resulting from the sampled continuous-time signal and their processing by discrete systems in general and switched-capacitor filters in particular.

## 8.3   SEQUENCES AND DISCRETE SYSTEMS

Consider a set of numbers $\{f(nT,\}$ defining a sequence. Let us now drop the reference to time in the sequence and replace $(nT)$ by a discrete variable $(n)$ which assumes integral values. Thus, a discrete system transforms a sequence $\{f(n)\}$ into another sequence $\{g(n)\}$ as shown in Figure 8.8. The following notation may be used to denote the operation performed by the system

$$\{g(n)\} = \Upsilon[\{f(n)\}]  \tag{8.28}$$

In the notation of (8.28), curly brackets are used to denote the entire sequence while $f(n)$ denotes a single sample. However, it is usually convenient to drop the brackets and use $f(n)$ to denote either the entire sequence or a single sample. We shall often do this, whenever no confusion may arise, and let the context define which one is meant.

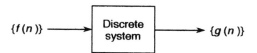

**Figure 8.8**   A discrete-time system.

The system is *linear* if

$$\Upsilon[a\{f_1(n)\} + b\{f_2(n)\}] = a\{g_1(n)\} + b\{g_2(n)\} \tag{8.29}$$

where $a$ and $b$ are constants. The system is time-invariant if for any integer $n_0$

$$\Upsilon[\{f(n - n_0)\}] = \{g(n - n_0)\} \tag{8.30}$$

Now, a sequence $\{f(n)\}$ can be represented by a set of lines as a function of the discrete integral variable $n$, as shown in Figure 8.9. Let us define a *discrete* unit impulse $u_0(n)$ as the *sequence*

$$u_0(n) = 1 \quad \text{for } n = 0$$
$$= 0 \quad \text{for } n \neq 0 \tag{8.31}$$

which is shown in Figure 8.10. It must be emphasized that this discrete impulse is physically realizable as just a *number* existing at $n = 0$ and is zero elsewhere. This is not to be confused with the delta function studied in relation to continuous signals, which must be defined as a distribution or generalized function [36].

Now, a sequence $\{f(n)\}$ can be expressed as the sum of shifted impulses

$$f(n) = \sum_{m=-\infty}^{\infty} f(m)u_0(n - m) \tag{8.32}$$

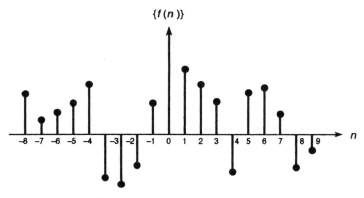

**Figure 8.9**   A discrete-time signal.

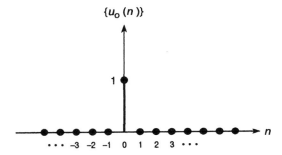

**Figure 8.10**   The discrete unit impulse.

where

$$u_0(n - m) = 1 \quad n = m$$
$$= 0 \quad n \neq m \tag{8.33}$$

which is the shifted impulse shown in Figure 8.11.

Consequently, suppose that the response of the system to a unit impulse is the sequence $\{h(n)\}$, i.e.

$$\{h(n)\} = \Upsilon[\{u_0(n)\}] \tag{8.34}$$

which is the *impulse response* of the system. Then, the shift-invariance property, expressed in (8.29), implies

$$\{h(n - m)\} = \Upsilon[\{u_0(n - m)\}] \tag{8.35}$$

Next, using the linearity constraint together with (8.31) we obtain for the output sequence

$$\{g(n)\} = \Upsilon[\{f(n)\}]$$
$$= \Upsilon\left( \sum_{m=-\infty}^{\infty} f(m)u_0(n - m) \right) \tag{8.36}$$

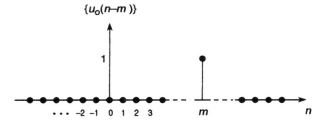

**Figure 8.11**   Shifted discrete impulse.

which upon use of (8.34) gives

$$g(n) = \sum_{m=-\infty}^{\infty} f(m)h(n-m)$$

$$= \sum_{m=-\infty}^{\infty} h(m)f(n-m) \qquad (8.37)$$

The above expression is written as

$$\{g(n)\} = \{h(n)\} * \{f(n)\} \qquad (8.38)$$

This is the *convolution* of the two sequences $\{h(n)\}$ and $\{f(n)\}$. Thus, a linear time-invariant system is completely defined by its impulse response. The output sequence $\{g(n)\}$ for any input sequence $\{f(n)\}$ is obtained from the convolution of $\{f(n)\}$ with the impulse response $\{h(n)\}$ as defined by (8.37).

Now, the system is said to be *causal* if the output for $n = n_0$ depends only on inputs with $n < n_0$. This implies

$$h(n) = 0 \quad \text{for } n < 0 \qquad (8.39)$$

and the output is

$$g(n) = \sum_{m=0}^{\infty} h(m)f(n-m) \qquad (8.40)$$

As in the case of continuous systems, we say that the system is *strictly stable* if any bounded input produces a bounded output. The necessary and sufficient condition for bounded input–bounded output (BIBO) stability is

$$\sum_{n} |h(n)| < \infty \qquad (8.41)$$

To prove the necessity of this condition, consider the input sequence

$$f(n) = 1 \qquad h(n) \geq 0$$
$$= -1 \qquad h(n) < 0 \qquad (8.42)$$

so that for $n = 0$, (8.37) gives

$$g(0) = \sum_{m} |h(m)| \qquad (8.43)$$

and if the inequality (8.41) is not satisfied, $g(0)$ is not bounded and the system is not (BIBO) strictly stable. Next, to prove sufficiency, we assume a bounded input

$$|f(n)| \leq M \quad \text{for all } n \qquad (8.44)$$

for which (8.37) gives

$$|g(n)| = \sum_m |h(m)| \, |f(n-m)|$$

$$\leqslant M \sum_m |h(m)|$$

(8.45)

which is bounded if (8.41) is satisfied.

## 8.4   THE $z$-TRANSFORMATION

The manipulation of sequences, and the description of discrete linear shift-invariant systems, can be conveniently accomplished using the $z$-transform. This is reviewed in this section.

### 8.4.1   Definition and General Considerations

Let $\{f(n)\}$ be a sequence of real numbers. Its *two-sided* $z$-transform is defined by

$$\hat{F}(z) = \hat{\mathcal{Z}}\{f(n)\}$$

$$\underset{=}{\triangle} \sum_{n=-\infty}^{\infty} f(n)z^{-n}$$

(8.46)

where $z$ is a complex variable. The *one-sided* $z$-transform of the sequence is defined by

$$F(z) = \mathcal{Z}\{f(n)\}$$

$$\underset{=}{\triangle} \sum_{n=0}^{\infty} f(n)z^{-n}$$

(8.47)

Clearly, if the sequence is causal, i.e.

$$f(n) = 0 \qquad n < 0$$

(8.48)

then the two-sided and one-sided $z$-transforms are identical. Henceforth, we shall restrict ourselves to causal sequences. Therefore, no distinction will be made between the two versions of the transform, and expression (8.47) is taken to form the definition of the $z$-transform of the sequence.

*Example 8.1*   Consider the sequence

$$\{f(n)\} = \{1, \, -1.5, 2, 0, 3\}$$

$$= 0 \quad \text{for } n > 4$$

Its z-transform is obtained from (8.46) as

$$F(z) = 1 - 1.5z^{-1} + 2z^{-2} + 3z^{-4}$$

*Example 8.2*   Consider the discrete impulse sequence defined by (8.30). Its z-transform is given by

$$\mathcal{Z}\{u_0(n)\} = \sum_{n=0}^{\infty} u_0(n)z^{-n}$$

$$= u_0(0)z^0$$

or

$$\mathcal{Z}\{u_0(n)\} = 1 \tag{8.49}$$

*Example 8.3*   Consider the discrete unit step sequence

$$\{u_1(n)\} = 1 \quad n > 0$$
$$= 0 \quad n < 0 \tag{8.50}$$

which is shown in Figure 8.12. By (8.47) the z-transform of the above sequence is obtained as

$$\mathcal{Z}\{u_1(n)\} = \sum_{n=0}^{\infty} z^{-n}$$

or

$$\mathcal{Z}\{u_1(n)\} = \frac{1}{1 - z^{-1}} \quad |z| > 1 \tag{8.51}$$

*Example 8.4*   Consider the shifted discrete impulse sequence defined by (8.33) and shown in Figure 8.11. Its z-transform is obtained by (8.47) as

$$\mathcal{Z}\{u_0(n - m)\} = \sum_{n=0}^{\infty} u_0(n - m)z^{-n} \tag{8.52}$$

$$= z^{-m}$$

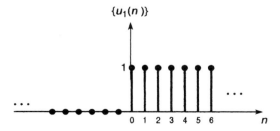

**Figure 8.12**   The discrete unit step.

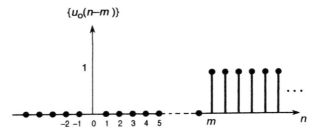

**Figure 8.13**   Shifted discrete unit step.

*Example 8.5*   Consider the shifted unit step sequence shown in Figure 8.13:

$$\{u_1(n-m)\} = 1 \quad n > m$$
$$= 0 \quad n > m$$

(8.53)

ts $z$-transform is

$$\mathcal{Z}\{u_1(n-m)\} = \sum_{n=m}^{\infty} z^{-n}$$

$$= z^{-m} + z^{-(m+1)} + z^{(m+2)} + \cdots$$
$$= z^{-m}(1 + z^{-1} + z^{-2} + \cdots)$$

i.e.

$$\mathcal{Z}\{u_1(n-m)\} = \frac{z^{-m}}{1 - z^{-1}} \quad |z| > 1$$

(8.54)

*Example 8.6*   Consider the exponential sequence

$$\{f(n)\} = \exp(-\alpha n) \quad \alpha > 0$$

Its $z$-transform is given by

$$\mathcal{Z}\{\exp(-\alpha n)\} = \sum_{n=0}^{\infty} \exp(-\alpha n)z^{-n}$$

$$= \sum_{n=0}^{\infty} [e^{-\alpha}z^{-1}]^n$$

i.e.

$$\mathcal{Z}\{\exp(-\alpha n)\} = \frac{1}{1 - e^{-\alpha}z^{-1}} \quad |z| < e^{\alpha}$$

(8.56)

*Example 8.7*   From the result of the previous example, with

$$b = e^{\alpha}$$

we obtain

$$\mathcal{Z}\{b^n\} = \frac{1}{1 - bz^{-1}} \quad |z| > b \tag{8.57}$$

## 8.4.2   Relationship Between the z-transform and the Laplace Transform

Now, suppose that the sequence $\{f(n)\}$ has been obtained as a result of sampling a causal continuous signal $f(t)$ as discussed in Section 8.2. Then we may write for the sample values

$$\{f(nT)\} \triangleq \{f(0), f(T), f(2T), \ldots\} \tag{8.58}$$

and the z-transform of the sequence is given by

$$F(z) = \sum_{n=0}^{\infty} f(nT)z^{-n} \tag{8.59}$$

We have seen, however, that the sampled signal can be written as the train of delta functions. Hence, (8.3) gives for a causal signal

$$f_s(t) = \sum_{n=0}^{\infty} f(nT)\delta(t - nT) \tag{8.60}$$

which upon taking the Laplace transform of both sides yields

$$\mathcal{L}[f_s(t)] = F(s)$$
$$= \sum_{n=0}^{\infty} f(nT)\exp(-nTs) \tag{8.61}$$

Comparison of (8.59) with (8.61) reveals that if we make the identification

$$z^{-1} \equiv \exp(-Ts) \tag{8.62}$$

then the z-transform and the Laplace transform of the causal sampled signal are identical. Noting that $s$ is the complex frequency

$$s = \sigma + j\omega$$

we can write

$$z = \exp(Ts)$$
$$= \exp(T\sigma)\exp(jT\omega) \tag{8.64}$$

The mapping between the $s$-plane and the $z$-plane is shown in Figure 8.14 and reveals the following features:

1. The left half of the complex s-plane maps onto the inside of the unit circle $z = 1$ in the $z$-plane. this is because the left half-plane is defined by

$$\text{Re } s(=\sigma) < 0 \qquad (8.65)$$

   which, from (8.64) corresponds to values of $z$ satisfying

$$|z| < 1 \qquad (8.66)$$

2. The right half of the $s$-plane defined by

$$\text{Re } s(=\sigma) > 0 \qquad (8.67)$$

   maps onto the outside of the unit circle

$$|z| > 1 \qquad (8.68)$$

   in the $z$-plane.

3. The $j\omega$-axis in the $s$-plane maps onto the unit circle $|z| = 1$ in the $z$-plane. The correspondence between the $s$-plane and $z$-plane allows us to exploit the theory of continuous systems to develop many parallel concepts and tools for the analysis and design of discrete systems, as we shall see throughout the rest of this book.

Now, if the sequence $\{f(nT)\}$ represents the samples of the signal $f(t)$ taken every $T$ seconds, then the Fourier transform of the sequence is obtained from (8.59) by letting $s \to j\omega$, so that

$$F(j\omega) = \sum_{n=0}^{\infty} f(nT) \exp(-jnT\omega) \qquad (8.69)$$

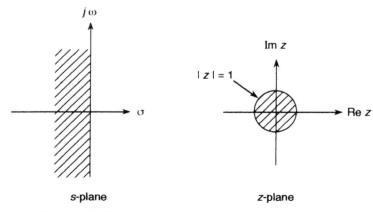

**Figure 8.14**   The mapping between the $s$-plane and the $z$-plane.

It follows that for

$$z = \exp(jT\omega) \tag{8.70}$$

the z-transform of the *causal sequence* $\{f(nT)\}$ is identical to its Fourier transform.

### 8.4.3 Properties of the z-Transform

The z-transform has the following properties which can be easily derived from the basic definition in (8.47).

#### *Linearity*

Let

$$F_1(z) = \mathcal{Z}\{f_1(n)\} \tag{8.71}$$

and

$$F_2(z) = \mathcal{Z}\{f_2(n)\} \tag{8.72}$$

then the z-transform is *linear*, i.e.

$$\mathcal{Z}[a\{f_1(n)\} + b\{f_2(n)\}] = aF_1(z) + bF_2(z) \tag{8.73}$$

where $a$ and $b$ are constants.

#### *Shifting*

If

$$\mathcal{Z}\{f(n)\} = F(z)$$

then

$$\mathcal{Z}\{f(n - m)\} = z^{-m}F(z) \tag{8.74}$$

with $m > 0$ and $\{f(n)\}$ assumed causal.

To prove this prroperty, write

$$\mathcal{Z}\{f(n - m)\} = \sum_{n=0}^{\infty} f(n - m)z^{-n}$$

and put

$$(n - m) = k$$

so that

$$\mathcal{Z}\{f(n - m)\} = \sum_{k=-m}^{\infty} f(k)z^{-(m+k)}$$

$$= z^{-m} \sum_{k=-m}^{-\infty} f(k)z^{-k}$$

i.e.

$$\mathcal{Z}\{f(n - m)\} = z^{-m} \sum_{k=0}^{\infty} f(k)z^{-k} + z^{-m} \sum_{k-1}^{m} f(-k)z^{k} \qquad (8.75)$$

Since $\{f(n)\}$ is assumed causal, then $f(-k) = 0$ for $k \geq 1$. Thus the second sum on the right vanishes, which proves (8.74). We note, however, that if the sequence is non-causal then the same analysis leading to (8.75) applies, but in this case

$$\mathcal{Z}\{f(n - m)\} = z^{-m}F(z) + z^{-m} \sum_{k=1}^{m} f(-k)z^{k} \qquad (8.76)$$

where the last sum on the right represents the initial conditions.

Clearly, from the above shifting property, delaying a sequence by 1 unit (or one sample period) amounts to multiplication by $z^{-1}$ in the z-domain. Therefore $z^{-1}$ is sometimes called the *unit delay operator*. Figure 8.15 illustrates the delay operation on a causal sequence while Figure 8.16 illustrates the same operation for a non-causal sequence.

If the sequence is shifted to the *left* by $m$, we obtain in a manner similar to that used in the derivation of (8.75),

$$\mathcal{Z}\{f(n + m)\} = z^{m} \sum_{k=0}^{\infty} f(k)z^{-k} - z^{m} \sum_{k=0}^{m-1} f(k)z^{-k}$$

$$(8.77)$$

$$= z^{m}F(z) - z^{m} \sum_{k=0}^{m-1} f(k)z^{-k}$$

and causality does *not* eliminate the sum in the above expression.

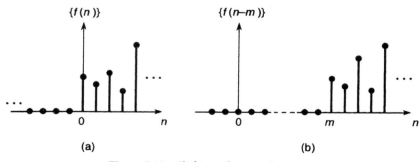

Figure 8.15   Shifting of a causal sequence.

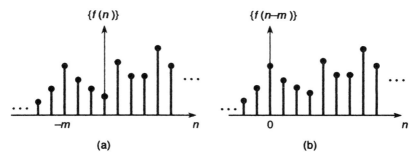

**Figure 8.16**  Shifting of a non-causal sequence.

## Convolution of sequences

The convolution of two sequences $\{f_1(n)\}$ and $\{f_1(n)\}$ is defined by (8.37) and (8.38). For two causal sequences, the convolution sequence is

$$\{f_1(n)\} * \{f_2(n)\} \triangleq \sum_{m=0}^{\infty} f_1(m) f_2(n-m)$$

$$= \sum_{m=0}^{n} f_1(m) f_2(n-m) \qquad (8.78)$$

We now show that the z-transform of the convolution sequence is the product of the ztransform of the individual sequences. Write

$$F_1(z) = f_1(0) + f_1(1)z^{-1} + \cdots + f_1(n)z^{-n} + \cdots$$
$$F_2(z) = f_2(0) + f_2(1)z^{-1} + \cdots + f_2(n)z^{-n} + \cdots \qquad (8.79)$$

Multiplying the two ztransform we obtain

$$F_1(z)F_2(z) = f_1(0)f_2(0) + [f_2(0)f_2(1) + f_1(1)f_2(0)]z^{-1} + \cdots$$

$$+ \left( \sum_{m=0}^{n} f_1(m)f_2(n-m) \right) z^{-n} + \cdots \qquad (8.80)$$

i.e.

$$F_1(z)F_2(z) = \sum_{n=0}^{\infty} \left( \sum_{m=0}^{n} f_1(m)f_2(n-m) \right) z^{-n} \qquad (8.81)$$

The right side is, however, the z-transform of the right-hand side of (8.78). Hence

$$\mathcal{Z}[\{f_1(n)\} * \{f_2(n)\}] = F_1(z)F_2(z)$$
$$= F_2(z)F_1(z) \qquad (8.82)$$

and it follows that

$$\mathcal{Z}\left(\sum_{m=0}^{n}f_1(m)f_2(n-m)\right)=\mathcal{F}\left(\sum_{m=0}^{n}f_1(n-m)f_2(m)\right) \tag{8.83}$$

$$=F_1(z)F_2(z)$$

## Convergence

The (one-sided) $z$-transform of the causal sequence $\{f(n)\}$ is

$$F(z)=\sum_{n=0}^{\infty}f(n)z^{-n} \tag{8.84}$$

which is defined only at all points in the $z$-plane where the right-hand side of (8.83) converges. Writing

$$z=\exp(x+jy)$$

and

$$z^{-1}=\exp(-x)\exp(-jy) \tag{8.85}$$

then the region of convergence of the series in (8.84) is defined by

$$x>0$$

or

$$|z|>1 \tag{8.86}$$

It follows that the $z$-transform of a causal sequence converges for all points outside the unit circle in the z-plane, as shown in Figure 8.17.

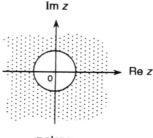

Figure 8.17   The regions of the $z$-plane.

## 8.5 DESCRIPTION OF DISCRETE SYSTEMS

### 8.5.1 The Difference Equation and Transfer Function

A very important class of linear shift-invariant systems is that for which the input and output are related by a linear *difference* equation with constant coefficients [36]. These provide a very accurate description of practical switched-capacitor filters. Consider such a system, shown in Figure 8.8 whose input sequence is $\{f(n)\}$ and the output is $\{g(n)\}$. The causal system is defined by the difference equation

$$g(n) = \sum_{r=0}^{M} a_r f(n-r) - \sum_{r=1}^{N} b_r g(n-r) \quad \text{with } M \leqslant N \tag{8.87}$$

which is simply a resursion formula allowing the present value of the output to be calculated from: the $N$ past output values, the $M$ past input values, and the present input value. In (8.87) $a_r$ and $b_r$ are real constants.

Taking the $z$-transform of both sides of (8.87) and making use of (8.74) we obtain

$$G(z) = F(z) \sum_{r=0}^{M} a_r z^{-r} - G(z) \sum_{r=0}^{N} b_r z^{-r}. \tag{8.88}$$

Thus, a *transfer function* $H(z)$ of the system may be formed as

$$H(z) = \frac{G(z)}{F(z)} \tag{8.89}$$

which upon use of (8.87) takes the general form

$$H(z) = \frac{\displaystyle\sum_{r=0}^{M} a_r z^{-r}}{1 + \displaystyle\sum_{r=1}^{N} b_r z^{-r}} \tag{8.90}$$

This is a real rational function of $z$ (or $z^{-1}$) relating the input and output in the z-domain.

### 8.5.2 The Frequency Response

If the input to the system is $\{f(nT)\}$, representing a sampled continuous signal, then the output sequence $\{g(nT)\}$ may also be thought of as a sampled continuous output signal. Since, however, $z$ and the complex frequency $s$ are related by (8.62), then the transfer function $H(z)$ of the system is such that: letting $z \to \exp(jTw)$ in (8.90) we obtain

$$H(\exp(jwT)) = \frac{\displaystyle\sum_{r=0}^{M} a_r \exp(-jrwT)}{1 + \displaystyle\sum_{r=1}^{N} b_r \exp(-jrTw)} \tag{8.91}$$

or

$$H(\exp(j\omega T)) = |H(\exp(j\omega T)| \exp(j\psi(\omega)) \tag{8.92}$$

where $|H(\exp(j\omega T))|$ is the *amplitude response*. Due to the fact that $H(z)$ is a rational function of $z$ and the periodicity of the exponential $\exp(-j\omega T)$, it follows that the amplitude and phase responses of the system are *periodic* with respect to $\omega$, the period being $\omega_N = 2\pi f_N$, where $f_N$ is the sampling frequency. The frequency response of the system is, therefore, obtained by considering the points on the $j\omega$-axis of the $s$-plane which, in the $z$-plane, is equivalent to considering the points on the unit circle $|z| = 1$.

### 8.5.3   Stability

Rewriting expression (8.89) in the form

$$G(z) = H(z)F(z) \tag{8.93}$$

and taking the inverse transform using the convolution relation (8.78) we find

$$g(n) = \sum_{m=0}^{n} h(m)f(n-m) \tag{8.94}$$

or

$$\{g(n)\} = \{h(n)\} * \{f(n)\} \tag{8.95}$$

where $\{h(n)\}$ is the *impulse response* of the (causal) system, related to the transfer function by

$$H(z) = \mathcal{Z}\{h(n)\} \tag{8.96}$$

$$\{h(n)\} = \mathcal{Z}^{-1}[H(z)] \tag{8.97}$$

Now we have seen that the system is strictly stable (bounded-input–bounded-output) only if its impulse response $\{h(n)\}$ satisfies (8.41), i.e. it is an absolutely summable sequence,

$$\sum_{n=0}^{\infty} |h(n)| < \infty \tag{8.98}$$

If, however, the transfer function is written as

$$H(z) = \frac{P(z^{-1})}{\displaystyle\prod_{r=1}^{N}(1 - p_r z^{-r})}$$

$$= \sum_{r=1}^{N} \frac{a_r}{(1 - p_r z^{-r})} \tag{8.99}$$

then using (8.57), the impulse response sequence is given by

$$\{h(n)\} = a_r p_r^n \tag{8.100}$$

which satisfies (8.98) if and only if

$$|p_r| < 1 \quad \text{for all } r \tag{8.101}$$

that is, the poles of the transfer function defined by $z = p_r$ must lie inside the unit circle in the $z$-plane. This, of course, corresponds to the left half of the complex frequency $s$-plane, as shown in Figure 8.14. But the stability condition has been obtained here without reference to time or complex frequency.

### 8.5.4 The Bilinear Variable: $\lambda$

Although the variable $z$ appears naturally in the transfer functions of sampled-data systems, it is often more convenient to use a different variable $\lambda$ [1], [36] defined by

$$\lambda = \frac{1 - z^{-1}}{1 + z^{-1}} = \frac{z - 1}{z + 1} \tag{8.102}$$

which is a bilinear function of $z$. This variable is particularly useful in two respects:

1. Testing the stability of the sampled-data system without resort to factorization of polynomials.

2. The solution of the approximation problem in filter design, i.e. the determination of the transfer function such that it possesses filtering characteristics meeting the required specifications.

Let us now study the mapping between the $z$-plane and the $\lambda$-plane as defined by (8.102). Write

$$z = x + jy \tag{8.103}$$

and

$$\lambda = \Sigma + j\Omega \tag{8.104}$$

so that

$$\Sigma + j\Omega = \frac{(x + jy) - 1}{(x + jy) + 1} \frac{[(x^2 + y^2) - 1]}{(1 + x)^2 + y^2} \tag{8.105}$$
$$+ j \frac{2y}{(1 + x)^2 + y^2}$$

or

$$\Sigma + j\Omega = \frac{(|z|^2 - 1)}{(1 + x)^2 + y^2} + j \frac{2y}{(1 + x)^2 + y^2} \tag{8.106}$$

From the above relation the following correspondence is evident.

1. Points inside the unit circle in the $z$-plane, map onto points in the left half of the $\lambda$-plane, i.e.

$$|z| < 1 \quad \text{corresponds to } \Sigma < 0$$

2. Points outside the unit circle in the $z$-plane, map onto points in the right half of the $\lambda$-plane, i.e.

$$|z| > 1 \quad \text{corresponds to } \Sigma > 0$$

3. Points on the circumference of the unit circle in the $z$-plane map onto points on the $j\Omega$-axis, in the $\lambda$-plane, i.e.

$$|z| = 1 \quad \text{corresponds to } \Sigma = 0$$

In particular, on the $j\Omega$-axis (8.106) shows that

$$z = 1 \quad \text{corresponds to } \Omega = 0$$

and

$$z = -1 \quad \text{corresponds to } \Omega = \pm\infty$$

Therefore, moving on the unit circle in the $z$-plane, corresponds to moving along the $j\Omega$-axis in the $\lambda$-plane. Figure 8.18 shows the mapping between the $\lambda$-plane and the $z$-plane. Now, the transfer function of a sampled-data system is of the form

$$H(z) = \frac{N(z^{-1})}{D(z^{-1})} \tag{8.107}$$

where $N(z^{-1})$ and $D(z^{-1})$ are polynomials with real coefficients. Using (8.102) in (8.107), the transfer function assumes the alternative form

$$H(\lambda) = \frac{P_M(\lambda)}{Q_N(\lambda)} \tag{8.108}$$

where $P(\lambda)$ and $Q(\lambda)$ are real polynomials of the new variable $\lambda$.

We have seen that strict stability of the system described by (8.107) requires that all the poles of $H(z)$ be inside the unit circle in the $z$-plane. However, this domain corresponds to the left half of the complex $\lambda$-plane. It follows that in its new form $H(\lambda)$ in (8.107) is stable if and only if

$$M \leqslant N \tag{8.109}$$

and

$$Q_N(\lambda) \neq 0 \quad \text{in Re } \lambda > 0 \tag{8.110}$$

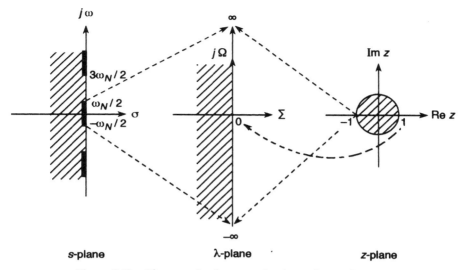

**Figure 8.18**   The mapping between the three planes of interest.

i.e. $Q_N(\lambda)$ has all its zeros in the open left half-plane. Condition (8.109) excludes poles at $\lambda = \pm\infty$, corresponding to the point $z = -1$ on the unit circle.

From the above discussion, it follows that for stability, the polynomial $Q_N(\lambda)$ must be a strictly Hurwitz polynomial in $\lambda$. Thus, we can use the simple test for strict Hurwitz character to check the stability of $H(\lambda)$ without the need for factorization of a polynomial to determine its roots. This test is now illustrated by an example.

*Example 8.8*   Test the following transfer function for stability

$$H(z) = \frac{(1 + z^{-1})^3}{37 + 51z^{-1} + 27z^{-2} + 5z^{-3}}$$

*Solution*   We transform into the $\lambda$-domain via the relation

$$z^{-1} \rightarrow \frac{1 - \lambda}{1 + \lambda}$$

to obtain

$$H(\lambda) = \frac{8/(1 + \lambda)^3}{37 + 51\left(\dfrac{1 - \lambda}{1 + \lambda}\right) + 27\left(\dfrac{1 - \lambda}{1 + \lambda}\right)^2 + 5\left(\dfrac{1 - \lambda}{1 + \lambda}\right)^3}$$

$$= \frac{8}{37(1 + \lambda)^3 + 51(1 - \lambda)(1 + \lambda)^3 + 27(1 - \lambda^2)(1 + \lambda) + 5(1 - \lambda)^3}$$

which, after some simplification, becomes

$$H(\lambda) = \frac{1}{15 + 15\lambda + 6\lambda^2 + \lambda^3}$$

It now remains to test the denominator for strict Hurwitz character. Separating the denominator into its odd and even parts

$$N(\lambda) = \lambda^3 + 15\lambda$$

$$M(\lambda) = 6\lambda^2 + 15$$

we perform the continued fraction expansion of $N(\lambda)/M(\lambda)$ to obtain

$$
\begin{array}{r|ll}
 & \frac{1}{6}\lambda & \\
6\lambda^2 + 15 & \lambda^3 + 15\lambda & \\
 & \underline{\lambda^3 + \frac{5}{2}\lambda} & \frac{12}{25}\lambda \\
 & \frac{25}{2}\lambda & 6\lambda^2 + 15 \\
 & & \underline{6\lambda^2} \qquad \frac{5}{3} \\
 & & 15 \mid 25\lambda \\
 & & \qquad \underline{25\lambda} \\
 & & \qquad 0
\end{array}
$$

which yields strictly positive quotients, demonstrating that all the poles of $H(\lambda)$ are in the open left half $\lambda$-plane corresponding to the inside of the unit circle in the $z$-plane. Hence, the function is (strictly) stable.

Now, if the sampled-data system processes a sampled time signal $\{f(nT)\}$, then we can be more specific about the variable $\lambda$ by relating it to the complex frequency variable $s$. This is achieved using $z = \exp(Ts)$, so that

$$\lambda = \frac{1 - z^{-1}}{1 + z^{-1}}$$

$$= \frac{1 - \exp(-Ts)}{1 + \exp(-Ts)}$$

$$= \frac{\exp[(T/2)s] - \exp[-(T/2)s]}{\exp[(T/2)s] + \exp[-(T/2)s]}$$

or

$$\lambda = \frac{\sinh(T/2)s}{\cosh(T/2)s}$$

i.e.

$$\lambda = \tanh(T/2)s \tag{8.111}$$

where $T$ is the sampling period, and $s$ is the usual complex frequency

$$s = \sigma + j\omega \tag{8.112}$$

Thus with $\lambda = \Sigma + j\Omega$ we have

$$\Sigma + j\omega = \tanh(T/2)(\sigma + j\omega)$$

$$= \frac{[1 + \tan^2(T/2)\omega]\tanh(T/2)\sigma + j[1 - \tanh^2(T/2\sigma)]\tan(T/2)\omega}{1 + \tan^2(T/2)\omega\tanh^2(T/2)\Sigma} \qquad (8.113)$$

This shows that points in the right half $s$-plane map on to points in the right half $\lambda$-plane.

Similarly the left half-planes map on to each other. The real and imaginary axes map on to the corresponding ones. In the case of the imaginary axes (of particular interest since they represent the sinusoidal steady state) $\sigma = 0$, $\Sigma = 0$ and (8.113) shows that the correspondence is periodic. Thus, for $s = j\omega$, $\lambda = j\Omega = j\tan(T/2)\omega$. With

$$T = \frac{1}{f_N} = \frac{2\pi}{\omega_N} \qquad (8.114)$$

where $f_N$ is the sampling (Nyquist) frequency, we have

$$\frac{T}{2}\omega = \left(\frac{\omega}{\omega_N}\right)\pi = \left(\frac{f}{f_N}\right)\pi \qquad (8.115)$$

so that the periodic correspondence between the $j\omega$-axis and the $j\Omega$-axis is described by

$$-\infty \leqslant \Omega \leqslant \infty \quad \text{for} \frac{-(2r+1)}{2} \leqslant \frac{\omega}{\omega_N} \leqslant \frac{(2r+1)}{2} \qquad (8.116)$$

$$r = 0, 1, 2, 3, \ldots$$

But the sampling theorem limits the useful frequency range by

$$0 \leqslant |\omega| < \frac{\omega_N}{2} \qquad (8.117)$$

so that the continuous signal is recoverable unambiguously from its samples if it is band-limited to the range $< \omega_N/2$, outside which aliasing occurs. The mapping between the $s$-plane, the $\lambda$-plane and the $z$-plane is shown in Figure 8.18.

### 8.5.5 Amplitude, Phase and Delay Functions

The transfer function of a general sampled-data filter is of the form

$$H(z) = \frac{\displaystyle\sum_{n=0}^{M} a_n z^{-n}}{1 + \displaystyle\sum_{n=1}^{N} b_n z^{-n}} \qquad (8.118)$$

$$\frac{A_M(z^{-1})}{B_N(z^{-1})}$$

The frequency response of the filter is obtained by letting

$$z \rightarrow \exp(j\omega T) \tag{8.119}$$

so that

$$H(\exp(j\omega T)) = \frac{\displaystyle\sum_{n=0}^{M} a_n \exp(-jn\omega T)}{1 + \displaystyle\sum_{n=1}^{N} b_n \exp(-jn\omega T)} \tag{8.120}$$

which may be put in the form

$$H(\exp(j\omega T)) = |H(\exp(j\omega T))| \exp[j\psi(\omega T)] \tag{8.121}$$

where $|H(\exp(j\omega T))|$ is the amplitude response and $\psi(\omega T)$ is the phase response of the filter. Clearly

$$|H(\exp(j\omega T))|^2 = H(z)H(z^{-1})|_{z=\exp(j\omega T)} \tag{8.122}$$

Taking logarithms of both sides of (8.121) we have

$$\ln H(\exp(j\omega T) = \ln|H(\exp(j\omega T)| + j\psi(\omega T)$$
$$= \tfrac{1}{2}\ln[H(\exp(j\omega T))H(\exp(-j\omega T))] + j\psi(\omega T) \tag{8.123}$$

If we let

$$\psi(z) \triangleq -\frac{1}{2}\ln\left(\frac{H(z)}{H(z^{-1})}\right) \tag{8.124}$$

then

$$\psi(\omega T) = -j\psi(z)|_{z=\exp(j\omega T)} \tag{8.125}$$

The group delay is given by

$$T_g(\omega T) = -\frac{d\psi(\omega T)}{d\omega}$$
$$= j\frac{d\psi(z)}{dz}\bigg|_{z=\exp(j\omega T)} \frac{d\exp(j\omega T)}{d\omega} \tag{8.126}$$
$$= T\left(z\frac{d\psi(z)}{dz}\right)_{\exp(j\omega T)}$$

However, (8.124) gives

$$-\frac{d\psi(z)}{dz} = \frac{1}{2}\frac{d}{dz}\left(\ln\frac{H(z)}{H(z^{-1})}\right)$$

$$= \frac{1}{2}\left(\frac{H'(z)}{H(z)} + \frac{1}{z^2}\frac{H'(z^{-1})}{H(z^{-1})}\right) \tag{8.127}$$

the group delay in (8.126) becomes

$$T_g(\omega T) = \frac{T}{2}\left(\frac{z\{H'(z)\}}{H(z)} + z^{-1}\frac{H'(z^{-1})}{H(z^{-1})}\right)_{z=\exp(j\omega T)} \tag{8.128}$$

or

$$T_g(\omega T) = T\,\mathrm{Re}\left(z\frac{H'(z)}{H(z)}\right)_{z=\exp(j\omega T)}$$

$$= T\,\mathrm{Re}\left(z\frac{d}{dz}\ln H(z)\right)_{z=\exp(j\omega T)} \tag{8.129}$$

It follows that expressions (8.122), (8.124) and (8.125)–(8.129) can be used for the analysis of any sampled-data filter by evaluating its amplitude, phase and delay responses.

In terms of the *bilinear variable* $\lambda$, the frequency response of the filter is obtained by letting $s \to j\omega$ so that

$$\lambda \to j\Omega \tag{8.130}$$

where

$$\Omega = \tan\tfrac{1}{2}T\omega$$

$$= \tan\pi\frac{\omega}{\omega_N} \tag{8.131}$$

and $\omega_N$ is the radian sampling frequency. If *critical sampling* is assumed then $\omega_N$ is the radian *Nyquist* frequency, which is chosen as twice the highest frequency of the band of interest, as explained in Section 8.2. On the $j\omega$-axis, (8.108) becomes

$$H(j\Omega) = \frac{P_m(j\Omega)}{Q_n(j\Omega)}$$

$$= |H(j\Omega)|\exp(j\psi(\Omega)) \tag{8.132}$$

For the group delay we can write

$$T_g(\omega T) = -\frac{d\psi(\lambda)}{ds}\bigg|_{s=j\omega}$$

$$= -\frac{d\psi(\lambda)}{d\lambda}\frac{d\lambda}{ds}\bigg|_{s=j\omega} \tag{8.133}$$

$$= \frac{T}{2}(1-\lambda^2)\frac{d\psi(\lambda)}{d\lambda}\bigg|_{s=j\omega}$$

or

$$T_g(\Omega) = -\frac{T}{2}(1 + \Omega^2)\frac{d\psi(\lambda)}{d\lambda}\bigg|_{\lambda=j\Omega} \tag{8.134}$$

which can be expressed as

$$T_g(\Omega) = \frac{T}{2}\operatorname{Re}(1 + \Omega^2)\left(\frac{Q_n'(\lambda)}{Q_n(\lambda)} - \frac{P_m'(\lambda)}{P_m(\lambda)}\right)_{\lambda=j\Omega} \tag{8.135}$$

Having disposed of the above preliminaries, we devote the remainder of this chapter to a discussion of the derivation of switched-capacitor filter transfer functions based on the bilinear variable. We concentrate on amplitude oriented filter design since the discussion of selective linear-phase filters will be undertaken in a separate chapter.

## 8.6  SWITCHED-CAPACITOR FILTER MODELS DERIVED FROM CONTINUOUS-TIME PROTOTYPES

The magnitude-squared function of a sampled-data filter can be written from (8.132) as

$$|H(j\Omega)|^2 = \frac{|P_m(j\Omega)|^2}{|Q_n(j\Omega)|^2}$$

$$= \frac{\displaystyle\sum_{r=0}^{m} c_r\Omega^{2r}}{\displaystyle\sum_{r=0}^{n} d_r\Omega^{2r}} \tag{8.136}$$

which, as explained earlier, is periodic in $\omega$ due to the periodicity of the variable $\Omega = \tan(\pi\omega/\omega_N)$. Thus

$$-\infty \leqslant \Omega \leqslant \infty \quad \text{for} \frac{-(2r+1)}{2} \leqslant \frac{\omega}{\omega_N} \leqslant \frac{2r+1}{2} \quad r = 0, 1, 2, \dots \tag{8.137}$$

however, the useful band is limited by the sampling theorem to the range

$$0 \leqslant \frac{|\omega|}{\omega_N} \leqslant 0.5 \tag{8.138}$$

The ideal low-pass amplitude characteristic is shown in Figure 8.19 where the dotted lines represent the periodic nature of the *hypothetical* response. However, the frequencies above $\omega_N/2$ are excluded if the input signal is bandlimited to below $\omega_N/2$. If not, aliasing will occur as explained in Section 8.2.

Now, consider a continuous-time filter low-pass prototype transfer function

$$H(s) = \frac{N_m(s)}{D_n(s)} \quad m \neq n \tag{8.139}$$

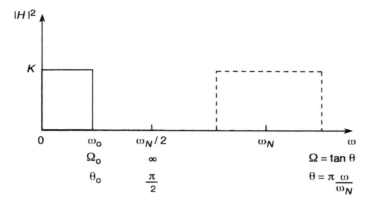

**Figure 8.19**   Ideal low-pass response of a discrete system.

whose amplitude response

$$|H(j\omega)|^2 = \frac{|N_m(j\omega)|^2}{|D_n(j\omega)|^2} \tag{8.140}$$

is obtained subject to a certain optimality criterion, as discussed in Chapter 7, with passband edge at $\omega = 1$. In the prototype function, let

$$s \rightarrow \frac{\lambda}{\Omega_0} \rightarrow \frac{1}{\Omega_0} \frac{1 - z^{-1}}{1 + z^{-1}} \tag{8.141}$$

where

$$\Omega_0 = \tan \pi \frac{\omega_0}{\omega_N} \tag{8.142}$$

and $\omega_0$ is the actual passband edge of the required sampled-data filter. Expression (8.141) transforms the continuous function into a discrete one of the form

$$H(\lambda) = \frac{P_m(\lambda)}{Q_n(\lambda)} \tag{8.143}$$

with amplitude response defined by

$$|H(j\Omega)|^2 = \frac{|P_m(j\Omega)|^2}{|Q_n(j\Omega)|^2} \tag{8.144}$$

The so-called *bilinear transformation* defined by (8.141) has the following basic features:

1. The point $\omega = 1$ in the continuous prototype is transformed to $\omega_0$ in the discrete domain. The point $\omega = 0$ is transformed to the same point $\omega = 0$, while $\omega = \infty$ is transformed to $\omega = \omega_N/2$ since $\Omega = \infty$ corresponds to half the sampling frequency.

2. The stability of the continuous function is preserved under the transformation, since the resulting denominator $Q_n(\lambda)$ is guaranteed strictly Hurwitz in $\lambda$ by virtue of (8.141) and stability is a result of the mapping in Figure 8.18.

3. The properties of the continuous filter along the $j\omega$-axis are transformed into analogous properties on the $j\omega$-axis, which corresponds to the unit circle in the $z$-plane. However, periodicity as stated in (8.137) applies in this case.

4. The resulting transfer function is rational in $z^{-1}$ with real coefficients.

Figure 8.20 shows the transformation of the specifications of a continuous low-pass prototype response into the discrete domain. Due to the mapping

$$\omega = 0 \rightarrow \omega = 0$$

$$\omega = 1 \rightarrow \omega = \omega_0 \qquad\qquad (8.145)$$

$$\omega = \infty \rightarrow \omega = \omega_N/2$$

the stopband now extends from some specified frequency $\omega_{s1}$ to half the sampling frequency $\omega_N/2$. It is assumed that the input to the filter is band-limited to $\omega_N/2$ as required by the sampling theorem.

It follows from the above discussion that we can derive the switched-capacitor counterparts of the maximally-flat, Chebyshev or elliptic filters by the transformation (8.141).

Given a lumped filter of the types dicussed in Chapter 7, applying the transfomation in (8.141) every inductor becomes an element of impedance $L/\Omega_o$ while every capacitor becomes an element of admittance $C/\Omega_o$ as shown in Figure 8.20. These new elements have to be implemented using discrete-time building blocks using switched-capacitor

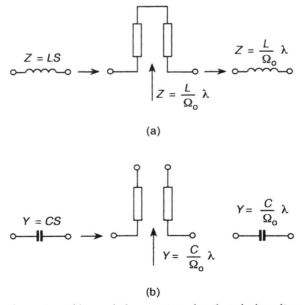

(a)

(b)

**Figure 8.20** Transformation of lumped elements into distributed, then discrete elements. (a) inductor and (b) capacitor.

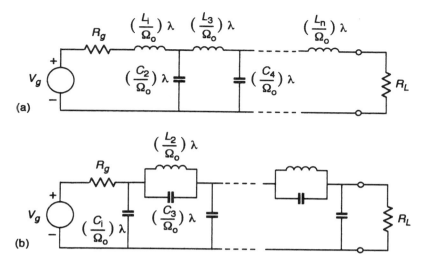

**Figure 8.21** Examples of the transformation of a lumped filter into discrete-time equivalents. (a) simple low-pass ladder and (b) mid-shunt ladder.

techniques; these will be discussed in a later chapter. Meanwhile let us make a number of observations. First we may call the elements with impedances of the form $k\lambda$ *discrete inductors* and those of admittances of the form $1/k\lambda$ are called *discrete capacitors*. More accurately, with reference to distributed network models using unit elements discussed in Section 7.7, we may reinterpret the parameter $\tau$ in the distributed domain as half the sampling period in the discrete domain. With this reinterpretation the distributed variable $\lambda = \tanh(\tau s)$ becomes the discrete variable $\lambda = \tanh(T/2)s$ and the short-circuited stub becomes a discrete short-circuited stub while the open-circuited stub becomes a discrete open-circuited stub. This is shown in Figure 8.20 which is the exact correct analogy. But since the origin of the network is, anyway, lumped we shall loosely speak of discrete

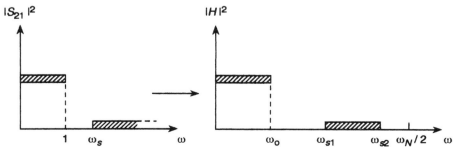

**Figure 8.22** Transformation of the low-pass continuous-time response into a switched-capacitor response using (8.141).

inductors and discrete capacitors and still sketch the discrete counterparts as inductors labelled in the $\lambda$ domain as illustrated in Figure 8.20. Figure 8.21 shows typical low-pass filters obtained by this transformation.

Another $\lambda$-domain distributed network which has a switched-capacitor equivalent is the cascade of unit elements shown in Figure 8.22. This will be discussed fully in a later chapter since it requires special treatment and has no lumped counterpart.

## 8.7   AMPLITUDE-ORIENTED FILTER TRANSFER FUNCTIONS

We now discuss the derivation of the various bilinearly-transformed lumped prototype functions giving rise to the standard amplitude responses.

### 8.7.1   Low-pass Filters

#### Maximally flat response in both bands

Letting $s \to \lambda/\Omega_0$ in (7.100) we obtain

$$S_{21}(s) \to H(\lambda) = \frac{K}{\displaystyle\prod_{r=1}^{n}\left(\frac{\lambda}{\Omega_0} - j\exp(j\theta_r)\right)} \tag{8.146}$$

where

$$\theta_r = \frac{(2r-1)}{2n} \quad r = 1, 2, \ldots n \tag{8.147}$$

and $K$ is chosen such that the dc gain is any prescribed value; $K = 1$ for $H(0) = 1$. $\omega_0$ is the point at which the gain falls by 3 dB. In the z-domain, use of (8.141) in (8.146) yields

$$H(z) = \frac{K(1 + z^{-1})^n}{\displaystyle\prod_{r=1}^{n}\left[\left(\frac{1}{\Omega_0} - j\exp(j\theta r)\right) - \left(\frac{1}{\Omega_0} + j\exp(j\theta r)\right)z^{-1}\right]} \tag{8.148}$$

The resulting response is obtained from (7.95) by letting

$$\omega \to \Omega/\Omega_0 \tag{8.149}$$

This gives

$$|H(j\Omega)|^2 = \frac{K^2}{1 + (\Omega/\Omega_0)^{2n}} \tag{8.150}$$

which has $(2n - 1)$ zero derivatives around $\Omega = 0$ corresponding to $2n - 1$ zero derivatives around $\omega = 0$. It also possesses $(2n - 1)$ zero derivatives around $\Omega = \infty$ corresponding to $(2n - 1)$ zero derivatives around $\omega_N/2$. Figure 8.23 shows the general appearance of a typical maximally-flat response.

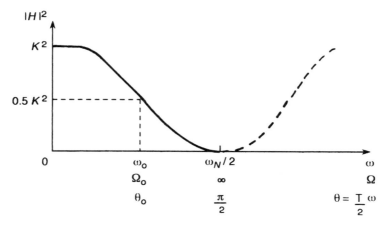

**Figure 8.23**   General appearance of a typical maximally-flat switched-capacitor filter response.

In order to determine the required degree of the filter, let the specifications be given as

$$\text{Passband } 0 \leqslant \omega \leqslant \omega_0, \quad \text{attenuation} \leqslant 3 \text{ dB}$$
$$\text{Stopband } \omega_{s1} \leqslant \omega \leqslant \omega_{s2}, \quad \text{attenuation} \geqslant \alpha_s \text{ dB}$$
$$\text{(8.151)}$$

We note that the stopband must be finite in the sampled-data domain, ending at a frequency less than half the sampling frequency. We begin by choosing a sampling frequency of at least twice the highest frequency of the band of interest. Thus, we must take

$$\omega_N \geqslant 2\omega_{s2} \qquad (8.152)$$

Let us assume that critical sampling is employed, so that (8.152) is satisfied with equality

$$\omega_N = 2\omega_{s2} \qquad (8.153)$$

Then, the passband is defined by

$$0 \leqslant \frac{\omega}{\omega_N} \leqslant \frac{\omega_0}{\omega_N} \qquad (8.154)$$

and the stopband is in the range

$$\frac{\omega_{s1}}{\omega_N} \leqslant \frac{\omega}{\omega_N} \leqslant 0.5 \qquad (8.155)$$

Next, the corresponding $\Omega$-values are obtained to define the bands in the $\Omega$-domain as

$$\text{Passband } 0 \leqslant \Omega \leqslant \Omega_0$$
$$\text{Stopband } \Omega_{s1} \leqslant \Omega \leqslant \Omega_{s2} (= \infty)$$
$$\text{(8.156)}$$

where

$$\Omega_0 = \tan \pi \frac{\omega_0}{\omega_N} \tag{8.157}$$

$$\Omega_{s1} = \tan \pi \frac{\omega_{s1}}{\omega_N} \tag{8.158}$$

$$\Omega_{s2} = \tan \pi \frac{\omega_{s2}}{\omega_N} = \infty \tag{8.159}$$

Of course, if the sampling frequency is chosen higher than $2\omega_{s2}$, then the stopband is defined by

$$\Omega_{s1} \leqslant \Omega \leqslant \Omega_{s2} \tag{8.160}$$

where, instead of (8.159) we have

$$\Omega_{s2} = \tan \pi \frac{\omega_{s2}}{\omega_N} \neq \infty \tag{8.161}$$

Finally, the degree of the required filter is obtained by substitution of (8.157) and (8.158) in (7.109) with $\omega_s \rightarrow \Omega_{s1}/\Omega_0$. This gives

$$n \geqslant \frac{\log(10^{0.1a_s} - 1)}{2\log(\Omega_{s1}/\Omega_0)} \tag{8.162}$$

An alternative format for the specifications may be given as

$$\begin{array}{ll} \text{Maximum passband attenuation} = \alpha_p & \omega \leqslant \omega_0 \\ \text{Minimum stopband attenuation} = \alpha_s & \omega \geqslant \omega_{s1} \end{array} \tag{8.163}$$

In this case (7.114) may be used to determine the required degree after transformation to the $\Omega$-domain by means of (8.141). This gives

$$n \geqslant \frac{\log\left(\dfrac{10^{0.1\alpha_s} - 1}{10^{0.1\alpha_p} - 1}\right)}{2\log(\Omega_{s1}/\Omega_0)} \tag{8.164}$$

*Example 8.9* Find the degree of a switched-capacitor maximally-flat filter, with the following specifications:

Passband 0 to 1 kHz with attenuation $\leqslant 1$ dB
Stopband 2 to 4 KHz with attenuation $\geqslant 20$ dB.

*Solution* Assuming critical sampling, the highest frequency in the band of interest is 4 kHz so that we may choose a sampling frequency of twice this value to give

$$\omega_N = (2\pi) \times 8 \times 10^3$$

Thus, in the $\Omega$-domain we have

$$\Omega_0 = \tan\tfrac{1}{8}\pi = 0.4142$$

$$\Omega_{s1} = \tan\tfrac{2}{8}\pi = 1.0$$

Using (8.164) the required filter degree is

$$n \geqslant 3.373$$

so that we may take

$$n = 4$$

### Chebyshev response

Letting $s \rightarrow \lambda/\Omega_0$ in the denominator of (7.136) we obtain

$$S_{21}(s) \rightarrow H(\lambda) = \frac{K}{\displaystyle\prod_{r=1}^{n}\left(\frac{\lambda}{\Omega_0} + [\eta\sin\theta_r + j(1+\eta^2)^{1/2}\cos\theta_r]\right)} \qquad (8.165)$$

where $\theta_r$ is given by (7.128) and

$$\eta = \sinh\left(\frac{1}{n}\sinh^{-1}\frac{1}{\varepsilon}\right) \qquad (8.166)$$

The numerator of $H(\lambda)$ is taken to be an arbitrary constant $K$, by contrast with (7.136), since the switched-capacitor filter gain can be adjusted arbitrarily, and is not restricted to have a maximum value of unity as in the case of passive filters. Again, using (8.141) in (8.165), the $z$-domain representation of the Chebyshev filter is obtained as

$$H(z) = \frac{K(1+z^{-1})^n}{\displaystyle\prod_{r=1}^{n}\left[\left(\frac{1}{\Omega_0}+jy_r\right) - \left(\frac{1}{\Omega_0}-jy_r\right)z^{-1}\right]} \qquad (8.167)$$

where

$$y_r = \cos(\sin^{-1}j\eta + \theta_r) \qquad (8.168)$$

Evidently, the resulting transfer function has a magnitude squared function given by (7.115) with $\omega \rightarrow (\Omega/\Omega_0)$ i.e.

$$|H(j\Omega)|^2 = \frac{K^2}{1 + \varepsilon^2 T_n^2(\Omega/\Omega_0)} \qquad (8.169)$$

where $T_n(\Omega/\Omega_0)$ is the Chebyshev polynomial defined by (7.120) and $\varepsilon$ is the ripple factor. $|H(j\Omega)|^2$ has an optimum equiripple response in the passband and $(2n - 1)$ zero derivatives around $\Omega = \infty$, i.e. around $w_N/2$ Figure 8.24 shows the general appearance of a typical Chebyshev response.

To determine the required degree of the filter, the specifications are transformed to the $\Omega$-domain as in (8.157)–(8.161), then use is made of (8.141) with $w \rightarrow (\Omega/\Omega_0)$. Let the specifications be given as

$$\text{Passband } 0 \leqslant w \leqslant w_0, \quad \text{attenuation} \leqslant \alpha_p$$
$$\text{Stopband } w_{s1} \leqslant w \leqslant w_{s2}, \quad \text{attenuation} \geqslant \alpha_s \tag{8.170}$$

Choosing $w_N \geqslant 2w_{s2}$, and evaluating $\Omega_0$ and $\Omega_{s1}$ from (8.157)–(8.159), we then use (7.141) to obtain for the degree

$$n \geqslant \frac{\cosh^{-1}[(10^{0.1\alpha_s} - 1(/(10^{0.1\alpha_p} - 1)]^{1/2}}{\cosh^{-1}(\Omega_{s1}/\Omega_0)} \tag{8.171}$$

*Example 8.10*  Find the transfer function of a low-pass Chebyshev switched-capacitor filter with the following specifications

Passband 0 to 0.5 kHz with 0.1 dB ripple.
Stopband edge: 0.7 kHz with attenuation $\geqslant$ 40 dB.
Sampling frequency: 2 kHz.

*Solution*  From (8.157)–(8.159) the $\Omega$-domain values are

$$\Omega_0 = \tan \pi \frac{0.5}{2} = 1$$
$$\Omega_{s1} = \tan \pi \frac{0.7}{2} = 1.963 \tag{8.172}$$

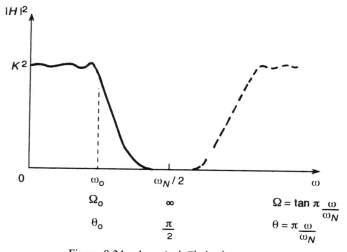

Figure 8.24   A typical Chebyshev response.

and (8.171) gives the required degree as $n = 6$. Also

$$\varepsilon = (10^{0.1\alpha_p} - 1)^{1/2} = (10^{0.01} - 1)^{1/2} = 0.1526$$

so that the auxiliary parameter is obtained from (8.166) as

$$\eta = \sinh\left(\frac{1}{6}\sinh^{-1}\frac{1}{0.1526}\right) = 0.443$$

Finally the filter transfer function is obtained from (8.167) and (8.168) as

$$H(z) = H_1(z)H_2(z)H_3(z)$$

$$H_1(z) = \frac{0.426 + 0.851z^{-1} + 0.426z^{-2}}{1 + 0.103z^{-1} + 0.805z^{-2}}$$

$$H_2(z) = \frac{0.431 + 0.863z^{-1} + 0.431z^{-2}}{1 - 0.266z^{-1} + 0.459z^{-2}} \qquad (8.173)$$

$$H_3(z) = \frac{0.472 + 0.944z^{-1} + 0.472z^{-2}}{1 - 0.696z^{-1} + 0.192z^{-2}}$$

### Elliptic function response

The Low-pass elliptic transfer function can be obtained using the same transformation (8.141) and the extensive expressions available for the analog elliptic transfer functions. The procedure is illustrated by the following example.

*Example 8.11* The transfer function of a third-order low-pass elliptic continuous-time filter is given by

$$H(s) = \frac{0.314(s^2 + 2.806)}{(s + 0.767)(s^2 + 0.453s + 1.149)}$$

which gives 0.5 dB passband ripple and a minimum stopband attenuation of 21 dB for $\omega_s/\omega_0 \geqslant 1.5$. Use this prototype function to design a switched-capacitor filter with pass-band edge at 500 Hz and a sampling frequency of 3 kHz.

*Solution* From the specifications and (8.157),

$$\Omega_0 = \tan\pi\frac{500}{3000} = 0.577$$

Thus, using the bilinear transformation (8.141) the transfer function becomes

$$H(z) = \left(\frac{0.126 + 0.126z^{-1}}{1 - 0.386z^{-1}}\right)\left(\frac{1.177 - 0.079z^{-1} + 1.1778z^{-2}}{1 - 0.75z^{-1} + 0.682z^{-2}}\right)$$

$$= H_1(z)H_2(z)$$

## 8.7.2  High-pass filters

The continuous-time low-pass prototype transfer function of Section 7.5 can also be used to derive high-pass switched-capacitor transfer functions. This can be achieved via the transformation

$$s \rightarrow \frac{\Omega_0}{\lambda} \rightarrow \Omega_0 \left( \frac{1 + z^{-1}}{1 - z^{-1}} \right) \tag{8.174}$$

where

$$\Omega_0 = \tan \pi \frac{\omega_0}{\omega_N} \tag{8.175}$$

in which $\omega_0$ is the high-pass filter passband edge. The transformation is illustrated in Figure 8.25. The point $\omega = 0$ is transformed to $\omega_N/2$ and $\omega = \infty$ is transformed to $\omega = 0$. The passband now occupies the range

$$\omega_0 \leqslant |\omega| \leqslant \omega_N/2 \tag{8.176}$$

The resulting transfer functions are obtained by means of the transformation in (8.174) and any of the expressions in Section 7.5. The maximally flat and Chebyshev amplitude functions are now

$$|H(j\Omega)|^2 = \frac{K^2}{1 + (\Omega_0/\Omega)^{2n}} \tag{8.177}$$

and

$$|H(j\Omega)|^2 = \frac{K^2}{1 + \varepsilon^2 T_n^2(\Omega_0/\Omega)} \tag{8.178}$$

To obtain the required degree of the filter, let the specifications be given as

$$\begin{aligned} \text{Passband } \omega_0 \leqslant \omega \leqslant \omega_N/2, \quad & \text{attenuation} \leqslant \alpha_p \, dB \\ \text{Stopband } 0 < \omega \leqslant \omega_{s1}, \quad & \text{attenuation} \geqslant \alpha_s \, dB \end{aligned} \tag{8.179}$$

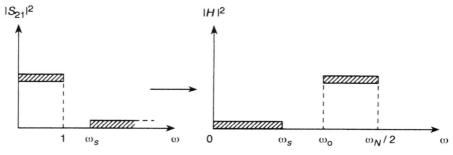

**Figure 8.25**   Low-pass prototype to switched-capacitor high-pass transformation.

with

$$\Omega_0 = \tan \pi \frac{\omega_0}{\omega_N}$$

$$\Omega_{s1} = \tan \pi \frac{\omega_{s1}}{\omega_N}$$

$$(8.180)$$

The required degree of the maximally-flat filter is obtained from (8.164) by letting $(\Omega_s/\Omega_0) \to (\Omega_0/\Omega_s)$ to give

$$n \geqslant \frac{\log[(10^{0.1\alpha_s} - 1)/(10^{0.1\alpha_p} - 1)]}{2\log(\Omega_0/\Omega_{s1})} \qquad (8.181)$$

Alternatively, if the passband edge is the 3 dB point then (8.162) gives

$$n \geqslant \log(10^{0.1\alpha_s} - 1)2\log(\Omega_0/\Omega_{s1}) \qquad (8.182)$$

For the Chebyshev response, the required degree is

$$n \geqslant \frac{\cosh^{-1}[(10^{0.1\alpha_s} - 1)/(10^{0.1\alpha_p} - 1)]^{1/2}}{\cosh^{-1}(\Omega_0/\Omega_s)} \qquad (8.183)$$

Having obtained the required degree, the analog low-pass prototype function is obtained, which upon use of the transformation (8.141) gives the switched-capacitor high-pass transfer function.

### 8.7.3   Band-pass Filters

Again, the continuous-time low-pass prototype transfer functions can be used to obtain discrete-time band-pass functions by means of a transformation given by

$$s \to \frac{\overline{\Omega}}{\Omega_2 - \Omega_1} \left( \frac{\lambda}{\overline{\Omega}} + \frac{\overline{\Omega}}{\lambda} \right) \qquad (8.184)$$

where

$$\overline{\Omega} = \tan \pi \frac{\overline{\omega}}{\omega_N} \qquad (8.185)$$

$$\Omega_{1,2} = \tan \pi \frac{\omega_{1,2}}{\omega_N} \qquad (8.186)$$

$$\overline{\Omega} = (\Omega_1\Omega_2)^{1/2} \qquad (8.187)$$

This transformation is illustrated in Figure 8.26. The band centre $\overline{\omega}$ is arbitrary, but the end of the upper stopband is at most equal to $\omega_N/2$. The passband extends from $\omega_1$ to $\omega_2$. Expression (8.141) may be used to transform to the $z$-domain for the purpose of realization of the transfer function.

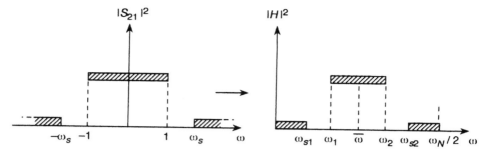

**Figure 8.26**  Low-pass prototype to switched-capacitor band-pass transformation.

The resulting maximally flat and Chebyshev responses are given, respectively, by

$$|H(j\Omega)|^2 = \frac{K^2}{1 + \left[\dfrac{\Omega}{\Omega_2 - \Omega_1}\left(\dfrac{\Omega}{\overline{\Omega}} - \dfrac{\overline{\Omega}}{\Omega}\right)\right]} \qquad (8.188)$$

and

$$|H(j\Omega)|^2 = \frac{K^2}{1 + \varepsilon^2 T_n^2\left[\dfrac{\Omega}{\Omega_2 - \Omega_1}\left(\dfrac{\Omega}{\overline{\Omega}} - \dfrac{\overline{\Omega}}{\Omega}\right)\right]} \qquad (8.189)$$

The actual filter transfer functions are obtained by the transformation (8.184) in the appropriate analog functions given, for example, by (7.102) and (7.135).

The required degree of the prototype filter is obtained by substituting in (8.188), (8.189) for the required attenuation values at the given frequencies, and forming the necessary equations. In particular, for the maximally flat response, the two 3 dB points occur at $\omega_1$ and $\omega_2$. Furthermore, there are two frequencies, $\omega_{s1}$ and $\omega_{s2}$ at which the attenuation is the same value $\alpha_s$, the two frequencies being related by

$$\overline{\Omega} = \left[\left(\tan \pi \frac{\omega_{s1}}{\omega_N}\right)\left(\tan \pi \frac{\omega_{s2}}{\omega_N}\right)\right] \qquad (8.190)$$

The filter is designed according to the more severe of the two requirements, and the response has geometric symmetry about $\overline{\Omega}$ and the $\Omega$ domain. These considerations also apply to Chebyshev and elliptic filters.

### 8.7.4  Band-stop Filters

Typical specifications in this case are shown in Figure 8.27. These can be met using the passive prototype low-pass function by the transformation

$$s \to \frac{(\Omega_2 - \Omega_1)}{\overline{\Omega}}\bigg/\left(\frac{\lambda}{\overline{\Omega}} + \frac{\overline{\Omega}}{\lambda}\right) \qquad (8.190)$$

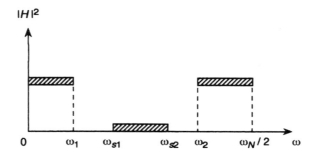

**Figure 8.27**   Switched-capacitor band-stop filter specifications.

where $\overline{\Omega}$, and $\Omega_{1,2}$ are also given by (8.185)–(8.187) but in this case $\omega_1$ is the lower passband edge and $\omega_2$ is the upper passband edge. Again, the resulting response exhibits geometric symmetry about $\overline{\Omega}$, so that (8.190) and the preceding discussion are also valid in this case.

## CONCLUSION

In this chapter, the methods of description of sampled-data filters in general and switched-capacitor filters in particular were discussed. Then, it was shown, in detail, how a switched-capacitor amplitude-oriented filter transfer function may be derived from a corresponding passive lumped-element prototype. We have drawn heavily on the material of Chapter 7.

## PROBLEMS

8.1  Find the transfer function of a low-pass maximally-flat switched-capacitor filter with the following specifications:

   Passband 0 to 1.5 kHz, attenuation $\leqslant 0.8$ dB
   Stopband 2.0 to 3.5 kHz, attenuation $\geqslant 30$ dB
   Clock frequency: 10 kHz

8.2  Find the transfer function of a low-pass Chebyshev switched-capacitor filter with the following specifications:

   Passband 0 to 3.4 kHz, attenuation $\leqslant 0.1$ dB
   Stopband 4.6 to 5 kHz, attenuation $\geqslant 40$ dB
   Clock frequency: 10 kHz

8.3  Find the transfer function of a maximally-flat band-pass switched-capacitor filter with the following specifications:

   Passband 300 to 3400 kHz, attenuation $\geqslant 3$ dB
   Stopband edge frequencies $= 100$ Hz and 4.6 kHz with minimum attenuation of 32 dB
   Clock frequency: 20 kHz

8.4   Find the transfer function of a Chebyshev band-pass switched-capacitor filter with the following specifications:

Passband 300 to 3400 kHz, 0.25 dB ripple
Stopband edge frequencies $= 100$ Hz and 4.6 kHz with minimum attenuation of 32 dB
Clock frequency: 10 kHz

8.5   Find the transfer function of a band-stop Chebyshev switched-capacitor filter with the following specifications

Passband edge frequencies: 1 kHz and 3 kHz with 0.5 dB ripple
Stopband 1.2–1.5 kHz
Clock frequency $= 8$ kHz

# 9

# Amplitude-oriented Filters of the Lossless Discrete Integrator Type

## 9.1 INTRODUCTION

In this chapter, a very useful class of switched-capacitor filters is studied in detail. This is characterized by its low-sensitivity properties relative to element value variations; a highly desirable attribute from the practical viewpoint. These filters have no lumped-element counterparts, but their transfer functions are of the same form as those of cascaded unit elements distributed filters introduced in Section 7.7. Hence a complete synthesis algorithm is required, which is implemented by a computer program given as part of the package in ISICAP. The basic ideas of the state-variable or leapfrog ladder filter are introduced then applied specifically to the design of maximally-flat and Chebyshev filters [1], [40]–[42]. We shall draw heavily upon the results of Chapter 7, in particular the use of Darlington synthesis technique as applied to $\lambda$-domain networks instead of the usual $s$-domain type. The procedure and basic principles are precisely the same in the manner indicated in Section 7.7 once we have incorporated the one-to-one correspondence between the $s$-plane and $\lambda$-plane studied in detail in 8.5.4.

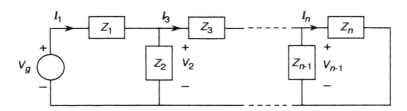

**Figure 9.1**   General passive ladder.

## 9.2   THE STATE-VARIABLE LADDER FILTER

Consider the general passive ladder shown in Figure 9.1 where the branches
are arbitrary impedances. Write the state equations of the ladder, relating the
series currents to the shunt voltages. For the sake of specificity, we assume that $n$ is
odd.

$$I_1 = Z_1^{-1}(V_g - V_2)$$

$$V_2 = Z_2(I_1 - I_3)$$

$$I_3 = Z_3^{-1}(V_2 - V_4) \tag{9.1}$$

$$\vdots \quad \vdots$$

$$I_n = Z_n^{-1}V_{n-1}$$

The transfer function of interest is given by

$$H_{21} = \frac{I_n}{V_g} \tag{9.2}$$

Regardless of the specific form of the branch impedances, any other circuit which
implements the same set of equations (9.1) will produce the same transfer function.
Note that we have deliberately refrained from using any specific frequency variable
in (9.1).

   Now, in seeking an active implementation of (9.1) all we need are some building
blocks capable of providing the frequency dependence of the branches, as well as
performing the mathematical operations of (9.1). A block diagram of the required
implementation is shown in Figure 9.2 and is known as the *state-variable leap-frog*
*ladder* realization. This consists of differential-input boxes which possess voltage
transfer functions $T_1, T_2, \ldots T_n$ connected such that:

$$\tilde{V}_1 = T_1(V_0 - V_2)$$

$$V_2 = T_2(\tilde{V}_1 - \tilde{V}_3)$$

$$\tilde{V}_3 = T_3(V_2 - V_4) \tag{9.3}$$

$$\vdots \quad \vdots$$

$$\tilde{V}_n = T_n V_{n-1}$$

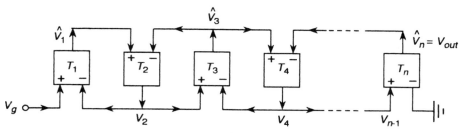

**Figure 9.2**   State-variable (leap-frog) equivalent of the ladder in Figure 9.1.

Let the currents $I_1, I_3, I_5, \ldots, I_n$ in (9.1) be simulated by the voltages $\tilde{V}_1, \tilde{V}_3, \tilde{V}_5, \ldots, \tilde{V}_n$ in (9.3). The internal structure of each box in Figure 9.2 is chosen such that

$$T_1 = \alpha Z_1^{-1}$$

$$T_2 = \alpha^{-1} Z_2$$

$$T_3 = \alpha Z_3^{-1} \tag{9.4}$$

$$\vdots \qquad \vdots$$

$$T_n = \alpha Z_n^{-1}$$

where $\alpha$ is a constant. Then the transfer function of the leap-frog ladder of Figure 9.2

$$\tilde{H}_{21} = \frac{V_{out}}{V_g} \tag{9.5}$$

only differs from $H_{21}$ of the passive ladder by a constant. Consequently, given a ladder of the general form in Figure 9.1, with the specific type and values of elements, it is possible to determine the state variable simulation provided we can find the necessary active building blocks with transfer functions satisfying (9.4). Conversely, if we decide on certain types of building blocks for the boxes in Figure 9.2 we can find the equivalent ladder. The use of this technique is now applied to the two main categories of filters separately.

## 9.3   LOW-PASS FILTERS

### 9.3.1   The Basic Generic Building Blocks

Starting from Figure 9.2, two basic types of building blocks are used as the boxes. These are shown in Figure 9.3 and are known as lossless discrete integrator (LDI) and damped discrete integrator (DDI). These may be considered *generic* building blocks, in the sense that any other set which perform the same functions would be acceptable.

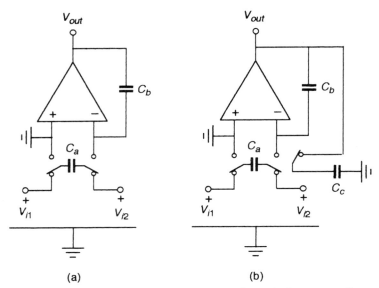

**Figure 9.3**   The basic **generic** building blocks for a class of switched-capacitor filters: (a) Type A: lossless discrete integrator (LDI); (b) Type B: damped discrete integrator (DDI).

*Type A*   This is the circuit of Figure 9.3(a) and is the LDI. The Op Amp is assumed ideal and the switches are driven by a biphase clock with non-overlapping pulses of period $T$ as shown in Figure 9.4. The capacitor $C_a$ is switched to the inputs $V_{i1}$ and $V_{i2}$ at $t = mT$ and to the Op Amp inputs at $t = (m + \frac{1}{2})T$. The output of the Op Amp is sampled at $t = (m + \frac{1}{2})T$. Thus

$$V_{out}(nT) = V_{out}\{(n-1)T\} + \frac{C_a}{C_b}\left(V_{i1}\left\{\left(n - \frac{1}{2}\right)T\right\} - V_{i2}\left\{\left(n - \frac{1}{2}\right)T\right\}\right) \quad (9.6)$$

Taking the $z$-transform $(z = e^{Ts})$

$$V_{out}(z) = z^{-1}V_{out}(z) + z^{-1/2}\frac{C_a}{C_b}\{V_{i1}(z) - V_{i2}(z)\} \quad (9.7)$$

Therefore, this building block has the transfer function

$$T_A = \frac{V_{out}(z)}{V_{i1}(z) - V_{i2}(z)}$$

$$= \frac{z^{-1/2}}{\left(\dfrac{C_b}{C_a}\right)(1 - z^{-1})} \quad (9.8)$$

$$= \frac{1}{2\left(\dfrac{C_b}{C_a}\right)\gamma}$$

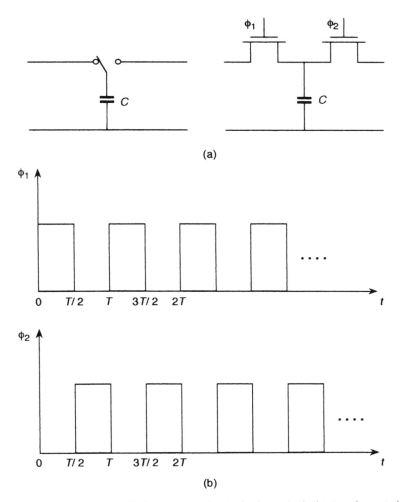

Figure 9.4   (a) A switched-capacitor, (b) the biphase clock driving the switches.

where

$$\gamma = \sinh(T/2)s \qquad (9.9)$$

*Type B*   This is the circuit of Figure 9.3(b) which is the same as that of the LDI but with a feedback capacitor $C_c$. This is known as a damped discrete integrator (DDI). With the same sequence of switching as Type A, we have:

$$V_{out}(nT) = V_{out}\{(n-1)T\} + \frac{C_a}{C_b}\left(V_{i1}\left\{\left(n-\frac{1}{2}\right)T\right\} - V_{i2}\left\{\left(n-\frac{1}{2}\right)T\right\}\right)$$

$$- \frac{C_c}{C_b}V_{out}\{(n-1)T\} \qquad (9.10)$$

and taking the $z$-transform

$$V_{out}(z) = z^{-1}V_{out}(z) + \frac{C_a}{C_b}z^{-1/2}\{V_{i1}(z) - V_{i2}(z)\} - \frac{C_c}{C_b}z^{-1}V_{out}(z) \qquad (9.11)$$

Therefore, this building block has the transfer function

$$T_B = \frac{V_{out}(z)}{V_{i1}(z) - V_{i2}(z)}$$

$$= \frac{z^{-1/2}}{\frac{C_b}{C_a}\left\{1 - z^{-1}\left(1 - \frac{C_c}{C_a}\right)\right\}} \qquad (9.12)$$

$$= \frac{1}{\left(\frac{2C_b - C_c}{C_a}\right)\gamma + \frac{C_c}{C_a}\mu}$$

where

$$\mu = \cosh(T/2)s \qquad (9.13)$$

Now let these building blocks be used as the boxes in Figure 9.2 such that the *first* and *last* boxes are *Type B* while all the *internal ones* are *Type A*. The resulting network takes the form shown in Figure 9.5 and is clearly of the analogue sampled-data type. It is called a *switched-capacitor state-variable ladder*. Examination of (9.1) and (9.4) shows that the switched-capacitor network has the equivalent network of Figure 9.6 where

$$L_k(\text{or } C_k) = 2\left(\frac{C_b}{C_a}\right)_k, \quad k = 2 \to (n-1) \qquad (9.14)$$

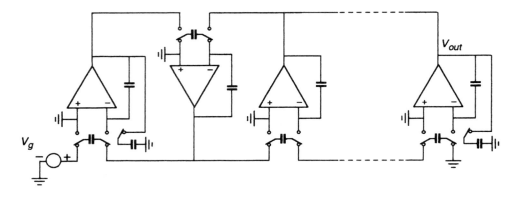

**Figure 9.5**  The switched-capacitor state variable (leap-frog) ladder filter using generic building blocks.

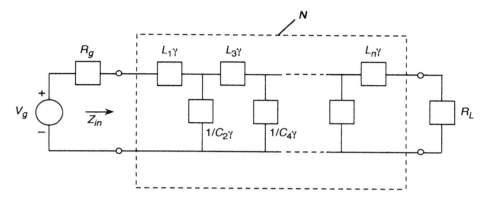

**Figure 9.6** Equivalent network of the ladder in Fig. 9.5.

$$L_{1,n} = \left(\frac{2C_b - C_c}{C_a}\right)_{1,n} \tag{9.15}$$

$$R_{g,L} = \left(\frac{C_c}{C_a}\right)_{1,n} \tag{9.16}$$

At first glance the equivalent network of Figure 9.6 appears rather peculiar. Viewed as a doubly terminated ladder two-port, the elements have frequency dependence $L_k\gamma$, $1/C_k\gamma$ and their terminations are also frequency dependent of the form $R_g\mu$, $R_L\mu$. We shall see in a moment that the transfer function of this network is very familiar!

Consider Figure 9.6 and let the two-port $N$ be described by a polynomial transmission matrix

$$[t(\gamma)] = \begin{bmatrix} a(\gamma) & b(\gamma) \\ c(\gamma) & d(\gamma) \end{bmatrix} \tag{9.17}$$

Viewed *abstractedly* as a polynomial transmission matrix of an odd degree ladder composed of elements of the indicated frequency dependence, we have from Section 7.3.2

$$a(\gamma) = a_{n-1}(\gamma) \quad \text{all even}$$

$$b(\gamma) = b_n(\gamma) \quad \text{all odd}$$

$$c(\gamma) = c_{n-2}(\gamma) \quad \text{all odd} \tag{9.18}$$

$$d(\gamma) = d_{n-1}(\gamma) \quad \text{all even}$$

Next, obtain the transfer function $I_n/V_g$ of the equivalent network as

$$H_{21} = \frac{I_n}{V_g} = \frac{1}{R_g R_L \mu a(\gamma) + b(\gamma) + R_g R_L \mu^2 c(\gamma) + R_g \mu d(\gamma)} \tag{9.19}$$

and using

$$\gamma = \lambda/\sqrt{1 - \lambda^2} \tag{9.20}$$

$$\mu = 1/\sqrt{1 - \lambda^2} \tag{9.21}$$

where

$$\lambda = \tanh\frac{T}{2}s \tag{9.22}$$

we have

$$H_{21}(\lambda) = \frac{(1 - \lambda^2)^{n/2}}{P_n(\lambda)} \tag{9.23}$$

which is the transfer function realized by the switched-capacitor network of Figure 9.5 as a voltage transfer function $V_{out}/V_g$.

But the transfer function in (9.23) is identical in form to the transfer function of a cascade of UEs, as shown in Section 7.7, if we make the identification $\tau = T/2$. Nevertheless the equivalent network of Figure 9.6 is *not a resistively terminated lossless two-port*, so we require a modified synthesis technique. To this end, consider the process of impedance-scaling the equivalent network of Figure 9.6 by the factor $1/\mu$. This results in the resistively terminated lossless two-port shown in Figure 9.7 which is called the *auxiliary network*. This is taken as the starting-point in performing the synthesis of the original network of Figure 9.6. The reason is that the standard synthesis technique (see Section 7.3.3) is directly applicable to the auxiliary network. Its transfer function is given by

$$\tilde{H}_{21} = \mu H_{21}$$

$$= \frac{(1 - \lambda^2)^{(n-1)/2}}{P_n(\lambda)} \tag{9.24}$$

describing a cascade of $(n - 1)$ UEs and one transmission zero at $\lambda = \infty$. If we denote the transducer power gain of the auxiliary network by $|S_{21}|^2$ then

$$|\tilde{H}_{21}|^2 = |S_{21}|^2/4R_g R_L \tag{9.25}$$

It also follows that if the input impedance $\tilde{Z}_{in}$ of the auxiliary network is calculated using Darlington's procedure of Section 7.3.3, for any prescribed $|S_{21}|^2$, then the input impedance of the original network of Figure 9.6 can be obtained from

$$Z_{in} = \mu \tilde{Z}_{in} \tag{9.26}$$

Before discussing the approximation and synthesis we note that the necessary and sufficient condition for stability is that $P_n(\lambda)$ in (9.23) be strictly Hurwitz.

### 9.3.2 Approximation and Synthesis

Comparing the transfer function (9.23) and that of (7.202) it follows that we already know the solutions to the amplitude approximation problem. Thus, using (7.209) for a maximally flat response around the origin and one zero-derivative at $(T/2)\omega = \pi/2$,

$$|H_{21}|^2 = K \left/ \left( 1 + \left\{ \frac{\sin \dfrac{T}{2}\omega}{\sin \dfrac{T}{2}\omega_0} \right\}^{2n} \right) \right. \tag{9.27}$$

where $\omega_0$ is the 3 dB point. Clearly

$$\frac{T}{2}\omega = \pi \frac{\omega}{\omega_N}, \qquad \frac{T}{2}\omega_0 = \pi \frac{\omega_0}{\omega_N} \tag{9.28}$$

where $\omega_N$ is the radian sampling frequency. For an optimum equiripple response up to $\omega_0$ we have by (7.210),

$$|H_{21}|^2 = K \left/ \left( 1 + \epsilon^2 T_n^2 \left\{ \frac{\sin \dfrac{T}{2}\omega}{\sin \dfrac{T}{2}\omega_0} \right\} \right) \right. \tag{9.29}$$

Thus, for the auxiliary network in Figure 9.7, assuming $R_g = 1\,\Omega$, we use (9.23) to (9.25) to obtain

$$|S_{21}|^2 = 4KR_L \cos^2 \frac{T}{2}\omega \left/ \left( 1 + \left\{ \frac{\sin \dfrac{T}{2}\omega}{\sin \dfrac{T}{2}\omega_0} \right\}^{2n} \right) \right. \tag{9.30}$$

for a maximally-flat response of the original network and

$$|S_{21}|^2 = 4KR_L \cos^2 \frac{T}{2}\omega \left/ \left( 1 + \epsilon^2 T_n^2 \left\{ \frac{\sin \dfrac{T}{2}\omega}{\sin \dfrac{T}{2}\omega_0} \right\} \right) \right. \tag{9.31}$$

for an equiripple passband response of the original network.

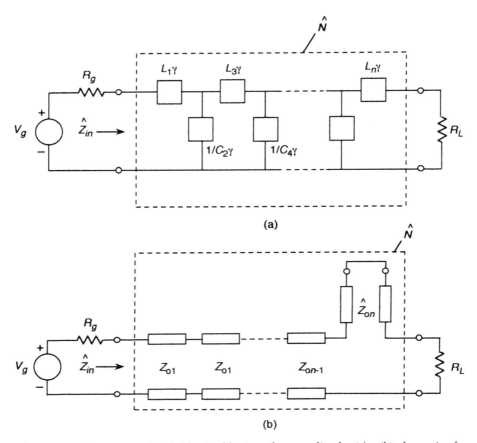

**Figure 9.7** Auxiliary neetwork: (a) Obtained by impedance-scaling by $1/\mu$; (b) alternative form as a cascade of $(n-1)$ UEs and a stub.

Now, in order to obtain the input impedance of the original network of Figure 9.6 we must first proceed along the lines of Darlington's synthesis, in Section 7.3.3 for the auxiliary network of Figure 9.7, since this technique applies to resistively terminated lossless two-ports. Having found the input impedance of this network, (9.26) can be used to determine the input impedance of the original network of Figure 9.6. Thus, from (9.30) and (9.31) we evaluate

$$|S_{11}|^2 = 1 - |S_{21}|^2 \tag{9.32}$$

and the input reflection coefficient of the auxiliary network $S_{11}(\lambda)$ is obtained by factorization of (9.32) and selecting the left half $\lambda$-plane poles, i.e.

$$S_{11}(\lambda) = \frac{N(\lambda)}{D(\lambda)} \tag{9.33}$$

where $D(\lambda)$ is a strictly Hurwitz polynomial in $\lambda$.

Once $S_{11}$ is determined the input impedance of the original (realized) network is obtained using (9.26) and (9.33). Thus

$$Z_{in} = \mu \tilde{Z}_{in} = \mu \frac{1 + S_{11}}{1 - S_{11}} \qquad (9.34)$$

Using

$$\mu^2 = (1 + \gamma^2), \qquad \lambda = \gamma/\mu \qquad (9.35)$$

$Z_{in}$ may be put in the form

$$Z_{in} = \frac{\mu a_{n-1}(\gamma) + b_n(\gamma)}{\mu c_{n-2}(\gamma) + d_{n-1}(\gamma)} \qquad (9.36)$$

where $a$, $b$, $c$, and $d$ are polynomials in $\gamma$ which can be identified with those in (9.18). The synthesis technique then suggests itself as performing the continued fraction expansion of $Z_{in}$ around $\gamma$, with $\mu$ treated as a constant. The coefficients of $\gamma$ in the expansion give $L_1$, $C_2$, $L_3, \dots$ the element values in the network of Figure 9.6. Subsequently the capacitance ratios can be obtained from (9.14)–(9.16)

Finally, the dc gain can be adjusted to any value by a suitable choice of K in (9.27). Alternatively, this can be accomplished after the synthesis has been completed by impedance-scaling the entire network by a suitable factor.

Now, **Program LDISCG** in Appendix A of ISICAP implements the synthesis technique for an arbitrary transfer function of the general form (9.23). Program **CHBYE** in Appendix C of ISICAP performs the synthesis for the specific case of Chebyshev response given arbitrary specifications.

*Example 9.1* Consider the calculation of the element values for a seventh-order low-pass Chebyshev filter with 0.05 dB passband ripple and passband edge at 3.4 kHz for a clock frequency of 28 kHz. In this case we have

$$f_0/f_N = \omega_0/\omega_N = 0.12$$

$$\sin(\pi f_0/f_N) = 0.37$$

Therefore from (9.29) and (9.30) with $\theta \equiv \pi\omega/\omega_N$, $K = 1/4$, and $R_L = 1$,

$$|H_{21}|^2 = \frac{1/4}{1 + 0.01 T_7^2 \left( \dfrac{\sin\theta}{0.37} \right)}$$

$$|S_{21}|^2 = \frac{\cos^2\theta}{1 + 0.01 T_7^2 \left( \dfrac{\sin\theta}{0.37} \right)}$$

$$|S_{11}|^2 = \frac{\sin^2\theta + 0.01T_7^2\left(\dfrac{\sin\theta}{0.37}\right)}{1 + 0.1T_7^2\left(\dfrac{\sin\theta}{0.37}\right)}$$

The above expression is factored in the $\lambda$-domain so that $\tilde{Z}_{in}$ is obtained from (9.28) as

$$\tilde{Z}_{in} = \frac{\begin{array}{c}1 + 16.55\lambda + 112.8\lambda^2 + 573.04\lambda^3 + 1660.94\lambda^4 \\ +4716.78\lambda^5 + 5854.26\lambda^6 + 10391.77\lambda^7\end{array}}{1 + 12.27\lambda + 84.28\lambda^2 + 301.87\lambda^3 + 948.69\lambda^4 + 1366.66\lambda^5 + 2425.928\lambda^6}$$

Then using (9.35) $Z_{in}(\lambda)$ is put in the form given by (9.36)

$$Z_{in}(\gamma,\mu) = \frac{\begin{array}{c}15074.13\gamma^7 + 7629.00\gamma^8\mu + 5924.49\gamma^5 + 1889.54\gamma^4 \\ \mu + 622.8\gamma^3 + 115.80\gamma^2\mu + 16.54\gamma + \mu\end{array}}{3549.9\gamma^6 + 1680.8\gamma^5\mu + 1120.25\gamma^4 + 326.41\gamma^3\mu + 87.27\gamma^2 + 12.17\gamma\mu + 1.0}$$

Performing the continued fraction expansion we obtain for the element values of Figure 9.6

$$L_1 = 4.53, \quad C_2 = 4.12, \quad L_3 = 5.18, \quad C_4 = 4.4$$

$$L_5 = 4.77, \quad C_6 = 3.74, \quad L_7 = 2.05, \quad R_L = 1$$

The actual capacitor ratios are then obtained from the above values and (9.14) to (9.16).
   The resulting network has a dc gain of 0.25 ($-6$ dB). To adjust the gain value to 1 (0 dB) the network is impedance-scaled by 0.5.

### 9.3.3   Strays-insensitive LDI Ladders

The real merit of these switched capacitor ladder filters is two-fold. First, the performance of the filter is determined by capacitance ratios and not by absolute element values. These ratios can be very accurately and conveniently realized using present-day silicon integrated circuit technologies; the absolute capacitor values can be reduced to increase the density of elements on a single chip. The other advantage is that the leap-frog ladder filter is modelled on a passive (distributed) prototype, and retains the good sensitivity property of passive filters.
   When designing switched-capacitor filters using silicon integrated circuits as discussed in Chapters 2–6 , the effect of stray parasitic capacitances must be taken into account. Attention has, therefore, been given to the use of building blocks which perform the same function as those of Figure 9.3 but are *insensitive* to parasitic capacitances. This objective can be easily achieved by modifying the building blocks. Before showing this modification, let us examine the problem of parasitics in some detail. Consider Figure 9.8 which shows a typical circuit which is sensitive to these parasitic capacitances. The capacitance $C_1$ represents all the parasitics associated with node 1. This includes the source-to-drain diffusion capacitances of the

transistors making up the switches. It also includes the capacitances of the leads connecting the top plate of $C_A$ to the two transistors. The total stray capacitance to ground could be of the order of 0.5 pF and is uncontrollable. Since $C_A$ may be of the order of 1pF, the error in this value could be as large as 50%. The rather obvious solution to take $C_A$ very large, e.g. as 50pF for an error of 1%, is impractical since this will result in inordinately large areas on the chip. Next consider the circuit of Figure 9.8(b). The parasitic capcitance $C_1$ is charged from $V_{in}$ then discharged to ground. Each of the parasitic capacitances $C_2$ and $C_3$ is grounded at both ends. Also $C_4$ is driven by the low impedance Op Amp output. It follows that none of these parasitic capacitances affects the performance of the circuit of Figure 9.8(b). However, the circuit is sensitive to the parasitic capacitance between the clock signals to node 4. Minimization of this effect will be discussed in a later chapter, and it will be shown that this can be achieved by a modification of the clocking scheme.

Now, using building blocks that are strays-insensitive in the manner discussed above, we can obtain parasitic-insensitive leap frog ladders according to the following procedure.

(a) Modify the state-variable leap-frog block diagram of Figure 9.2 as shown in Figure 9.9. Clearly, this new structure still implements the same state equations (9.3). Each building block together with the summer at its input implements a

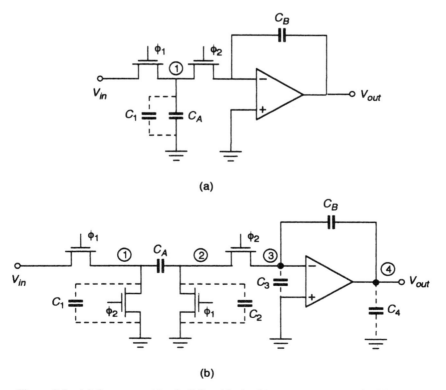

(a)

(b)

**Figure 9.8** (a) Strays-sensitive building block, (b) strays-insensitive building block.

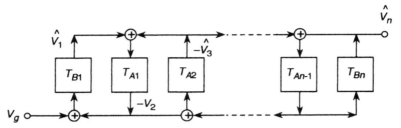

**Figure 9.9** Alternative active simulation of the ladder in Figure 9.1 using single-input circuits.

typical equation in the set of state equations (9.3) which simulate the same ladder in Figure 9.1.

(b) Instead of the building blocks of Figure 9.3, use the modified ones of Figures 9.10 and 9.11. These are as follows:

*Modified Type A*   These are the circuits of Figure 9.10, with transfer functions

$$T_A \frac{z^{-1}}{\dfrac{C_b}{C_a}(1 - z^{-1})} = \frac{z^{-1/2}}{2\dfrac{C_b}{C_a}\gamma} \tag{9.37}$$

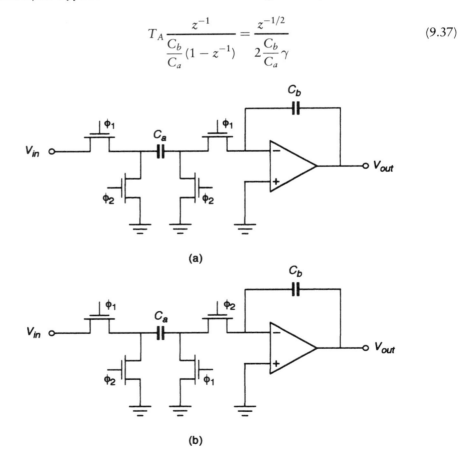

(a)

(b)

**Figure 9.10**   Strays-insensitive building blocks of type A: (a) positive; (b) negative.

for the circuit of Figure 9.10(a) and

$$\tilde{T}_A = \frac{-1}{\dfrac{C_b}{C_a}(1 - z^{-1})} = \frac{-z^{1/2}}{2\dfrac{C_b}{C_a}\gamma}$$                                      (9.38)

for the network of Figure 9.10(b).

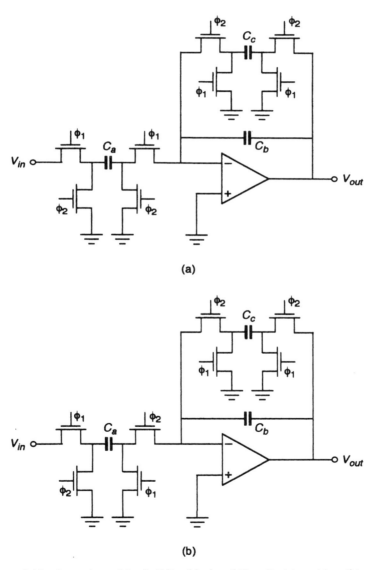

(a)

(b)

**Figure 9.11**  Strays-insensitive building blocks of Type B: (a) positive; (b) negative.

*Modified Type B*     These are the circuits of Figure 9.11 with transfer functions

$$T_B = z^{-1} \bigg/ \left( \frac{C_b}{C_a} \left\{ 1 - z^{-1} + \frac{C_c}{C_b} \right\} \right)$$

$$= z^{-1/2} \bigg/ \left[ \left( \frac{2C_b + C_c}{C_a} \right) \gamma + \frac{C_c}{C_a} \mu \right]$$

(9.39)

for the circuit of Figure 9.11(a); and

$$\tilde{T}_b = -1 \bigg/ \left( \frac{C_b}{C_a} \left\{ 1 - z^{-1} + \frac{C_c}{C_b} \right\} \right)$$

$$= z^{-1/2} \bigg/ \left[ \left( \frac{2C_b + C_c}{C_a} \right) \gamma + \frac{C_c}{C_a} \mu \right]$$

(9.40)

for the circuit of Figure 9.11(b).

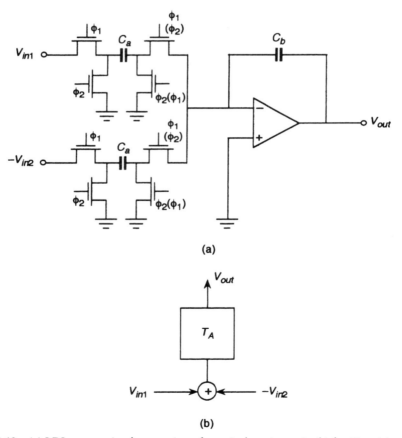

**(a)**

**(b)**

**Figure 9.12**     (a) LDI-summer implementation of a typical section as in (b) for $T$ positive. The clock phases for $T$ negative are between brackets. The same input arrangement is used with Type B circuits.

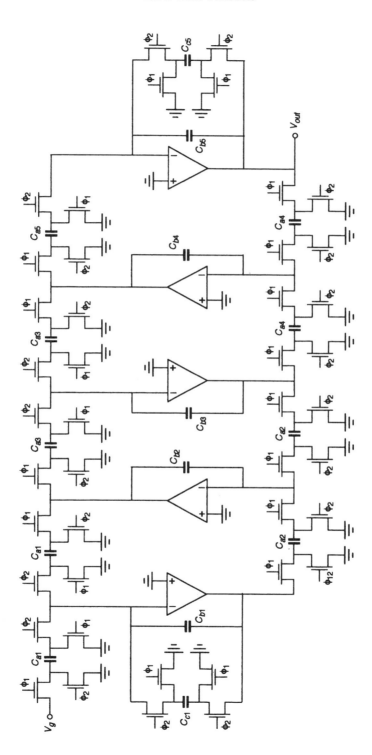

**Figure 9.13** The circuit of fifth order LDI low-pass filter with strays-insensitive building blocks.

(c) In the realization of Figure 9.9, using the modified building blocks, two points must be observed. First, positive and negative building blocks alternate. Second, the *first* and *last* building blocks are of *Type B* while the rest are of *Type A*. Clearly the summing operation is easily realized by adding another capacitor of the same value $C_A$ to every building block, i.e. between the voltage to be summed and the node where the original $C_a$ is connected as shown in Figure 9.12 for Type-A blocks. The clock phases for negative Type-A are shown between brackets. The same input arrangement is used with a Type-B building block together with the summer. A fifth order example is shown in Figure 9.13.

Having modified the procedure as outlined above, it is a simple matter to show that the resulting filter has the equivalent network shown in Figure 9.14, which is the same network as Figure 9.6 except that all impedances are scaled by $z^{1/2}$. This multiplies the transfer function in (9.23) by a factor $z^{-1/2}$, which of course has no effect on the amplitude response of the filter. The resulting structure, however, has the advantage of being completely insensitive to stray parasitic capacitances inherent in the IC fabrication process.

*Henceforth we shall use inductor and capacitor symbols to denote elements of impedance and admittance frequency dependence kx, respectively whether $x = \gamma$ or s or $\lambda$. We shall therefore speak of a $\gamma$-domain inductor and capacitor, etc. This should not result in any confusion since the context is very clear and explicit.*

*Example 9.2*   A strays-insensitive low-pass LDI ladder filter of the Chebyshev type is required to have a passband ripple of 0.1 dB, a ratio of passband to clock frequency of 0.12, a minimum stopband attenuation of 25 dB and a ratio of passband edge to clock frequency of 0.18. The dc gain is required to be 0 dB.

**Program CHBYE** in Appendix C of ISICAP is used to perform the design which requires a fifth-order filter. The circuit is shown in Figure 9.12 and the response is shown on Figure 9.15. The program prompts the user to specify the source resistor; this is the impedance level factor determining the dc gain. For a dc gain of 1 (0 dB) we specify the source resistor to be 0.5. The program gives the capacitor ratios as:

$$[C_c/C_a]_1 = 0.5$$
$$[C_b/C_a]_1 = 0.16\,037, \quad [C_b/C_a]_2 = 0.288\,72$$
$$[C_b/C_a]_3 = 0.198\,23, \quad [C_b/C_a]_4 = 0.314\,79$$
$$[C_b/C_a]_5 = 0.914\,37$$
$$[C_c/C_a]_5 = 0.5$$

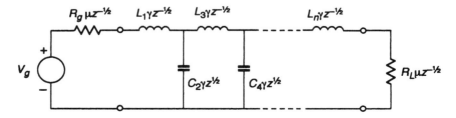

**Figure 9.14**   Equivalent network of strays-insensitive LDI ladder of Figure 9.13.

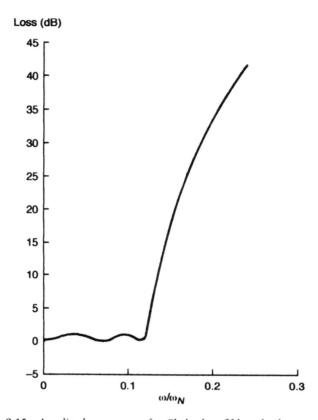

**Figure 9.15** Amplitude resopnse of a Chebyshev fifth-order low-pass LDF filter.

We note that with the choice of source resistor value as 0.5, the program automatically incorporates this value in the design as a scaling factor, thus producing the required dc gain value of 0 dB.

### 9.3.4 An Approximate Design Technique

The earliest design techniques of switched-capacitor filters relied on the assumption of a very high sampling rate as compared with the bandwidth of the signal. This leads to a design technique which gives a filter directly related to a lumped prototype. To see this, let the sampling frequency be such that

$$\omega_N \gg \omega \qquad (9.41)$$

for all $\omega$ in the frequency band of interest. Then on the $j\omega$-axis

$$\sinh(j\pi\omega/\omega_N) \approx j\pi\omega/\omega_N \qquad (9.42)$$

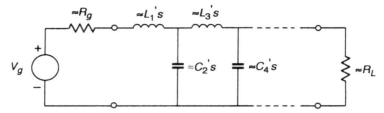

**Figure 9.16** Approximate equivalent circuit of LDI filter under the assumption of very high clock frequency.

and

$$\cosh(j\pi\omega/\omega_N) \approx 1 \tag{9.43}$$

and we arrive at an approximate equivalent circuit of the filter of the form given in Figure 9.16 which is a true resistively-terminated LC ladder. The element values can be obtained from those of the lumped designs of Chapter 7. If the latter are given by $g_r = L_r$ or $C_r$, then from the approximation in (9.42) and (9.43) we have the element values of the switched-capacitor equivalent network in Figure 9.16 as

$$g'_r = g_r\pi/\omega_N = g_r T/2 \tag{9.44}$$

However, this method is inexact with unacceptable disadvantages. First, it forces the use of an excessively high sampling rate, so that the useful bandwidth of applications of switched-capacitor filters becomes severely limited. Second, the element values depend on the sampling frequency and they lead to large capacitor ratios. The flexibility of programming the filter using a variable clock frequency is lost. For these reasons, this method has been very rarely used since the exact design techniques of this chapter were understood.

## 9.4  BAND-PASS FILTERS

The basic building blocks employed for the design of band-pass filters are of two types. The first includes precisely those used for the design of low-pass filters and are shown in Figure 9.10 and 9.11. These will be referred to collectively, as Type I building blocks. The second type, comprises the composite building blocks shown in Figure 9.17 and are referred to as Type II building blocks. The transfer functions of these are as follows

$$T_{IIA} = \frac{1}{[(2C_b/C_a)\gamma + (2C_d/C_a\gamma)^{-1}](\gamma + \mu)} \tag{9.45}$$

for the circuit of Figure 9.17(a) or

$$\tilde{T}_{IIA} = -(\gamma + \mu)^2 T_{IIA} \tag{9.46}$$

for the circuit of Figure 9.17(b)

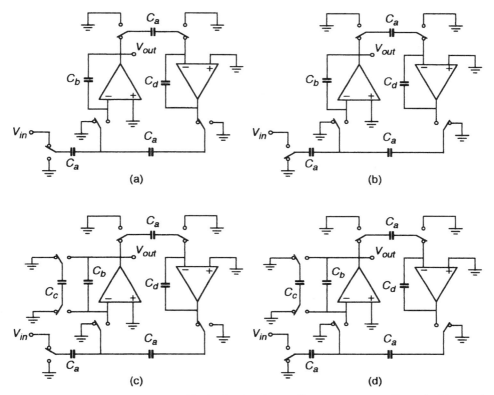

**Figure 9.17**   Second-order building blocks for band-pass filters: (a) positive Type-A; (b) negative Type-A; (c) positive Type-B; (d) negative Type-B.

The network of Figure 9.17(c) has the transfer function

$$T_{IIB} = \frac{1}{[(2C_b/C_a)\gamma + (2C_d/C_a)^{-1} + (C_c/C_b)\mu](\gamma + \mu)} \quad (9.47)$$

while that of Figure 9.17(d) has

$$\tilde{T}_{IIB} = -(\gamma + \mu)^2 T_{IIB} \quad (9.47)$$

It is now shown that both Type I and Type II building blocks can be employed in the same state variable ladder structure shown in Figure 9.9 to obtain bandpass filters. This allows greater flexibility in shaping the response in the three bands. First, direct analysis is used to obtain the general form of the transfer function resulting from the use of the building blocks in Figures 9.10, 9.11 and 9.17. Then, the solution to the approximation problem is given using the theory of generalized Chebyshev functions. The maximally flat response case is also given. Next, an exact synthesis algorithm is developed which allows the capacitor ratios in the filter to be determined in a straightforward manner. Finally, the synthesis technique is illustrated by an example.

## 9.4.1   General Form of Band-pass Transfer Functions

Consider the use of the building blocks of Figures 9.10, 9.11 and 9.17 to construct the leapfrog ladder structure shown in Figure 9.9. As in the low-pass case, the summers are simply realized by the addition of a switched capacitor connected between the input of one building block and the output of another. In the block diagram of Figure 9.9, the circuits of either Type I or Type II may be employed quite freely provided that inverting and non-inverting building blocks are used alternately. Moreover, the first and last ones are of the type $T_{IB}$ or $T_{IIB}$, i.e., those employing feedback switched capacitors. For reasons to be explained later, we shall also assume that at least one Type II building block is employed.

Our present task consists in finding the general form of the transfer function of such band-pass structures. First, remember that the active leapfrog network of Figures 9.9 is a realization of the state equations describing a ladder filter of the general form shown in Figure 9.1, where the same relations hold true.

Let the number of Type I building blocks be $k$ and that of Type II be $m$, where $k$ must be even for a band-pass filter. Furthermore, we deal first with the case where $m$ is odd, and the opposite case is discussed later. For the purpose of determining the degree and general form (i.e., location of transmission zeros, etc.) of the transfer function, the order in which the two types are connected does not affect the conclusions. Thus for convenience, we let the first $k$ building blocks be of Type I and the next $m$ be of Type II.

With the above assumptions, the structures in Figures 9.10, 9.11 and 9.17 realize the equivalent ladder in Figure 9.18. Let its transfer function be

$$H_{21} = I_L/V_g \tag{9.49}$$

To simplify the analysis, we scale all impedances in the equivalent ladder of Figure 9.18 by the factor $(\mu + \gamma) = z^{1/2}$. This gives the network of Figure 9.19 whose transfer function $H_{21}$ is related to that of the original network by

$$H'_{21} = \frac{1}{(\mu + \gamma)} H_{21}$$

$$= z^{-1/2} H_{21} \tag{9.50}$$

$$= \left(\frac{1 - \lambda}{1 + \lambda}\right)^{1/2} H_{21}$$

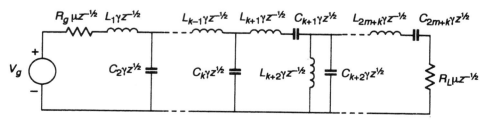

**Figure 9.18**   Equivalent network of LDI band-pass filter ($m$ odd).

It follows that, as in the low-pass case, this scaling is trivial since the factor $z^{1/2}$ has unity magnitude at all frequencies, and, therefore, does not affect the amplitude response of the filter. Thus we can work with the scaled network of Figure 9.19 to obtain the general form of $H'_{21}$ and the prime on $H'_{21}$ is henceforth dropped for convenience.

Now, the network of Figure 9.19 may be viewed as a two-port $N$ with elements of frequency dependence $L_i\gamma$, $1/C_i\gamma$ which is terminated at both ends in elements of frequency dependence $R_g\mu$ and $R_L\mu$. The two-port $N$ is a cascade of two subnetworks $N_1$ and $N_2$ as shown in Figure 9.17. Thus the transmission matrix of $N$ is given by

$$[T] = [T'] \times [T''] \qquad (9.51)$$

where $[T']$ and $[T'']$ are the transmission matrices of the networks $N_1$ and $N_2$, respectively. Direct analysis of these sub-networks reveals the degrees of the entries of their transmission matrices. Hence

$$[T'] = \begin{bmatrix} a'_k(\gamma) & b'_{k-1}(\gamma) \\ c'_{k-1}(\gamma) & d'_{k-2}(\gamma) \end{bmatrix} \qquad (9.52)$$

where $a'_k(\gamma)$ and $d'_{k-2}(\gamma)$ are even, whereas $b'_{k-1}(\gamma)$ and $c'_{k-1}(\gamma)$ are odd real polynomials. Similarly, for $N_2$,

$$[T''] = \frac{1}{\gamma^m} \begin{bmatrix} a''_{2m-1}(\gamma) & b''_{2m}(\gamma) \\ c''_{m-2}(\gamma) & d''_{2m-1}(\gamma) \end{bmatrix} \qquad (9.53)$$

where $a''_{2m-1}(\gamma)$ and $d''_{2m=1}(\gamma)$ are odd, while $b''_{2m}(\gamma)$ and $c''_{2m-2}(\gamma)$ are even real polynomials. Using (9.52) and (9.53) in (9.51), the transmission matrix of the overall network $N$ is obtained in the form

$$[T] = \frac{1}{\gamma^m} \begin{bmatrix} a_{2m+k-1}(\gamma) & b_{2m+k}(\gamma) \\ c_{2m+k-2}(\gamma) & d_{2m+k-1}(\gamma) \end{bmatrix} \qquad (9.54)$$

where $a(\gamma)$ and $d(\gamma)$ are odd while $b(\gamma)$ and $c(\gamma)$ are even real polynomials.

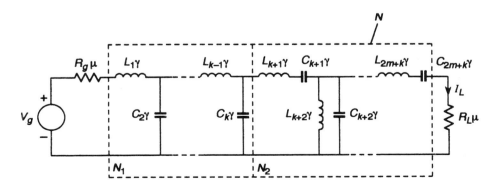

**Figure 9.19** The network of Figure 9.18 after impedance scaling by $z^{\frac{1}{2}}$.

Now, the general properties of $[T]$ in (9.54) may be used to obtain the form of the transfer function of the filter as

$$H_{21}(\gamma) = \frac{I_L}{V_g}$$

$$= \frac{\gamma^m}{R_L \mu a(\gamma) + b(\gamma) + R_g \mu d(\gamma) + R_L R_g \mu^2 c(\gamma)}$$

(9.55)

Using (9.20) and (9.21), we finally obtain

$$H_{21}(\lambda) = \frac{K\lambda^m (1 - \lambda^2)^{\frac{m+k}{2}}}{P_{2m+k}(\lambda)}$$

(9.56)

where $K$ is constant.

Due to the mapping between the $\lambda$ plane and the usual complex frequency $s$ plane, the general properties of the transfer function of the switched-capacitor filter are best studied in the $\lambda$ domain; hence, the importance of expressing $H_{21}$ as a function of $\lambda$ as in (9.56). From this expression of $H_{21}(\lambda)$, the following conclusions may be reached:

a)   Every Type I building block contributes a factor $\sqrt{1 - \lambda^2}$ in the numerator of the transfer function;

b)   Every Type II building block contributes a transmission zero at $\lambda = 0$ (corresponding to $s = 0$) as well as a factor $\sqrt{1 - \lambda^2}$ in the numerator of the transfer function;

c)   The necessary and sufficient condition for the stability of the bandpass filter is that $P_{2m+k}(\lambda)$ in (9.56) be a strictly Hurwitz polynomial in the variable $\lambda = \tanh(T/2)s$.

d)   At real frequencies $s = j\omega$, $\lambda = j\Omega$ where $\Omega = \tan(\omega T/2)$ and (9.56) gives

$$|H_{21}(j\Omega)|^2 = \frac{K^2 \Omega^{2m}(1 + \Omega^2)^{m+k}}{D_{2m+k}(\Omega^2)}$$

(9.57)

So far, we have only considered the case where $m$ is odd. The equivalent network for $m$ even is shown in Figure 9.20 which is somewhat different from that for $m$ odd in Figure 9.19. The basic difference is that the load termination becomes an element of impedance $Z_L = 1/G_L\mu$. Following very closely the steps which have been used to obtain (9.56), it may be readily shown that this case has a transfer function

$$H_{21} = \frac{V_{out}}{V_g} = \frac{K\lambda^m (1 - \lambda^2)^{\frac{m+k}{2}}}{P_{2m+k}(\lambda)}$$

(9.58)

which is identical to that in (9.56) for $m$ odd.

Thus the conclusions regarding the location of the transmission zeros, stability, etc.,

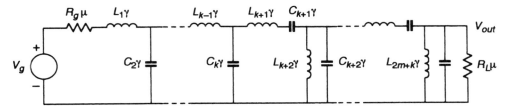

Figure 9.20   Equivalent network of the LDI band-pass filter ($m$ even).

are equally valid in this case.

Having established the general properties of the filter transfer function, the next step is to solve the corresponding approximation problem.

### 9.4.2   The Approximation Problem

From (9.57), the magnitude-squared function of the bandpass filter may be written as

$$|H_{21}|^2 = K^2 \Big/ \left(1 + \frac{h_{2m+k}(\Omega)^2}{\Omega^{2m}(1+\Omega^2)^{m+k}}\right) \qquad (9.59)$$

Let

$$x = \sin \omega \frac{T}{2}$$
$$= \sin \pi \frac{\omega}{\omega_N} \qquad (9.60)$$

where $\omega_N$ is the radian sampling (Nyquist) frequency. Thus from (9.20)

$$\Omega^2 = \frac{x^2}{1 - x^2} \qquad (9.61)$$

Substituting for $\Omega^2$ from (9.61) into (9.57), the expression for $|H_{21}|^2$ can be put in the form

$$|H_{21}|^2 = \frac{K^2}{1 + \epsilon^2 F_{2m+k}^2(x)} \qquad (9.62)$$

where

$$F_{2m+k} = \frac{Q_{m+k/2}(x^2)}{x^m} \qquad (9.63)$$

with $Q_{m+k/2}(x^2)$ being an even polynomial in $x$. Denoting by $2n$, the degree of the transfer function in (9.56) we have

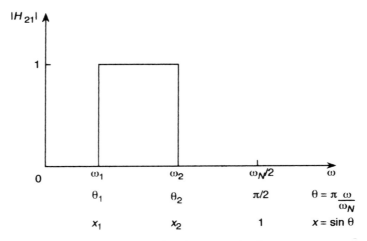

**Figure 9.21** Ideal band-pass amplitude response.

$$F_{2n}(x) = \frac{Q_n(x^2)}{x^m} \qquad (9.64)$$

where

$$2n = 2m + k \qquad (9.65)$$

Our present task is to find suitable functions $F_{2n}(x)$ which approximate the ideal bandpass characteristics shown in Figure 9.21 in which

$$x_1 = \sin \pi \, \frac{\omega_1}{\omega_N}$$

$$x_2 = \sin \pi \, \frac{\omega_2}{\omega_N} \qquad (9.66)$$

where $\omega_1$ and $\omega_2$ are the passband edge frequencies.

Two types of approximating functions are now derived. The first is optimum equiripple in passband and monotonic in both stopbands. The second type is maximally flat about passband-center while maintaining the monotonic response in the stopbands.

### Transfer functions with optimum equiripple passband

Examination of (9.64) shows that the function $F_{2n}(x)$ has 'prescribed' poles. Clearly, this is a result of the specification of the types of building blocks employed in the realization. Thus the derivation of $F_{2n}(x)$ reduces to the determination of $Q_n(x^2)$ in (9.64) such that $|H_{21}|^2$ in (9.62) is optimum equiripple in the passband. This can be achieved using a generalized Chebyshev function as follows. Define the variable

$$y = \left( \frac{x_2^2 - x^2}{x_1^2 - x^2} \right)^{1/2} \tag{9.67}$$

and re-express $F_{2n}(x)$ as

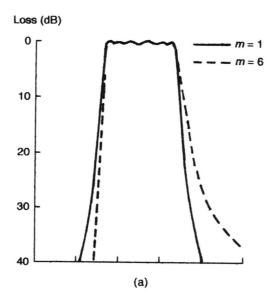

(a)

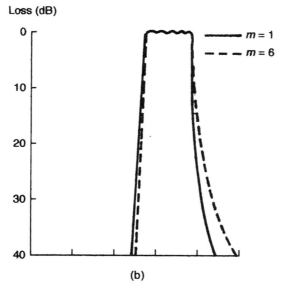

(b)

**Figure 9.22**  Amplitude response for $2n = 12$, $\epsilon^2 = 0.1$ and different values of m: (a) $\theta_1 = 30°$; $\theta_2 = 60°$; and (b) $\theta_1 = 50°$; $\theta_2 = 70°$.

$$F(y) = \frac{f(y)}{g(y)} \tag{9.68}$$

From (9.63) and (9.67), $g(y)$ can be obtained as

$$g(y) = \left[ y^2 - \left( \frac{x_2}{x_1} \right)^2 \right]^{m/2} (y^2 - 1)^n \tag{9.69}$$

The theory of approximation by generalized Chebyshev functions allows us to express $f(y)$ as

$$f(y)) = Ev \left\{ \left( y + \frac{x_2}{x_1} \right)^m (y+1)^{2n-m} \right\} \tag{9.70}$$

We also have

$$f(x) \overset{\Delta}{=} f(y) \cdot (x_1^2 - x^2)^n$$
$$\tag{9.71}$$
$$g(x) \overset{\Delta}{=} g(y) \cdot (x_1^2 - x^2)^n$$

Thus transforming back to the variable $x = \sin \pi(\omega/\omega_N)$, we finally obtain

$$F_{2n}(x) = \frac{(x_1/x)^m}{(x_2^2 - x_1^2)^n} \sum_{\ell=0}^{n} \left\{ \sum_{i=0}^{2\ell} \binom{m}{i} \binom{2n-m}{2\ell-1} \left( \frac{x_2}{x_1} \right)^{m-1} \right\} \cdot (x_2^2 - x^2)^\ell (x_1^2 - x^2)^{n-\ell} \tag{9.72}$$

It is to be noted that, regarding the realization of the above function, the required number of Type I building blocks is $2(n - m)$ and that of Type II is $m$. Of course, the degree of the transfer function is $2n$.

Since $m$ is the number of transmission zeros at $\lambda = 0$, then increasing $m$ will increase the achieved attenuation and improve the selectivity in the lower stopband. On the other hand, the upper stopband performance can be improved by increasing $n$. The passband ripple is determined by suitable choice of $\epsilon$ in (9.62). Thus a combination of $m$, $n$, and $\epsilon$ can be found to meet the filter specifications. Figure 9.22 shows typical response curves.

### Transfer functions with maximally flat response in the passband

In this case $|H_{21}|^2$ in (9.62) is required to have the maximum possible number of zero derivatives at a point $x_0$ lying between $x_1$ and $x_2$. For $\epsilon = 1$ in (9.62), $x_1$ and $x_2$ define the 3 dB points. The required function $F^2(x)$ can be easily derived by noting that for a maximally flat response around $x = x_0$ and $m$ transmission zeros at $x = 0$, $F_{2n}(x)$

must be of the form

$$F_{2n}(x) = \frac{h(x_1, x_2)(x^2 - x_0^2)^n}{x^m}$$ (9.73)

where $h$ is a constant depending on the specified passband edges $x_1$ and $x_2$. For such specifications, it is now required to find $x_0$ and $h$. From (9.73), we require

$$\left[\frac{h^2(x^2 - x_0^2)^{2n}}{x^{2m}}\right]_{\substack{x=x_1 \\ x=x_2}} = 1$$ (9.74)

From which

$$x_0^2 = \frac{x_2^2 + \left(\dfrac{x_2}{x_1}\right)^{m/n} x_1^2}{1 + \left(\dfrac{x_2}{x_1}\right)^{m/n}}$$ (9.75)

and

$$h = \frac{x_1^m \left[1 + \left(\dfrac{x_2}{x_1}\right)^{m/n}\right]^n}{(x_2^2 - x_1^2)^n}$$ (9.76)

Substituting from (9.75) and (9.76) into (9.73), we finally obtain

$$F_{2n}(x) = \frac{(x_1/x)^m}{(x_2^2 - x_1^2)^n} \left[(x_2^2 - x^2) + \left(\frac{x_2}{x_1}\right)^{m/n} (x_1^2 - x^2)\right]^n$$ (9.77)

Again, the required performance of the filter in the lower stopband can be achieved by selecting a suitable value of $m$. On the other hand, increasing $n$ results in increased passband flatness as well as greater attenuation in the upper stopband.

*A Special Case* In certain situations, it may be either necessary or convenient to use Type II building blocks alone for the design of the filter. Such a choice may be due to the specifications on the relative widths and the attenuation levels in the lower and upper stopbands. These may require the use of as many transmission zeros at $x = 0$ as possible, leading in some cases to the choice of $m = n$ in (9.63), i.e., $k = 0$ in (9.68). Therefore, only Type II building blocks are needed. In these cases, expressions (9.72) and (9.77) reduce to the following simpler functions:

$$F_{2n}(x) = T_n \left[\frac{x^2 - x_1 x_2}{x(x_2 - x_1)}\right]$$ (9.78)

for the equiripple passband response, and

$$F_{2n}(x) = \left[\frac{x^2 - x_1 x_2}{x(x_2 - x_1)}\right]^n \tag{9.79}$$

for the maximally flat passband response. In (9.78) $T_n$ is the Chebyshev polynomial.

### 9.4.3    The Synthesis Technique

In order to determine the element values in the equivalent network of Figures 9.19 and 9.20, and subsequently the capacitor ratios in the switched-capacitor filter, an exact synthesis algorithm is now derived. This applies to networks which realize the transfer functions of the types derived in the previous section. The technique is a generalization of that discussed earlier for low-pass filters.

Consider the process of impedance-scaling the equivalent network of Figures 9.19 by the factor $1/\mu$. This results in the network of Figure 9.23 which, as in the low-pass case, we shall call the *auxiliary network*. This can be viewed abstractedly as a lossless two-port terminated in resistors at both ports. The auxiliary network has a transfer function $\tilde{H}_{21}$ related to that of the original network by

$$\tilde{H}_{21} = \mu H_{21}. \tag{9.80}$$

Moreover, the network $\tilde{N}$ of Fig. 9.23 is describable by a scattering parameter $S_{21}$ given by

$$\begin{aligned}
S_{21} &= 2\sqrt{R_g R_L} \hat{H}_{21} \\
&= 2\sqrt{R_g R_L} \mu H_{21}
\end{aligned} \tag{9.81}$$

Substituting from (9.56) into (9.81), we obtain the general form of $S_{21}$ as

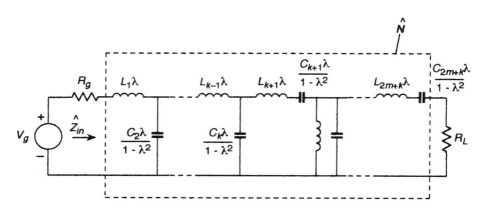

**Figure 9.23**   Auxiliary network obtained by impedance scaling.

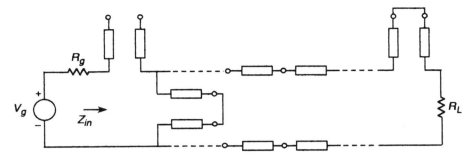

**Figure 9.24** Alternative form of auxiliary network containing unit elements and stubs.

$$S_{21}(\lambda) = 2K\sqrt{R_g R_L} \, \frac{\lambda^m (1 - \lambda^2)^{(2n-m-1)/2}}{P_{2n}(\lambda)} \qquad (9.82)$$

Thus the auxiliary two-port $\tilde{N}$ is formally equivalent to $(2n - m - 1)$ unit elements, $m$ high-pass stubs and a single low-pass stub, all of the same one-way propagation delay $T/2$. This structure is depicted in Figure 9.24.

The reason for forming the auxiliary network is the same as for the case of low-pass filters studied earlier, namely : that it is a resistively terminated lossless two-port to which the standard Darlington synthesis technique may be directly applied. It is also to be noted that if the input impedance $\tilde{Z}_{in}$ of the resistively terminated auxiliary two-port is obtained, the input impedance $Z_{in}$ of the original (realized) equivalent network can be obtained from

$$Z_{in} = \mu \tilde{Z}_{in} \qquad (9.83)$$

We now proceed with the derivation of the synthesis algorithm. Starting from either expression (9.72) or (9.77), we substitute in (9.62) to obtain the required magnitude-squared function. Next, we obtain the corresponding transducer power gain of the auxiliary network from (9.81) as

$$|S_{21}|^2 = 4R_g R_L \cos^2 \theta |H_{21}|^2 \qquad (9.84)$$

Choosing $R_g = R_L = 1$ and selecting $K$ in (9.56) accordingly, we obtain for the auxiliary network

$$|S_{11}|^2 = 1 - |S_{21}|^2 \qquad (9.85)$$

using

$$x^2 = \sin^2 \theta = \frac{\Omega^2}{1 + \Omega^2} \qquad (9.86)$$

and letting $\Omega^2 \to -\lambda^2$ in (9.85), we obtain $S_{11}(\lambda)S_{11}(-\lambda)$. This is factorized to construct a bounded real $S_{11}(\lambda)$ with a strictly Hurwitz denominator. Next, the input impedance of the (resistor-terminated) auxiliary network is calculated from

$$\tilde{Z}_{in}(\lambda) = \frac{1 + S_{11}(\lambda)}{1 - S_{11}(\lambda)} \tag{9.87}$$

Finally, the input impedance of the original equivalent network of Figure 9.18 is obtained from (9.83) as

$$Z_{in} = \mu \frac{1 + S_{11}(\lambda)}{1 - S_{11}(\lambda)} \tag{9.88}$$

It is now required to find a method for determining the element values in the equivalent network. Starting from (9.88), we separate the numerator and denominator polynomials into their even and odd parts. Then, relations (9.20) and (9.21) are used to express $Z_{in}$ in the form

$$Z_{in}(\gamma, \mu) = \frac{\mu a_{2m+k-1}(\gamma) + b_{2m+k}(\gamma)}{\mu c_{2m+k-2}(\gamma) + d_{2m+k-1}(\gamma)} \tag{9.89}$$

where $a$, $b$, $c$, and $d$ can be identified as the entries of the polynomial transmission matrix in (9.54). Then, the synthesis algorithm suggests itself as the successive extraction of poles at $\gamma = \infty$ and $\gamma = 0$ from $Z_{in}(\gamma, \mu)$ or $Y_{in}(\gamma, \mu)$ until the termination of impedance $R_L \mu$ is reached. This gives the element values $L_1, C_1, L_2, C_2, \ldots L_n, C_n$ in Figure 9.18 from which the exact capacitor ratios can be calculated using (9.45) to (9.48).

The case where $m$ is even can be treated in a similar manner. Starting from the equivalent network of Figure 9.20, we construct an auxiliary network by impedance scaling by the factor $1/\mu$. This will have a transmission coefficient of the same general form given by (9.82). The same steps leading to equation (9.88) are then followed. The input impedance of the equivalent network of Figure 9.20 is then put in the form

$$Z_{in}(\gamma, \mu) = \frac{\dfrac{1}{\mu} a + b}{\dfrac{1}{\mu} c + d} \tag{9.90}$$

The synthesis algorithm is the same as before, except that the load terminator will have the frequency dependence $Z_L = 1/G_L \mu$.

*Example 9.3*    Consider the calculation of the element values for an eighth-order filter of the maximally flat type of expression (33) with $m = 2$ and $n = 4$. The 3 dB points are at $f_1 = f_N/18$ and $f_2 = f_N/6$, i.e., $\theta_1 = 10°$ and $\theta_2 = 30°$. From (9.74), we have for the given values

$$F_{2n} = \frac{12.897}{x^2}(0.301 - 2.697x^2)^4$$

Therefore, from (9.62), (9.84) and (9.85)

$$|H_{21}|^2 = \frac{x^4}{x^4 + 166.332(0.301 - 2.697x^2)^8}$$

$$|S_{21}|^2 = \frac{(1 - x^2)x^4}{x^4 + 166.332(0.301 - 2.697x^2)^8}$$

$$|S_{11}|^2 = 1 - |S_{21}|^2 = \frac{x^6 + 166.332(0.301 - 2.697x^2)^8}{x^4 + 166.332(0.301 - 2.697x^2)^8}$$

The above expression is factored, poles in the left-half $\lambda$ plane are chosen and from (9.87), we calculate the input impedance of the auxiliary network. This gives

$$\tilde{Z}_{in}(\lambda) = \frac{\begin{array}{c} 2\lambda^8 + 1.858\lambda^7 + 1.885\lambda^6 + 0.960\lambda^5 + 0.458\lambda^4 + \\ 0.127\lambda^3 + 0.0317\lambda^2 + 0.0039\lambda + 0.0005 \end{array}}{0.247\lambda^7 + 0.230\lambda^6 + 0.197\lambda^5 + 0.085\lambda^4 + 0.03\lambda^3 + 0.005\lambda^2 + 0.00066\lambda}$$

(9.91)

Then, using (9.89), we have for the input impedance of the original equivalent network

$$Z_{in}(\lambda) = -\frac{\begin{array}{c} [437.517\gamma^8 + 289.847\gamma^6 + 55.648\gamma + 3.368\gamma^2 + 0.0499]\dfrac{1}{\mu} \\ +295.024\gamma^7 + 122.667\gamma^5 + 13.896\gamma^3 + 0.3916\gamma \end{array}}{[47.253\gamma^7 + 25.935\gamma^6 + 3.212\gamma^3 + 0.666\gamma]\dfrac{1}{\mu} + 32.048\gamma^6 + 9.581\gamma^4 + 0.522\gamma^2}$$

(9.92)

It is to be noted that the above impedance may now be scaled by an arbitrary constant to adjust the maximum passband amplitude and improve the spread of element values. Let this constant be 0.5.

We may now extract the poles at $\gamma = \infty$ by performing the first five steps of the continued fraction expansion of $Z_{in}(\gamma)$. This gives (see Figure 9.25) for the element values

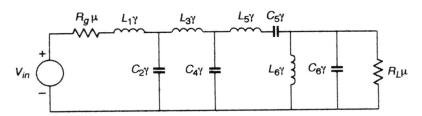

**Figure 9.25**  Equivalent network of the filter of Example 9.3.

$$L_1 = 4.602$$

$$C_2 = 1.860$$

$$L_3 = 15.219$$

$$C_4 = 0.530$$

$$L_5 = 39.415$$

The remainder impedance is then given by

$$Z'_{in} = \frac{(24.974 + 800.046\gamma^2)\dfrac{1}{\mu} + 195.80\gamma}{(6.882\gamma + 80.215\gamma^3)\dfrac{1}{\mu} + 53.798\gamma^2}$$

Extracting the pole at $\gamma = 0$ we have

$$C_5 = 0.275.$$

The remainder admittance is then given by

$$Y''_{in} = \frac{0.0135}{\gamma} + 1.487\gamma + \mu$$

from which

$$L_6 = 73.95$$

$$C_6 = 0.158$$

$$R_L = 0.105$$

Finally, the actual capacitor ratios can be calculated using the expressions for the transfer functions of the individual building blocks together with the above element values.

## 9.5   HIGH-PASS FILTERS

High-pass designs may be obtained using the same building blocks as those used for the band-pass case of the previous section. Thus, with the same notation, we use $k$ building blocks of Type I and $m$ of Type II. The structure has the same equivalent circuit as a corresponding band-pass design, except that the solution to the approximation problem under the constraints of a high-pass design is different. In this case we have

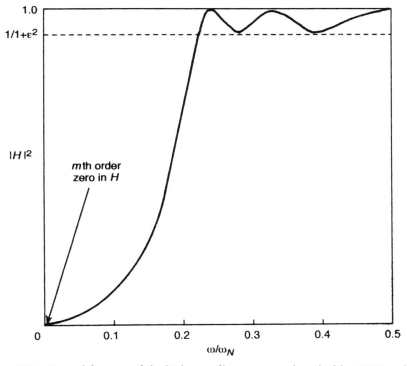

**Figure 9.26** General features of the high-pass filter response described by (9.93) and (9.94).

$$|H_{21}|^2 = \frac{1}{1 + F_n^2} \tag{9.93}$$

where, for an optimum equiripple response in the pass band and m zeros at $\omega = 0$ as depicted in Figure 9.26 we have

$$F_n = \epsilon \left[ (m + k) \cosh^{-1} \left( \frac{\cos \pi\omega/\omega_N}{\cos \pi\omega_o/\omega_N} \right) + m \cosh^{-1} \left( \frac{\cot \pi\omega/\omega_N}{\cot \pi\omega_o/\omega_N} \right) \right] \tag{9.94}$$

The synthesis of these functions is precisely the same as the band-pass ones.

## CONCLUSION

This chapter dealt with the amplitude-oriented design techniques of switched-capacitor filters employing the lossless discrete integrator. We have seen that these filters have no lumped-element counterparts; instead they correspond to classes of distributed filters. The use of the variable $\lambda$ was seen to establish this analogy, and the approximation and synthesis techniques were formulated accordingly. These are easily implemented as computer programs which form part of the package **ISCAP** supplied with this book. The classes of filter discussed in this chapter have achieved considerable popularity due to the simplicity of their structure, and low sensitivity properties with respect to element variations. In many ways, they epitomize all the fine attributes of switched-capacitor filters in general.

# PROBLEMS

9.1  Use Program **LDISCG** to realize the following transfer function:

$$H(\lambda) = \frac{(1 - \lambda^2)^{5/2}}{1 + 26.15\lambda + 294.10\lambda^2 + 2441.87\lambda^3 + 10018.2\lambda^4 + 44783.6\lambda^5}$$

9.2  Realize the following transfer function:

$$H(\lambda) = \frac{\lambda(1 - \lambda^2)^{3/2}}{4225.46 + 930.264\lambda + 186.943\lambda^2 + 19.739\lambda^3 + \lambda^4}$$

9.3  Design a maximally-flat LDI ladder filter with the following specifications:

Passband 0–1 kHz, with attenuation $\leqslant 0.1$ dB
Stopband edge at 3 kHz, with attenuation $\geqslant 30$ dB.

[Derive the transfer function in the $\lambda$ domain using Subroutine ROOT in Appendix K, then use **Program LDISCG** for the synthesis]

9.4  The following standard set of specifications are to be met by the receiver low-pass filter employed in a codec for PCM telephony:

Passband 0–3.4 kHz, with 0.25 dB ripple
Stopband edge at 4.6 kHz, with attenuation $\geqslant 32$ dB.

Design an LDI Chebyshev filter to meet the above set of specifications. [Use **Program CHBYE**]

9.5  Realize the following transfer function:

$$H(\lambda) = \frac{\lambda^2(1 - \lambda^2)^2}{0.1178 + 0.6173\lambda + 8.709\lambda^2 + 24.66\lambda^3 + 137.28\lambda^4 + 160.83\lambda^5 + 511.53\lambda^6}$$

9.6  Design a band-pass LDI filter with a monotonic passband having the following specifications:

Passband 10–16 KHz, with attenuation $\geqslant 3$ dB
Stopband edge frequencies at 5 kHz and 20 kHz with attenuation $\geqslant 20$ dB in both stopbands.

9.7  Design a band-pass LDI filter with optimum equiripple response in the passband to meet the following specifications:

Passband 10–16 KHz, with attenuation $\leqslant 3$ dB
Stopband edge frequencies at 5 kHz and 20 kHz with attenuation $\geqslant 20$ dB in both stopbands.

9.8  Design a maximally-flat high-pass filter, using LDI building blocks, with the following specifications:

   Passband edge at 2 kHz, with attenuation $\leqslant 3$ dB
   Stopband edge at 0.5 kHz, with attenuation $\geqslant 20$ dB.

9.9  Design a Chebyshev high-pass filter, using LDI building blocks, with the following specifications:

   Passband edge at 1 kHz, with 0.1 dB ripple
   Stopband edge at 0.25 kHz, with attenuation $\geqslant 20$ dB.

9.10 The following standard set of specifications are to be met by the transmit band-pass filter employed in a codec for PCM telephony:

   Passband: 300–3400 Hz, 0.25 dB ripple
   Lower stopband edge at 50 Hz, with 25 dB minimum attenuation
   Upper stopband edge at 4600 Hz, with 32 dB minimum attenuation.

   Design a filter which meets the above specifications using three different approaches:

   (a)  A band-pass LDI filter with optimum equiripple passband and monotonic stopbands, having two transmission zeros at the origin.

   (b)  The same type of filter as in (a) but with four zeros at the origin.

   (c)  The filter is to be designed as a cascade of two filters. The first is a low-pass LDI Chebyshev filter while the second is a high-pass maximally-flat LDI filter.

   Compare the three filters which meet the same specifications, on the basis of the degree. If two approaches yield the same degree, comparison should be made on the basis of the total capcitance required for the integrated circuit implementation.

# 10

# Amplitude-oriented Filters Derived from Passive Lumped Prototypes

## 10.1  INTRODUCTION

In this chapter, we present the techniques for designing switched-capacitor filters which are related to their continuous-time counterparts by the bi-linear transformation discussed in Chapter 8. There are two fundamental approaches to this design technique. The first is based on the exact simulation of the operation of the LC ladder prototypes discussed in Chapter 7. The advantage of this approach is that the resulting switched-capacitor filter retains the optimum sensitivity properties of the resistively-terminated passive ladder prototype. This is a highly desirable attribute from the practical view-point, and makes this approach , perhaps, the most desirable of all the approaches. The second approach consists in obtaining the required switched-capacitor filter *transfer function* from a continuous-time prototype transfer function by the bi-linear transformation given in Chapter 8. Next, the transfer function is factorized into second-order factors, and a possible first-order factor. Each factor is realized by a simple switched-capacitor circuit and the resulting networks are connected in cascade to form the entire filter. This is perhaps the simplest approach to the problem, and is particularly useful for relatively low-order filters, since higher order cases have much inferior sensitivity properties to the corresponding structures obtained by simulation of passive ladders. We shall consider both techniques and give illustrative examples. Both techniques discussed here allow the introduction of finite transmission zeros on the imaginary axis as required, for example, in the case of elliptic filters.

## 10.2  LOW-PASS LADDER FILTERS WITH FINITE TRANSMISSION ZEROS ON THE IMAGINARY AXIS

### 10.2.1  Derivation of the Switched-capacitor Circuit

In this approach, LC ladder prototypes are first designed completely in the s-domain, such that when the transformation in (8.141) is applied, the filter meets the required specifications. This technique was discussed fully in Section 8.7. Thus in the low-pass LC prototype, the following transformation is used

$$s \rightarrow \lambda/\Omega_0 \qquad (10.1)$$

this amounts to replacing every inductor in the prototype by an element with impedance $L/\Omega_0$ while every capacitor is replaced by an element with admittance $C/\Omega_0$. Here

$$\Omega_0 = \tan(\pi\omega_0/\omega_N) \qquad (10.2)$$

Thus, we obtain the $\lambda$-domain ladder shown in Figure 10.1. Its transfer function is that of the corresponding lumped prototype with the transformation (8.140) applied. The problem now is to find the switched-capacitor building blocks for the realization of the $\lambda$-plane ladder. Actually the required building blocks are basically the same ones used in the construction of the LDI ladder filters discussed in Chapter 9. There are some differences, however in the addition of some extra capacitors for the intro-duction of the finite zeros of transmission on the imaginary axis, and in the input section . Thus the circuits of Figures 9.10 and 9.11 are used , together with the more elaborate section [43] shown in Figure 10.2 in the input stage. This will now be explained in detail. First, we note that direct analysis of the circuit in Figure 10.2 shows that its transfer function is given by

$$\frac{V_{\text{out}}}{V_{\text{in}}} = \frac{-C_{a1}z^{1/2} + C_{a2}z^{-1/2} - 2C_{a3}\gamma - 2C_s\mu}{(2C_b + C_f)\gamma + C_f\mu} \qquad (10.3)$$

The additional Op Amp is required for the buffer of the sample-and-hold circuit.

Suppose that, based on the specifications of the filter, we already have an $n$th order elliptic lowpass $\lambda$-domain ladder as shown in Figure 10.1 in which the order of filter is assumed odd.

We begin, by some manipulations of the circuit, to convert it to a form suitable for simulation using the switched-capacitor building blocks. First, we extract from each of the mid-shunt capacitors, a negative capacitance of a value that resonates with the respective parallel inductor at $\lambda = \pm 1$. Now the parallel combination of $L_i\lambda$ and $-L_i/\lambda$ gives an impedance

$$\frac{L_i\lambda(-L_i/\lambda)}{L_i\lambda - (Li/\lambda)} = L_i\frac{\lambda}{1 - \lambda^2} = L_i\gamma\mu \qquad (10.4)$$

Then, the network of Figure 10.3 is obtained.

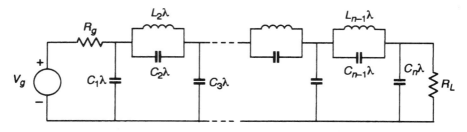

**Figure 10.1**   $\lambda$-domain ladder obtained from lumped passive prototype.

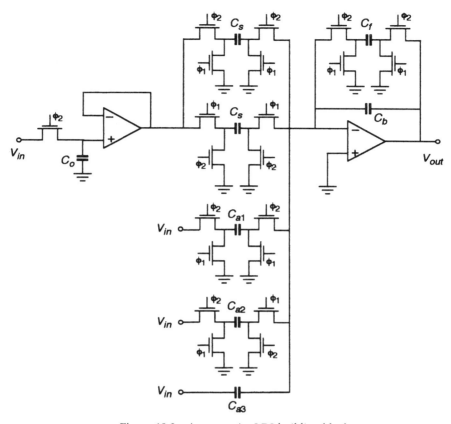

**Figure 10.2** A composite LDI building block.

Next we employ Norton's theorem to replace each of the bridging capacitors $C_2'$, $C_4'$, etc., by two voltage-controlled current sources (VCCS's). Finally, we divide all impedances by $\mu$ (including the transimpedances of the VCCS's). This scaling does not affect the voltage transfer function of the network which is, in this case, a voltage transfer ratio. The resulting network is shown in Figure 10.4.

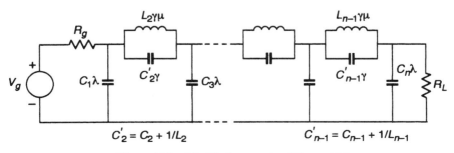

**Figure 10.3** Modified network of Figure 10.1.

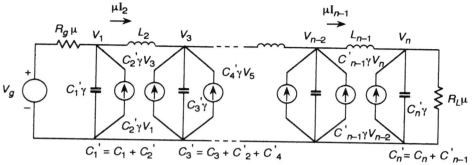

**Figure 10.4**   Network of Figure 10.3 after impedance-scaling by $1/\mu$ and modification using VCCS-equivalents.

Now the operation of the ladder network of Figure 10.4 can be described by the following state equations

$$V_j = \frac{(\mu I_{j-1}) - (\mu I_{j+1}) + \gamma C'_{j-1} V_{j-2} + \gamma C'_{j+1} V_{j+2}}{\gamma C'_j} \quad \text{for } j = 3, 5, 7, \ldots \quad (10.5)$$

$$\mu I_j = \frac{V_{j-1} - V_{j+1}}{\gamma L_j} \quad \text{for } j = 2, 4, 6, \ldots \quad (10.6)$$

$$V_1 = \frac{(\mu/R_g)V_g - (\mu I_2) + C'_2 \gamma V_3}{C'_1 \gamma + (\mu/R_g)} \quad (10.7)$$

$$V_n = \frac{(\mu I_{n-1}) + C'_{n-1} \gamma V_{n-2}}{C'_n \gamma + (\mu/R_L)} \quad (10.8)$$

Comparing the transfer functions of the the LDI building blocks in Figure 9.10, with equations (10.5) and (10.6), it is clear that the latter can be realized using these building blocks. On the other hand, equations (10.7) and (10.8) of the terminating sections (at the input and output) can be realized using the building block of Figure 10.2 with transfer function of the form given by (10.3). Let us examine the procedure in detail in relation to a fifth-order filter. In this case, the five equations are given by

$$V_1 = \frac{(\mu/R_g)V_g - (\mu I_2) + C'_2 \gamma V_3}{C'_1 \gamma + (\mu/R_L)}$$

$$(\mu I_2) = \frac{V_1 - V_3}{L_2 \gamma}$$

$$V_3 = \frac{(\mu I_2) - (\mu I_4) + C'_2 \gamma V_1 + C'_4 \gamma V_5}{C'_3 \gamma} \quad (10.9)$$

$$(\mu I_4) = \frac{V_3 - V_5}{L_4 \gamma}$$

$$V_5 = \frac{(\mu I_4) + C_4' \gamma V_3}{C_5' \gamma + (\mu/R_L)}$$

(10.9)

Figure 10.5 shows the switched-capacitor circuit of a fifth-order elliptic lowpass filter. To determine the capacitor ratios, the state equations of the SC circuit are written as follows:

$$\tilde{V}_1 = \frac{-2C_{s1}\mu V_g - C_{21}(z^{1/2}V_2) - 2C_{31}\gamma V_3}{(2C_{11} + C_s)\gamma + C_s\mu}$$

$$\tilde{V}_2 = \frac{C_{21}\tilde{V}_1 + C_{32}\tilde{V}_3}{2C_{22}}$$

$$\tilde{V}_3 = \frac{-2C_{13}\gamma\tilde{V}_1 - C_{23}(z^{1/2}\tilde{V}_4) - 2C_{53}\gamma\tilde{V}_5}{2C_{33}\gamma}$$

(10.10)

$$\tilde{V}_4 = \frac{C_{34}\tilde{V}_3 + C_{54}\tilde{V}_5}{2C_{44}\gamma}$$

$$\tilde{V}_5 = \frac{-2C_{35}\gamma\tilde{V}_3 - C_{45}(z^{1/2}\tilde{V}_4)}{(2C_{55} + C_\ell)\gamma + C_\ell\mu}$$

Next, all the variables appearing in (10.9) are simulated by voltages in expressions (10.10) and we establish the following correspondences

$$\tilde{V}_g \Leftrightarrow -V_g$$

$$\tilde{V}_1 \Leftrightarrow V_1$$

$$(z^{1/2}\tilde{V}_2) \Leftrightarrow (\mu I_2)$$

$$\tilde{V}_3 \Leftrightarrow -V_3$$

(10.11)

$$(z^{1/2}\tilde{V}_4) \Leftrightarrow -(\mu I_4)$$

$$\tilde{V}_5 \Leftrightarrow V_5$$

Then, the capacitor ratios are obtained by equating the coefficients of the transfer function of the building block to the coefficients in the state equations of the ladder

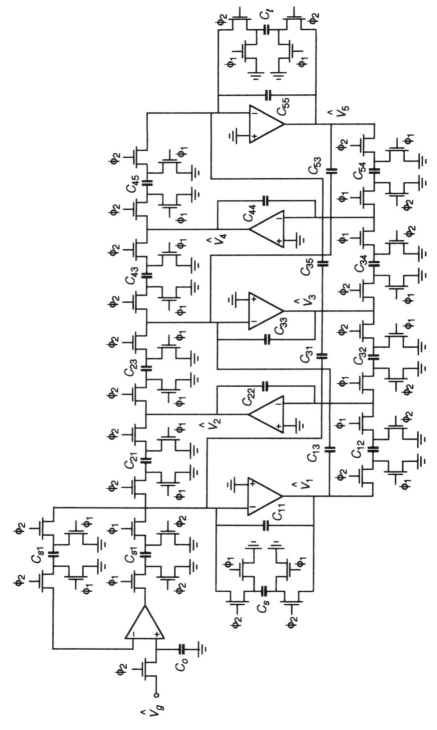

**Figure 10.5** Fifth-order ladder filter with finite $j\omega$-axis zeros.

prototype. This gives:

$$\frac{2C_{11} + C_s}{C_s} = R_g C_i' \qquad\qquad \frac{2C_{s1}}{C_s} = K$$

$$\frac{2C_{11} + C_s}{C_{21}} + C_1' = C_1' \qquad\qquad \frac{2C_{11} + C_s}{2C_{31}} = \frac{C_1'}{C_2'}$$

$$\frac{2C_{11}}{C_{21}} + \frac{2C_{22}}{C_{32}} = L_2 \qquad\qquad \frac{2C_{33}}{C_{23}} = \frac{2C_{33}}{C_{32}} = C_3'$$

$$\frac{C_{33}}{C_{13}} = \frac{C_3'}{C_2'} \qquad\qquad\qquad \frac{C_{33}}{C_{53}} = \frac{C_3'}{C_4'}$$

$$\frac{2C_{44}}{C_{34}} = \frac{2C_{44}}{C_{54}} = L_4 \qquad\qquad \frac{2C_{55} + C_\ell}{C_\ell} = C_5' R_L$$

$$\frac{2C_{55} + C_\ell}{C_{45}} = C_5' \qquad\qquad \frac{2C_{55} + C_\ell}{2C_{35}} = \frac{C_5'}{C_4'}$$

(10.12)

where $K$ is the ratio of the required gain to the ladder gain.

### 10.2.2 Computer-aided Implementation of the Design Procedure

The above procedure is implemented using **Program CMCRL** in Appendix F of ISICAP. The program only requires an input file containing the element values of the elliptic lumped prototype, or any lumped filter realized in the same form. The format of the input file with reference to Figure 10.1 is given in file 'ellp.dat' in directory ISICAPS and is as follows:

degree, ratio of pass-band edge to clock frequency

$$C_1$$

$$C_2$$

$$L_2$$

$$C_3$$

$$.. C_n$$

With reference to Figure 10.5, the program produces the capacitor ratios, in the following order:

For building block 1:

$$ratio\ 1 = \frac{C_s}{2C_{11} + C_s}$$

$$ratio\ 2 = \frac{2C_{s1}}{C_s}$$

$$ratio\ 3 = \frac{C_{21}}{2C_{11} + C_s} \tag{10.13}$$

$$ratio\ 4 = \frac{2C_{31}}{2C_{11} + C_s}$$

For building block $n$:

$$ratio\ 1 = \frac{C_{n-1,m}}{2C_{n,n} + C_\ell}$$

$$ratio\ 2 = \frac{C_\ell}{2C_{n,n} + C_\ell} \tag{10.15}$$

$$ratio\ 3 = \frac{2C_{n-2,n}}{2C_{n,n} + C_\ell}$$

For building block $i$ ($i$ odd)

$$ratio\ 1 = \frac{C_{i-1,i}}{2C_{i,i}}$$

$$ratio\ 2 = \frac{C_{i+1,i}}{2C_{i,i}}$$

$$ratio\ 3 = \frac{C_{i-2,i}}{2C_{i,i}} \tag{10.15}$$

$$ratio\ 4 = \frac{C_{i+2,i}}{2C_{i,i}}$$

For building block $i$ ($i$ even)

$$ratio\ 1 = \frac{C_{i-1,i}}{2C_{i,i}}$$

$$ratio\ 2 = \frac{C_{i+1,i}}{2C_{i,i}} \tag{10.16}$$

Another program **CMPCL1** uses the results of the previous program to perform scaling of the capacitance values for minimum total capacitance and maximum dynamic range. These ideas will be discussed in a later chapter. But this program can be bypassed in the design. Then the output of the program is the actual capacitance values in the following order

$$C_{s1}, C_{11}, C_{21}, C_{31}, C_s$$
$$C_{i-1,i}, C_{i,i}, C_{i+1,i} \quad \text{for } i \text{ even}$$
$$C_{i-1,1}, C_{i,i}, C_{i+1,i}, C_{i-2,i}, C_{i+2,i} \quad \text{for } i \text{ odd} \tag{10.17}$$
$$C_{n-1,n}, C_{n,n}, C_{n-2,n}, C_\ell \quad (\text{with } C_\ell = 1)$$

*Example 10.1* Consider the design of an elliptic filter with passband edge at 0.144 of the sampling frequency with 0.044 dB ripple, a stopband edge at 0.2 of the sampling frequency with minimum attenuation of 39 dB. The required degree is 5, and normalized prototype values are obtained from elliptic filter tables as

$$C_1 = 0.855\,35, \quad C_2 = 0.153\,67 \quad L_2 = 1.207\,63, \quad C_3 = 1.484\,38, \quad C_4 = 0.462\,65$$

$$L_4 = 0.897\,94, \quad C_5 = 0.637\,02, \quad R_g = R_L = 1$$

The program then divides these values by $\tan(\pi\omega_o/\omega_N)$ producing the $\lambda$-domain values, then performs the calculations giving as output the capacitor ratios. Figure 10.5 shows the circuit and the output of the program gives in the order of (10.13)–(10.16)

Building block 1   0.403 65, 2.0000, 0.403 65, 0.290 11.

Building block 2   0.402 74, 0.402 74

Building block 3   0.189 98, 0.1998, 0.136 54, 0.283 63

Building block 4   0.541 64, 0.541 64

Building block 5   0.356 81, 0.356 81, 0.532 67

If the minimum allowed capacitance value is 1 pF, then **Program CMPCL1** can be used with minimum capacitance scaling to obtain the capacitance values in the order given in (10.17) as follows

Block 1   2.782 81, 2.055 61, 2.782 81, 1.000 00, 2.782 81

Block 2   1.000 00, 1.241 49, 1.000 00

Block 3   2.782 81, 7.323 76, 2.782 81, 1.000 00, 2.077 21

Block 4   1.083 29, 1.000 00, 1.083 29

Block 5   1.000 00, 0.901 32, 0.746 44, 1.000 00

and the total capacitance is 37.426 44 pF where the final block is scaled such that $C_\ell = 1$, this can be further scaled at will. Figure 10.6 shows the response of the filter for a 10 kHz clock frequency.

## 10.3   BAND-PASS LADDER FILTERS WITH FINITE TRANSMISSION ZEROS ON THE IMAGINARY AXIS

Now consider the procedure for obtaining a band-pass filter using the frequency transformation on the lowpass prototype in the manner discussed in Chapter 8.

### 10.3.1   Low-pass to Band-pass Transformation

Suppose that the network of Figure 10.1 is the low-pass prototype designed based on the specifications for the band-pass filter. All the element values are normalized in the $\lambda$-domain, the centre frequency is also normalised to unity and the centre-to-bandwidth ratio is $r$. Then, the corresponding frequency transformation from low-pass to band-pass is

$$\lambda \rightarrow r\frac{\lambda^2 + 1}{\lambda} \tag{10.18}$$

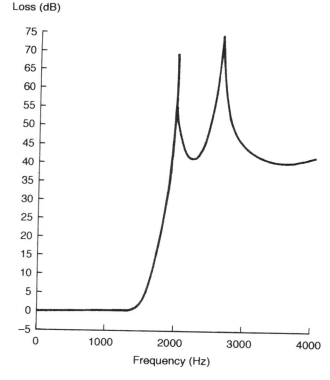

Figure 10.6   Amplitude response of the fifth-order elliptic filter of Example 10.1 with sampling frequency of 10 kHz.

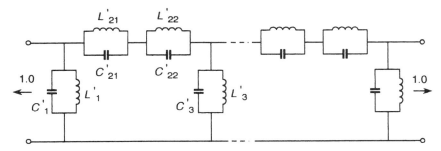

**Figure 10.7** Band-pass structure obtained from Figure 10.1 by frequency transformation.

Thus, a shunt capacitance in the lowpass prototype is transformed into a shunt LC-tank, and a series LC-tank is transformed into two LC-tanks in series as shown in Figure 10.7 with the element values given by

$$C_i' = rC_i \qquad\qquad L_i' = \frac{1}{rC_i} \qquad\qquad \text{for } i = 1, 3, 5, \ldots$$

$$C_{j1}' = rC_j(\beta_{-\infty}^2 + 1) \quad L_{j1}' = \frac{1}{rC_j}\frac{1}{\beta_{+\infty}^2 + 1} \qquad\qquad (10.19)$$

$$C_{j2}' = \frac{1}{L_{j1}'} \qquad\qquad L_{j2}' = \frac{1}{C_{j1}'} \qquad\qquad \text{for } j = 2, 4, 6, \ldots$$

where

$$\beta_{\pm\infty} = \frac{\Omega_\infty}{2r} \pm \sqrt{1 + (\Omega/2r)^2}, \quad \Omega_\infty = \frac{1}{\sqrt{C_j L_j}}$$

Now, consider Figure 10.8 which shows a network with an ideal transformer and its uncoupled Norton's equivalent. Then insert an ideal transformer with a turns-ratio of $t_i$ at the front of every shunt LC-tank except the first in the circuit of Figure 10.7. The result is shown in Figure 10.9. Furthermore, replacing every sub-network inside the dashed box of the network in Figure 10.9 by its Norton's equivalent, the network of Figure 10.10 is obtained.

Next, consider the sixth-order network of Figure 10.11 for illustration of the process of establishing the relations between the element values of the network of

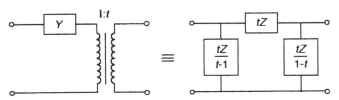

**Figure 10.8** Two equivalent network forms.

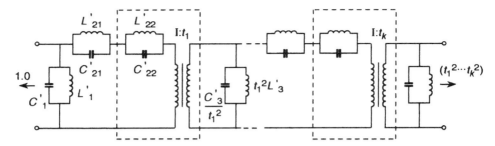

**Figure 10.9**  Network of Figure 10.7 with the addition of ideal transformers.

Figure 10.7 and that of Figure 10.10 . For Figure 10.11, these are as follows:

$$L''_1 = L_1 \qquad\qquad C''_1 = C_1$$

$$L''_2 = L_{21} \qquad\qquad C''_2 = C_{21}$$

$$L''_3 = \frac{L_{22}}{\left(1 - \dfrac{1}{t}\right)} \qquad C''_3 = \left(1 - \frac{1}{t}\right)C_{22} \qquad\qquad (10.20)$$

$$L''_4 = tL_{22} \qquad\qquad C''_4 = \frac{1}{t}C_{22}$$

$$\frac{1}{L''_5} = \frac{1}{t^2 L_3} + \frac{(1-t)}{t^2 L_{22}} \quad C''_5 = \frac{C_3}{t^2} + \frac{(1-t)C_{22}}{t^2}$$

To guarantee that $C''_3$, and $L''_3$ are positive, $t$ must be greater than 1. But, in order to have $L''_5 > 0$ and $C''_5 > 0$, the following inequalities must be satisfied:

$$t < 1 + \frac{L_{22}}{L_3}$$

$$t < 1 + \frac{C_3}{C_{22}}$$

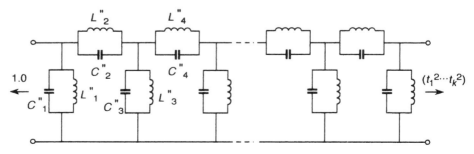

**Figure 10.10**  Network derived from Figure 10.7 by application of Figures 10.8 and 10.9.

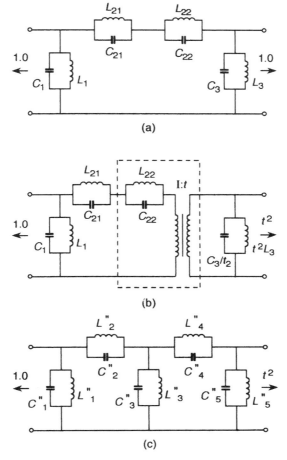

**Figure 10.11** Derivation of a sixth-order band-pass equivalent circuit for switched-capacitor realization: (a) network obtained by a low-pass to band-pass transformation; (b) application of Figure 10.9 to (a); (c) after application of Figure 10.8.

Thus, the condition for all the element values in the band-pass prototype to be positive is

$$1 < t < \min\left(1 + \frac{L_{22}}{L_3}, 1 + \frac{C_3}{C_{22}}\right) \qquad (10.16)$$

The value of $t$ could be chosen to obtain the smallest spread of the capacitor ratios of the filter.

It should be pointed out that whenever an ideal transformer is introduced in the network, the element values following the transformer are changed accordingly; so is the load impedance.

**Program LPTBP**, (available on disk in ISICAP), implements the transformation from low-pass to band-pass discussed above.

### 10.3.2   Complete Synthesis

We now consider the synthesis procedure. Figure 10.12 shows a band-pass ladder prototype obtained from the low-pass prototype by the technique discussed above. Now we divide all impedances by $\mu$ and note that

$$\lambda\mu = \gamma, \qquad \mu^2 = 1 + \gamma^2$$

and

$$\frac{L_i\lambda}{\mu} = L_i\frac{\gamma}{\mu^2} = L_i\frac{1}{\dfrac{1+\gamma^2}{\gamma}} = \frac{1}{\dfrac{1}{L_i\gamma} + \dfrac{1}{L_i}\gamma} \tag{10.22}$$

The resulting network is shown in Figure 10.13. Employing Norton's theorem to replace each of the bridging capacitors $C_2', C_4', \ldots$ by two voltage-controlled current sources as in the case of low-pass filters, we obtain the network shown in Figure 10.14. The operation of this network can be described by the following state equations.

$$V_1 = \frac{(\mu/R_g)V_g - I_1 - I_2 + C_2'\gamma V_3}{C_1''\gamma + (\mu/R_g)} \tag{10.23}$$

$$V_i = \frac{-I_1 + I_{i-1} - I_{i+1} + C_{i-1}'\gamma V_{i-2} + C_{i+1}'\gamma V_{i+2}}{C_i''\gamma} \quad \text{for } i = 3,5,7,\ldots \tag{10.24}$$

$$V_n = \frac{I_{n-1} - I_n + C_{n-1}'\gamma V_{n-2}}{C_n''\gamma + \mu/R_L} \tag{10.25}$$

However, from the network of Figure 10.14 we can see that

$$I_2 = \frac{V_1 - V_3}{L_2\gamma} = \frac{V_1}{L_1\gamma}\left(\frac{L_1}{L_2}\right) - \frac{V_3}{L_3\gamma}\left(\frac{L_3}{L_2}\right) = I_1\left(\frac{L_1}{L_2}\right) - I_3\left(\frac{L_3}{L_2}\right) \tag{10.26}$$

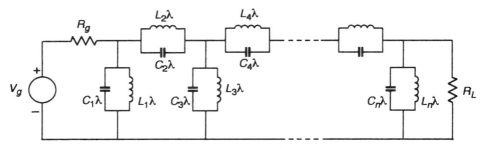

**Figure 10.12**   $\lambda$-domain band-pass structure.

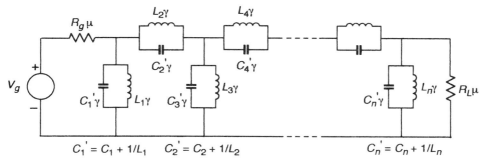

$C_1' = C_1 + 1/L_1 \quad C_2' = C_2 + 1/L_2 \qquad\qquad C_n' = C_n + 1/L_n$

**Figure 10.13** Modification of the element values of Figure 10.12 with impedance-scaling by $1/\mu$.

This is true for all series inductance currents. Thus, substituting this relation into the above state equations, the following equations result

$$V_1 = \frac{(\mu/R_g)V_s - I_1\left(1 + \dfrac{L_1}{L_2}\right) + I_3\left(\dfrac{L_3}{L_2}\right) + C_2'\gamma V_3}{C_1''\gamma + (\mu/R_g)} \tag{10.27}$$

$$V_i = \frac{-I_i\left(1 + \dfrac{L_i}{L_{i-1}}\dfrac{L_i}{L_{i+1}}\right) + I_{i-2}\left(\dfrac{L_{i-2}}{L_{i-1}}\right)}{+ I_{i+2}\left(\dfrac{L_{i+3}}{L_{i+1}}\right) + C_{i-1}'\gamma V_{i-2} + C_{i+1}'\gamma V_{i+2}}{C_i''\gamma} \tag{10.28}$$

$$I_i = \frac{V_i}{L_i\gamma} \quad \text{for } i = 3, 5, 7, \ldots \tag{10.29}$$

$$V_n = \frac{-I_n\left(1 + \dfrac{L_n}{L_{n-1}}\right) + I_{n-2}\left(\dfrac{L_{n-2}}{L_{n-1}}\right) + C_{n-1}'\gamma V_{n-2}}{C_n''\gamma + (\mu/R_L)} \tag{10.30}$$

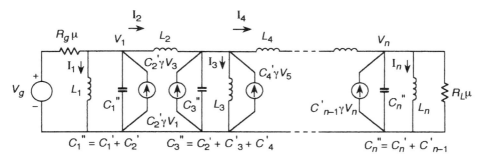

$C_1'' = C_1' + C_2' \qquad C_3'' = C_2' + C_3' + C_4' \qquad\qquad C_n'' = C_n' + C_{n-1}'$

**Figure 10.14** Network equivalent to that in Figure 10.13.

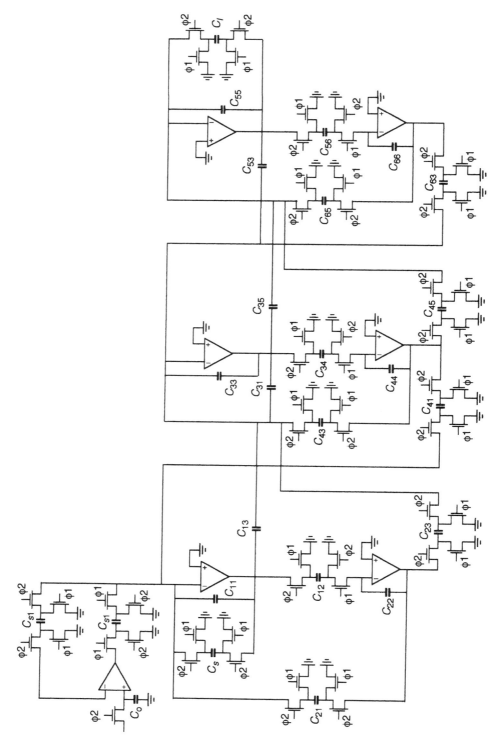

**Figure 10.15**    Sixth-order band-pass filter.

Now as in the case of low-pass filters, these equations can be easily simulated by the LDI building blocks as well as the composite building block of Figure 10.2. As an illustration, Figure 10.15 gives a switched-capacitor circuit of a sixth-order band-pass elliptic filter.

### 10.3.3 Computer-aided Implementation of the Design Procedure

**Program CMCRB** implements the design technique detailed above. The required input to the program contains the degree of the low-pass prototype and its element values. Also the ratio of centre frequency to bandwidth in the $\Omega$-*domain* must be given together with the ratio of centre frequency to clock frequency. The program will give the capacitor ratios as output. With reference to Figure 10.15, the order and format are as follows

For building block 1:

$$ratio\,1 = \frac{C_s}{2C_{11} + C_s}$$

$$ratio\,2 = \frac{2C_{s1}}{C_s}$$

$$ratio\,3 = \frac{C_{31}}{2C_{11} + C_s} \tag{10.31}$$

$$ratio\,4 = \frac{2C_{21}}{2C_{11} + C_s}$$

$$ratio\,5 = \frac{2C_{51}}{2C_{11} + C_s}$$

For building block $i$ ($i$ odd)

$$ratio\,1 = \frac{C_{i-1,i1}}{2C_{i,i}}$$

$$ratio\,2 = \frac{C_{i+3,i}}{2C_{i,i}}$$

$$ratio\,3 = \frac{C_{i-2,i}}{2C_{i,i}} \tag{10.32}$$

$$ratio\,4 = \frac{C_{i+2,i}}{2C_{ii}}$$

$$ratio\,5 = \frac{C_{i+1,i}}{2C_{ii}}$$

Building block $i$ ($i$ even)

$$ratio\ 1 = \frac{C_{i-1,i}}{2C_{i,i}} \tag{10.33}$$

For building block $(n-1)$

$$ratio\ 1 = \frac{C_\ell}{2C_{n-1,n-1} + C_\ell}$$

$$ratio\ 2 = \frac{2C_{n-3,n-1}}{2C_{n-1,n-1} + C_\ell}$$

$$ratio\ 3 = \frac{C_{n,n-1}}{2C_{n-1,n-1} + C_\ell} \tag{10.34}$$

$$ratio\ 4 = \frac{C_{n-2,n-1}}{2C_{n-1,n-1} + C_\ell}$$

Building block $n$

$$ratio\ 1 = \frac{C_{n-1,i}}{2C_{n,n}} \tag{10.35}$$

**Program CMCPB** gives the capacitance values with scaling for minimum capacitance and maximum dynamic range. It can also be bypassed. The output of the program gives the capacitance values in the following order

$$C_{s1}, C_s, C_{11}, C_{21}, C_{31}, C_{41}$$

$$C_{i-1,i}, C_{i,i} \quad \text{for } i \text{ even}$$

$$C_{i-2,i}, C_{i-1,i}, C_{i,i}, C_{i+1,i}, C_{i+2,i}, C_{i+3,i} \quad \text{for } i \text{ odd} \tag{10.36}$$

$$C_{n-3,n-1}, C_{n-2,n-1}, C_{n-1,n-1}, C_{n,n-1}, C_\ell$$

$$C_{n-1,n}, C_{n,n}$$

In the following, a design example is given to illustrate the use of the technique.

*Example 10.2*   The specifications on the response of the band-pass filter are:

Passband ripple $= 0.008$ dB for $0.12 < f/f_N < 0.14$
Stopband attenuation $= 50$ dB for $0.0 < f/f_N < 0.07$ and $0.222 < f/f_N < 0.5$

Then the centre-to-bandwidth ratio can be calculated from

$$r = \frac{\sqrt{\tan 0.14\pi \tan 0.12\pi}}{\tan 0.14\pi - \tan 0.12\pi} = 5.78$$

The center frequency $= 1300$ Hz.

Loss (dB)

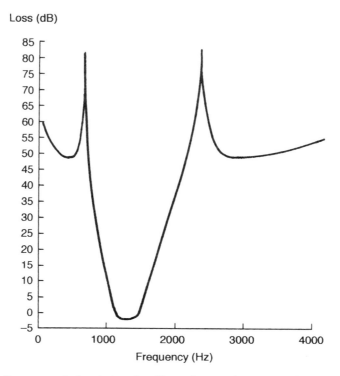

Figure 10.16 Response of the sixth-order filter of Example 10.2, with sampling frequency of 10 kHz.

The specifications require a third-order low-pass elliptic prototype with the following normalized element values in the $\lambda$-domain (which are in file elpbp.dat on the disk):

$$C_1 = 0.5776$$

$$C_2 = 0.0122 \quad L_2 = 0.9159$$

$$C_3 = 0.5776$$

Hence, a sixth-order band-pass filter is required.

Using **Program LPTBP** with $t_1 = 1.1734$, we obtain the element values of the band-pass filter normalized to the centre frequency. Then, **Program CMCRB** is used to obtain the capacitor ratios. These are given by

Building block 1   0.104 98, 2.0000, 0.038 42, 0.117 07, 3.161 17

Building block 2   1.447 10

Building block 3   0.089 16, 0.023 60, 0.028 323, 0.610 74, 28.533 596

Building block 4    0.005 272 873

Building block 5    0.099 15, 0.107 77, 0.140 73, 0.786 64

Building block 6    1.044 198

Figure 10.15 shows the circuit while Figure 10.16 shows the response of the filter.

## 10.4   CASCADE DESIGN AND ISICAPC

If the sensitivity properties of the filter are not of primary concern, as in the case of relatively low-order filters, the simplest method of realization is to decompose the transfer function into second-order factors(and a possible first-order one for odd degree cases), realize each factor by a simple network, then connect the resulting networks in cascade. Thus, the transfer function is obtained in the $\lambda$ or $z$-domain according to the specifications as explained earlier. Then it is written in the form

$$H(Z) = \prod_{k=1}^{m} H_k(z) \tag{10.37}$$

where a typical quadratic factor is of the form

$$H_k(z) = \frac{a_{ok} + a_{1k}z^{-1} + a_{2k}z^{-2}}{1 + b_{1k}z^{-1} + b_{2k}z^{-2}} \tag{10.38}$$

and its switched-capacitor realization is shown in Figure 10.17. The capacitor ratios can be obtained by comparing the above expression with the transfer function of the shown section as given by

$$Hz = \frac{I + (G - I - J)z^{-1} + (J - H)z^{-2}}{(1 + F) + (C + E - F - 2)z^{-1} + (1 - E)z^{-2}} \tag{10.39}$$

where it is assumed that the capacitances $A = B = D = 1$.

A first-order factor is of the form

$$H_k(z) = \frac{a_{ok} + a_{1k}z^{-1}}{1 + b_{1k}z^{-1}} \tag{10.40}$$

which can be realized as one of the four circuits shown in Figure 10.18, depending on the location of the pole-zero pair. Thus, for the circuit of Figure 10.18(a)

$$H_k(z) = \frac{(C_1 + C_3) - C_1 z - 1}{(1 + C_2) - z^{-1}} \tag{10.41}$$

while for the circuit of Figure 10.18(b)

$$H_k(z) = \frac{(C_1 + C_3) - C_1 z^{-1}}{(C_2 - 1)z^{-1} - 1} \tag{10.42}$$

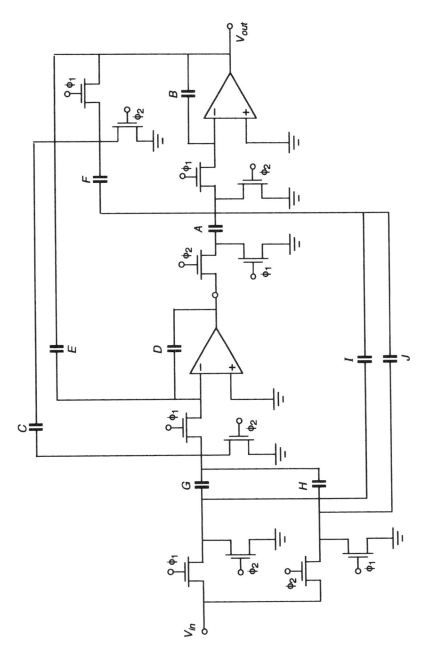

Figure 10.17 Second-order section.

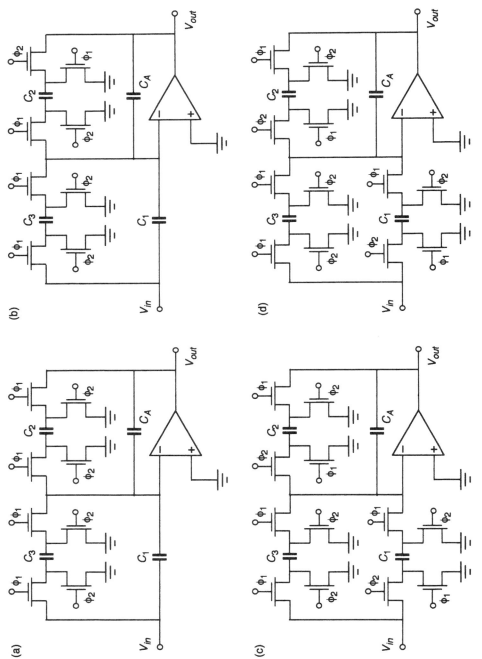

**Figure 10.18**  Linear section:  (a) type I; (b) type II; (c) type III and (d) type IV.

and the circuit of Figure 10.18(c) has

$$H_k(z) = \frac{C_3 + C_1 z^{-1}}{z^{-1} - (1 + C_2)} \qquad (10.43)$$

while for the circuit of Figure 10.18(d) we have

$$H_k(z) = \frac{C_3 + C_1 z^{-1}}{1 + (C_2 - 1)z^{-1}} \qquad (10.44)$$

*Example 10.3* Let the same filter transfer function of Example 10.1 be realized in cascade form. The elliptic transfer function is obtained from its pole zero locations available in tables, then the bilinear transformation is applied to obtain the z-domain function. The sections have the transfer functions

$$H_1(z) = \frac{0.519\,057 + 0.124\,971z + 0.519\,057z^2}{1 - 2.097\,346z + 2.260\,430z^2} \qquad (10.45)$$

$$H_2(z) = \frac{0.685\,511 - 0.376\,703z + 0.685\,511z^2}{1 - 1.219\,085z + 1.213\,404z^2} \qquad (10.46)$$

$$H_3(z) = \frac{0.610\,451 + 0.610\,451z}{1 - 2.220\,901z} \qquad (10.47)$$

By reference to Figures 10.17 and 10.18, and expressions (10.38)–(10.44) we obtain the capacitor values of the sections as

*Section 1*

$$E = 0, \quad F = 1.260\,430, \quad C = 1.163\,084, \quad I = 0.519\,057$$
$$H = 0.0000, \quad J = 0.519\,057, \quad G = 1.163\,084$$

All with reference to $A = B = D = 1$

*Section 2*

$$E = 0, \quad F = 0.213\,404, \quad C = 0.994\,319, \quad I = 0.685\,511$$
$$H = 0.0000, \quad J = 0.685\,511, \quad G = 0.994\,319$$

All with reference to $A = B = D = 1$.

*Section 3*

Linear section Type III:

$$C_2 = 1.220\,901, \quad C_1 = 0.610\,454, \quad C_3 = 0.610\,454$$

all with reference to $C_A = 1$.

The cascade design is given in ISICAPC on the disk.

## 10.5  GENERALIZED CHEBYSHEV FILTERS AND ISICAP2

As we have mentioned at several points, a small spread of element values and symmetry of structures are desirable attributes in integrated circuit implementation since they result in reduced area and easier matching of components. A particular type of filter response leads to both attributes, and at the same time gives very nearly the same selectivity as an elliptic filter for the same degree. In the continuous-time domain, the transducer power gain is of the same form given by

$$|S_{21}|^2 = \frac{1}{1 + \epsilon^2 F_n^2} \tag{10.48}$$

with

$$F_n(\omega^2) = \cosh\left\{(n-1)\cosh^{-1}\left[\omega\left(\frac{\omega_o^2 - 1}{\omega_o^2 - \omega^2}\right)^{1/2}\right] + \cosh^{-1}\omega\right\} \tag{10.49}$$

The general features of the response are shown in Figure 10.19, and reveals that it has a multiple transmission zero at $\omega_o$ only as well as an equiripple passband. The realization of this filter is the same as that of the elliptic filter in one of the forms of Figure 7.14. Thus, its $\lambda$-domain equivalent is the same as that of Figure 10.1 with the added property of symmetry.

Due to the importance of this filter, its switched-capacitor implementation is given separately in ISICAP as the executable program **ISICAP2.EXE** in the directory **ISICAPS**. It can be run by typing 'ISICAP2' then pressing return, and following the instructions in the menu. The program directly gives the capacitor ratios when the user enters the specifications. Both low-pass and band-pass designs are given. The format of the output is the same as that of the elliptic filter. It is felt that this filter should replace the elliptic filter in switched-capacitor filter design and, therefore, its

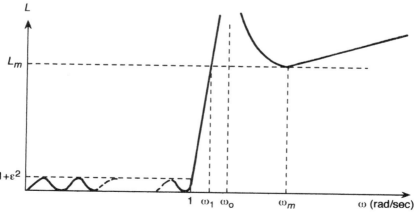

**Figure 10.19**  Response of the Generalized Chobyshev filter in the continuous domain $L = 1 + \epsilon^2 F_n^2$.

use is strongly recommended. In any event, in the present context the program obviates the need for using the tables of elliptic filter prototypes and the design is given directly from the specifications.

## CONCLUSION

This chapter dealt with the techniques of switched-capacitor filter design based directly on lumped passive lossless ladder filters. The main objective being to add the ability of introducing finite imaginary axis zeros of transmission. This allows the realization of responses such as the elliptic function response providing the optimum amplitude characteristic. Then, it was shown that the realization of the same transfer functions is possible in simple cascade form, at the expense of losing the excellent low-sensitivity-properties of the ladder structure. Throughout the chapter, extensive use was made of ISICAP to give several examples and facilitate the computational aspect of the design method.

## PROBLEMS

10.1 The filter shown in Figure P10.1 is a third-order normalized elliptic low-pass prototype with cutoff at $\omega = 1$ with 0.25 dB ripple, and the ratio of stopband edge to passband edge $= 1.5$.

(a) Calculate the minimum stopband attenuation achieved with this filter.

(b) Transform the prototype into a switched-capacitor ladder with cutoff at 3 kHz and using a clock frequency of 20 kHz.

[Use Program CMCRL]

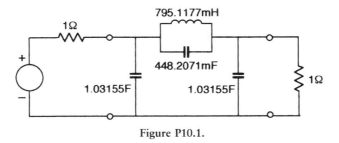

**Figure P10.1.**

10.2 The zeros and poles of the filter of Figure P10.1 are given by

Zeros at $s = \pm j1.675\,116$

Poles at $s = -2.593\,404 \pm j1.112\,388$ and $s = -9.694\,113$

Use the bilinear transformation to obtain the transfer function of the filter in the z-domain, in factored form, to have the same cutoff of 3.0 kHz and hence find the realization in cascade form. Compare the total required capacitance with that of the filter designed in Problem 10.1.

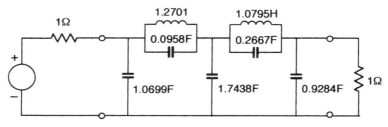

Figure P10.3.

10.3 The filter shown in Figure P10.3 is a fifth-order normalized elliptic low-pass prototype with cutoff at $\omega = 1$ with 0.1 dB ripple, and the ratio of stop band edge to passband edge $= 2$. In the stopband, the minimum attenuation provided is 40 dB.

   Transform the prototype into a switched-capacitor ladder with cutoff at 3.4 kHz and using a clock frequency of 12 kHz.

   [Use **Program CMCRL**]

10.4 The zeros and poles of the filter of Figure P10.3 are given by:

   Zeros at s$= \pm$j1.863 738 and s $= \pm$ j2.866 791

   Poles at s$= -4.260\,778 \pm$ j7.374 493, s $= -1.31259 \pm 1.071416$ and s $= \pm 6.076\,273$.

   Use the bilinear transformation to obtain the transfer function of the filter in the $z$-domain, in factored form, with the same cutoff.
   Hence, find a realization in cascade form. Compare the total required capacitance with that of the filter designed in Problem 10.3.

10.5 Transform the filter of Problem 10.1 into a band-pass switched-capacitor ladder filter with passband edge frequencies of 1.5 kHz and 2.5 kHz. Choose a suitable clock frequency and find the resulting stopband edge frequencies.

   [Use **Programs LPTBP, CMCRB**]

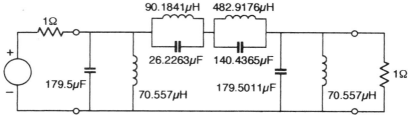

Figure P10.5.

10.6 Figure 10.3 shows an elliptic band-pass filter designed to meet the following specifications:

   Passband 1.0 to 2.0 kHz, with 0.2 dB ripple
   Stopband edges at 0.5 kHz and 3.0 kHz, with minimum attenuation of 30 dB.

(a) Find the low-pass prototype from which the filter was obtained.

(b) For the given band-pass filter, use the bilinear transformation to obtain a switched-capacitor ladder filter having the same passband edge frequencies with a clock frequency of 8 kHz. Find the resulting stopband edge frequencies in the switched-capacitor filter.

[Use **Programs LPTBP, CMCRB**]

10.7  Evaluate the transfer function of the filter of Problem 10.6 and realize it in cascade form.

10.8  Design a generalized Chebyshev low-pass ladder filter to meet the specifications on the receive filter of a speech CODEC as given by:

Passband 0–3.4 kHz with 0.25 dB maximum attenuation
Stopband edge at 4.6 kHz with 32 dB minimum attenuation.

[Use Program ISICAP2].

10.9  Design a generalized Chebyshev band-pass ladder filter to meet the specifications on the transmit filter of a speech CODEC as given by:

Passband 300–3.4 kHz with 0.25 dB maximum attenuation
Lower stopband edge at 60 Hz with 25 dB minimum attenuation.
Upper stopband edge at 4.6 kHz with 32 dB minimum attenuation.

[Use Program ISICAP2].

10.10 Realize the filters of Problems 10.6–10.9 in cascade form. Check the results using the Program ISICAPC.

# 11

# Selective Linear-phase and Data Transmission Filters

## 11.1 INTRODUCTION

In some applications, filters are designed on amplitude basis alone. For example, in the audio range, it is claimed that the human ear is insensitive to phase distortion, and the optimum solution to the filter design problem is usually taken to be the elliptic filter giving the lowest degree filter for a given set of specifications on the amplitude response. Other applications place more importance on the phase response while tolerating a moderate amount of amplitude distortion. In this case certain phase-oriented approximations are acceptable. However, in modern high-capacity communication systems, filters are required to meet stringent specifications on both the amplitude and phase simultaneously. Furthermore, for data transmission using impulses (or pulses), the impulse (or pulse) response of the filter is as important as its frequency response. This is also true in the case of video signal processing.

The use of switched-capacitor filters was, for some time, confined to audio frequencies. Recently, however, the operating range of these filters has been extended to frequency ranges, such as video signals, where phase linearity in the passbands becomes a major design consideration and for data transmission the time response is also an added constraint which is actually closely related to the phase response of the filter. Thus, the phase response of switched-capacitor filters is becoming of increasing importance.

The traditional approach to satisfying a given set of specifications on the amplitude and phase responses of the filter is to first design a filter which meets the amplitude specifications, then design an all-pass phase equalizer to correct for the deviation from phase linearity of the amplitude-oriented filter. Such an approach is, however, unacceptable due to the non- optimum nature of the design. This means that one obtains a filter with a higher degree than necessary to meet the given set of specifications. For switched-capacitor filters, this leads to an unnecessarily large number of active components including operational amplifiers and switches, as well as an increase in the number of capacitors. This leads to an increase in the chip area, power consumption, noise and all other associated non-ideal effects.

It follows that the traditional approach, which is the one usually adopted in all available computer-aided filter design programs, is unacceptable in many applications. This is particularly true since recent advances in approximation theory make it possible to design filters which satisfy the amplitude and phase specifications simultaneously at the outset in an optimum manner. That is, the transfer function is derived to possess the required properties without any need for subsequent equalization of either the phase or amplitude response; thus exploiting all the available degrees of freedom. This is the subject of this chapter and the concepts employed are extended to cover the design of filters for data transmission applications.

## 11.2    PHASE AND DELAY FUNCTIONS OF SWITCHED-CAPACITOR FILTERS

The transfer function of a sampled-data filter is of the form

$$H(z) = \frac{\displaystyle\sum_{i=0}^{M} a_i z^{-i}}{1 + \displaystyle\sum_{i=1}^{N} b_i z^{-i}} \tag{11.1}$$

or

$$H(\lambda) = \frac{P_m(\lambda)}{Q_n(\lambda)}, \quad m \leqslant n \tag{11.2}$$

The frequency response of the filter is obtained by letting

$$z \to \exp(j\omega T) \tag{11.3}$$

or

$$\lambda \to j\Omega \tag{11.4}$$

so that

$$H(j\Omega) = |H(j\Omega)| \exp[j\Psi(\Omega)] \tag{11.5}$$

where $|H(j\Omega)|$ is the amplitude response and $\Psi(j\Omega)$ is the phase response of the filter. The group delay is given by

$$\begin{aligned}
T_g(\omega T) &= -\left.\frac{d\Psi(\lambda)}{ds}\right|_{s=j\omega} \\
&= -\left.\frac{d\Psi(\lambda)}{d\lambda}\frac{d\lambda}{ds}\right|_{s=j\omega} \\
&= -\left.\frac{T}{2}(1-\lambda^2)\frac{d\Psi(\lambda)}{d\lambda}\right|_{s=j\omega}
\end{aligned} \tag{11.6}$$

or

$$T_g(\Omega) = -\frac{T}{2}(1+\Omega^2)\frac{d\Psi(\lambda)}{d\lambda}\bigg|_{\lambda=j\Omega} \tag{11.7}$$

Now, taking logarithm of both sides of (11.5) we obtain

$$\ln H(j\Omega) = \ln|H(j\Omega)| + \ln\Psi(\Omega) \tag{11.8}$$

and using

$$|H(j\Omega)|^2 = H(j\Omega)H(-j\Omega)$$

we have

$$\Psi(\Omega) = \frac{1}{2j}\ln\frac{H(j\Omega)}{H(-j\Omega)} \tag{11.10}$$

so that (11.7) can be expressed as

$$T_g(\Omega) = \frac{T}{2}(1+\Omega^2)\,\mathrm{Re}\left[\frac{Q'_n(\lambda)}{Q_n(\lambda)} - \frac{P'_m(\lambda)}{P_n(\lambda)}\right]_{\lambda=j\Omega} \tag{11.11}$$

where the prime denotes the derivative. Alternatively, in the $z$-domain we have for the group delay

$$T_g(\omega T) = \frac{T}{2}\left[z\frac{H'(z)}{H(z)} + z^{-1}\frac{H'(z^{-1})}{H(z^{-1})}\right]_{z=e^{j\omega T}} \tag{11.12}$$

or

$$T_g(\omega T) = T\,\mathrm{Re}\left[z\frac{H(z)}{H(z)}\right]_{z=e^{j\omega T}}$$

$$= T\,\mathrm{Re}\left[z\frac{d}{dz}\ln H(z)\right]_{z=e^{j\omega T}} \tag{11.13}$$

Now, if we attempt to obtain a sampled-data transfer function from a continuous-time one which approximates a linear phase in the passband using the familiar bi-linear transformation

$$s \to k\lambda \tag{11.14}$$

where $k$ is a constant, we see that the delay properties of the continuous-time filter are not preserved by this transformation. This is due to the factor $(1+\Omega^2)$ in (11.7).

Hence, this non-linear factor renders continuous-time prototypes such as the Bessel filter, of no use in the phase-oriented design of sampled-data filters. Moreover, frequency scaling by a factor $(\Omega \to k\Omega)$ is not possible either since the same factor $(1 + \Omega^2)$ does not scale. This necessitates the incorporation of a band-width scaling parameter in the expressions of transfer functions at the outset.

## 11.3   PHASE-ORIENTED DESIGN

In this section, phase linearity as a major consideration [45], [46] is introduced into the design of switched capacitors of the categories defined by the classes of transfer functions presented in Chapters 9 to 10. Here emphasis is laid entirely on the phase responses of the filters, but the resulting filters will also possess moderate amplitude selectivity. These will give rise to filters which can be acceptable for some applications in which the phase response is the main design objective while not requiring high selectivity. In the sections to follow the combined phase and amplitude approximation problem is considered.

### 11.3.1   Low-pass Maximally-flat Delay Filters

The filters considered here are of the lossless discrete integrator type discussed in Chapter 9, as shown in Figure 11.1, where it has been shown that the transfer function of the resulting structure (without the dashed-line capacitors) is of the general form

$$H_{21}(\lambda) = \frac{(1 - \lambda^2)^{n/2}}{P_n(\lambda)} \tag{11.15}$$

Furthermore, the addition of the extra dashed-line capacitor in Figure 11.1 allows the introduction of finite zeros of transmission at $\omega_N/2$ or any set of finite real frequencies. For example, if $m$ zeros are at $\omega_N/2$ we have a transfer function of the form

$$H_{21}(\lambda) = \frac{(1 - \lambda^2)^{(n-m)/2}}{P_n(\lambda)} \tag{11.16}$$

In all cases (11.15) and (11.16), stability requires $P_n(\lambda)$ to be strictly Hurwitz polynomial in $\lambda$.

Noting that the numerators of $H_{21}(\lambda)$ in (11.15)–(11.16) do not contribute to the delay variation, we obtain $P_n(\lambda)$ such that the filter approximates a linear phase response (constant group delay) in the passband, in a number of ways as follows.

Consider the polynomial

$$Q_n(\lambda, \alpha) = M(\lambda, \alpha) + N(\lambda, \alpha) \tag{11.17}$$

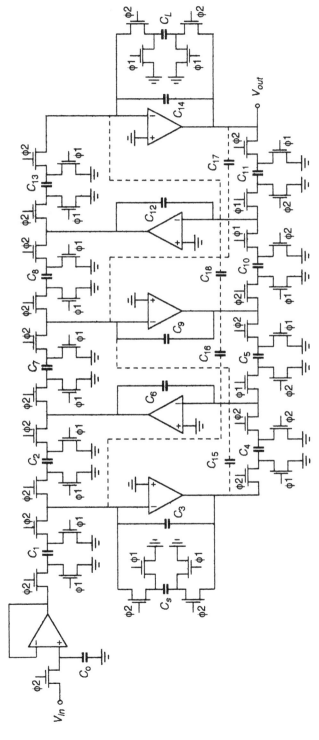

**Figure 11.1** Typical fifth-order LDI ladder filter. The dashed-line elements are used to realize finite $j\omega$-axis zeros.

Where $M$ and $N$ are the even and odd parts of $Q_n$, respectively, and $\alpha$ is a parameter whose significance will be discussed later. Let us identify the function

$$\tanh \Psi(\lambda, \alpha) = \frac{N(\lambda, \alpha)}{M(\lambda, \alpha)} \tag{11.18}$$

with the $n$th approximant in the continued fraction expansion of

$$\phi(\lambda, \alpha) = \tanh(\alpha \tanh^{-1}\lambda) = \sinh(\alpha \tanh^{-1}\lambda)/\cosh(\alpha \tanh^{-1}\lambda) \tag{11.19}$$

as

$$\frac{N(\lambda, \alpha)}{M(\lambda, \alpha)} \approx \cfrac{\alpha\lambda}{1 + \cfrac{(\alpha^2 - 1)\lambda^2}{3 + \cfrac{(\alpha^2 - 4)\lambda^2}{5 + \cfrac{(\alpha^2 - 9)\lambda^2}{\cdots}}}} \tag{11.20}$$

This means that we truncate the expansion at the $n$th step, and remultiply to obtain $N$ and $M$ which define the polynomial $Q_n(\lambda, \alpha)$ in (11.17). Subsequently, if we use (11.15)–(11.16) with

$$P_n(\lambda) \triangleq Q_n(\lambda, \alpha) \tag{11.21}$$

the resulting functions will have delay responses which are maximally flat around $\Omega = 0$. For stability we require

$$\alpha > (n - 1) \tag{11.22}$$

Using the properties of continued fractions, we may readily obtain a recurrence formula for $Q_n(\lambda, \alpha)$ as

$$Q_{n+1}(\lambda, \alpha) = Q_n(\lambda, \alpha) + \frac{\alpha^2 - n^2}{4n^2 - 1}\lambda^2 Q_{n-1}(\lambda, \alpha) \tag{11.23}$$

with

$$Q_0 = 1, \quad Q_1 = 1 + \lambda\alpha$$

and the maximally-flat delay character of this polynomial can be ascertained by substitution in (11.8).

Program MXLP, (available on disk in ISICAP), can be used to generate the maximally-flat delay polynomial defined by (11.23) while Program LDISCG in Appendix A can be used to perform the synthesis of the resulting function of the form (11.15).

Figure 11.2 shows typical amplitude and delay responses for $n = 5$ and the function of the type in (11.15). The parameter $\alpha$ has a value equal to the delay at $\Omega = 0$. However, its main significance is its role as a scaling parameter for adjusting the

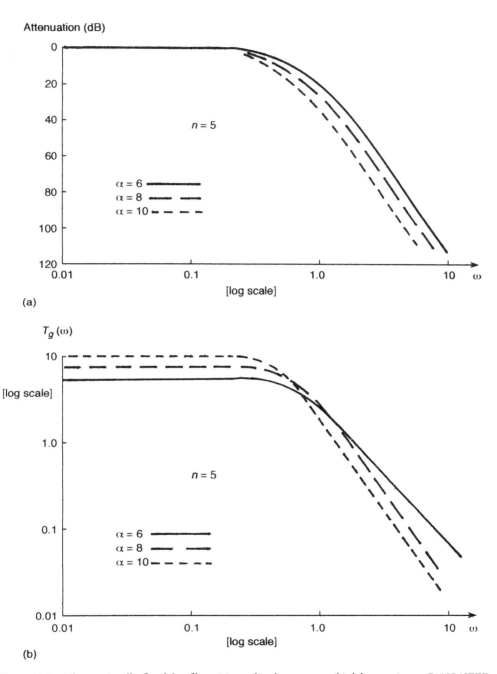

**Figure 11.2** The maximally-flat delay filter: (a) amplitude response; (b) delay response. © 1986 IEEE.

fractional bandwidth over which the delay is to be held constant. A combination of $n$ and $\alpha$ must be found in order to satisfy any set of specifications. Therefore, Figure 11.2 must be repeated for $n=3, 4, 5, 6, \ldots$, and the lowest order, together with a value of $\alpha$, are chosen. However, as all the degrees of freedom have been placed on the delay response, one should expect the amplitude selectivity of the filters to be rather poor, in a manner similar to Bessel filters in the lumped domain. This situation is improved later, but first an example of the use of Figure 11.2 is given.

*Example 11.1*   It is required to design a low-pass filter with the following specifications:

Passband 0 to 3 kHz
passband attenuation $\leqslant 1$ dB
passband delay variation $\leqslant 11$ ns
Stopband 6 kHz–21.4 kHz
Stopband attenuation $\geqslant 30$ dB.

To obtain the degree n, together with a value of $\alpha$, which meet the given set of specifications, we must use an extensive set of curves (or tables) of the type shown in Figure 11.2 for different $n$. For the sake of brevity, the specifications in the above example are chosen such that they can be met using Figure 11.2. From the given set of specifications, we may choose a sampling frequency of twice the highest frequency of 21.4 kHz, i.e.

$$f_N = 42.8 \text{ kHz}$$

so that we require

1 dB at $0.07 f_N$
30 dB at $0.14 f_N$
Passband delay variation $< 11 \text{ ns}/(T/2)$
$$< 9 \times 10^{-4}$$

Using the **Program MXLP**, we find that the amplitude specifications are met with $n = 5$, $\alpha = 9$. Checking the percentage delay variation we find that for $\alpha = 9$, $\Delta T_g = 0.01\%$, which is relative to a dc delay of 9. Thus, $\Delta T_g = 9 \times 10^{-4}$ corresponds to an actual delay variation of $9 \times 10^{-4} \times (T/2) = 10.5$ ns which meets the delay specifications. Program LDISCG can be used for the synthesis.

### 11.3.2   Low-pass Equidistant Linear-phase Response

An equidistant interpolation to a linear phase response in the passband can be obtained by setting $P_n(\lambda)$ in (11.15)–(1.16) as

$$P_n(\lambda) \overset{\Delta}{=} A_n(\lambda, \phi_0, \beta) \tag{11.24}$$

where $A_n(\lambda, \Phi_0, \beta)$ is the Rhodes polynomial which may be obtained by the recurrence relationship

$$A_{n+1} = A_n + \gamma_n(\lambda^2 + \tan^2 n\Phi_0)A_{n-1} \qquad (11.25)$$

with

$$A_0 = 1, \qquad A_1 = 1 + \frac{\tan \beta\phi_0}{\tan \phi_0}\lambda$$

where

$$\gamma_n = \frac{\cos(n-1)\phi_0 \cos(n+1)\phi_0 \sin(\beta+n)\phi_0 \sin(\beta-n)\phi_0 \cos^2 n\phi_0}{\sin(2n-1)\phi_0 \sin(2n+1)\phi_0 \cos^2 \beta\phi_0} \qquad (11.26)$$

The resulting function of (11.15)–(11.16) is such that

$$\arg H_{21}(\pm j \tan r\phi_0) = \pm\beta r\phi_0, \qquad r = 0, 1, \ldots n \qquad (11.27)$$

i.e. the phase is exactly linear at a set of points $\omega T/2 = 0$, $\phi_0$, $2\phi_0, \ldots, n\phi_0$ in the passband. **Program ELPP**, in Appendix E in ISICAP, generates this equidistant linear

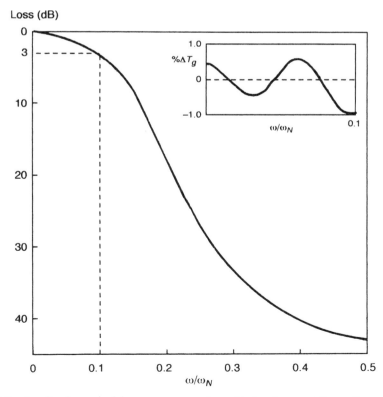

**Figure 11.3** Amplitude and delay responses of a third-order equidistant-linear-phase filter.
© 1986 IEEE.

phase polynomial. The choice of $\phi_0$ and $\beta$ determines the bandwidth together with the magnitude of the delay variation. Figure 11.3 shows typical amplitude and delay responses of a filter obtained using (11.23) in (11.15). The designer must try various combinations of $\phi_o$ and $\beta$ such that a given set of specifications may be met. There is no guarantee, however, that arbitrary amplitude selectivity would be possible as the polynomial in (11.24) is obtained on phase basis alone.

The conditions for stability of these filters are

$$\beta > n - 1, \qquad n\phi_0 < \frac{\pi}{2}, \qquad (\beta + n + 1)\phi_0 < \pi \qquad (11.28)$$

Having obtained the transfer function, **Program LDISCG** in Appendix A can be set for the synthesis.

## 11.4   OPTIMAL LOW-PASS LADDER DESIGN

In this section, a fundamental network theoretic idea [46], [47] is generalized and extended to accommodate switched-capacitor filters to be designed with the following properties:

(a) constant group delay in the specified portion of the passband within a given tolerance;

(b) prescribed constancy of passband amplitude, with arbitrarily selective response.

This selectivity is to be achieved by introducing a number of multiple finite zeros of transmission in the stopband. A comprehensive algorithm is also discussed which can be easily programmed on a computer, for the realization of an optimum compromise between delay constancy and amplitude selectivity.

The switched-capacitor filters considered in this section are of the typical LDI structure shown in Figure 11.1. Specifically, the transfer function of the filter is of the general form

$$H(\lambda) = \frac{(1 - \lambda^2)^{l/2}(\lambda^2 + \Omega_x^2)^m}{P_n(\lambda)} \qquad (11.29)$$

where $l$ is the number of strict LDI building blocks and $m$ is the number of modified ones (with dashes elements) used to introduce a multiple zero of transmission at $\lambda = \pm j\Omega_x$. Thus the zeros of $H(\lambda)$ are restricted to lie either on the $j\Omega$-axis or of the form $(1 - \lambda^2)^{1/2}$.

The high-pass solution to the same problem can be obtained from the low-pass case and the transformation

$$\lambda \rightarrow \frac{1}{\lambda} \qquad (11.30)$$

Then, it is only necessary to use the building blocks given in Chapter 9 for the high-pass case. Consequently, the high-pass design is covered by implication once the design technique for low-pass filters is presented.

### 11.4.1 Preliminaries

Strictly speaking, the transfer function in (11.29) is of the non-minimum-phase type, due to the half-order zeros at $\lambda = 1$. However, if $(1 - \lambda^2)^{1/2}$ is replaced by $(1 + \lambda)$ the resulting minimum-phase transfer function has the same amplitude $|H(j\Omega)|$ as the original one, and a group delay which differs from that of (11.29) by a constant $l$. For this reason, we shall call the transfer function in (11.29) *quasi-minimum-phase*. This class of function has the property that its amplitude and phase are related by a modified Hilbert transform.

The technique presented achieves an optimum compromise between amplitude selectivity and phase linearity for the classes of function under consideration, This is because the method meets the amplitude requirements then proceeds by maximizing the portion of the passband over which phase linearity may be maintained, for a given degree.

Consider the transfer function $H(\lambda)$ written as

$$H(\lambda) = e^{-W(\lambda)} \tag{11.31}$$

with

$$H(j\Omega) = |H(j\Omega)|\, e^{j\phi(\Omega)} \tag{11.32}$$

$$W(j\Omega) = \ln\frac{1}{H(j\Omega)} \tag{11.33}$$

so that

$$W(j\Omega) = A(\Omega^2) + j\phi(\Omega) \tag{11.34}$$

Let

$$
\begin{aligned}
A(\Omega^2) + j\phi(\Omega) &= A_1(\Omega^2) + j\phi_1(\Omega), \quad 0 \leqslant \Omega \leqslant \Omega_0 \\
&= A_2(\Omega^2) + j\phi_2(\Omega), \quad \Omega_0 \leqslant \Omega \leqslant \infty
\end{aligned}
\tag{11.35}
$$

with

$$\Omega_0 = \tan\left(\pi\frac{\omega_0}{\omega_N}\right) \tag{11.36}$$

where $\omega_0$ is the nominal passband edge. Also $A_1(\Omega^2)$ and $\phi_1(\Omega)$ are the loss and phase in the nominal passband while $A_2(\Omega^2)$ and $\phi_2(\Omega)$ are the corresponding functions outside the nominal passband.

Due to the quasi-minimum-phase nature of the transfer function, use of a modified Hilbert transform allows the real and imaginary parts of $W(j\Omega)$ to be related. However, the present problem is to construct a transfer function approximating a linear-phase in the passband together with high amplitude selectivity. Therefore, the Hilbert transform will be used to compute $A_1(\Omega^2)$ when $A_2(\Omega^2)$ and $\phi_1(\Omega)$ are specified. Thus if $\phi(\Omega)$ is specified in the passband as $\phi_1(\Omega)$ and $A(\Omega^2)$ is specified outside the passband as $A_2(\Omega^2)$, then we can calculate the passband amplitude $A_1(\Omega^2)$ by

$$
\frac{A_1(\Omega^2)}{\sqrt{(1 - \Omega^2/\Omega_0^2)}} = \frac{-2}{\pi} \int_0^{\Omega_0} \frac{x\phi_1(x)}{\sqrt{(1 - x^2/\Omega_0^2)}} \frac{dx}{(x^2 - \Omega^2)}
$$
$$
+ \frac{2}{\pi} \int_{\Omega_0}^{\infty} \frac{xA_2(x^2)}{\sqrt{(x^2/\Omega_0^2 - 1)}} \frac{dx}{(x^2 - \Omega^2)}, \qquad 0 \leqslant \Omega \leqslant \Omega_0 \tag{11.37}
$$

## 11.4.2  The Design Technique

The derivation of the transfer function $H(\lambda)$ satisfying the required properties (a) and (b) of Section 11.4.3, is accomplished in two stages.

### Stage 1
A prototype transfer function $\hat{H}(\lambda)$ (which may be unrealizable) is constructed such that it possesses an exact linear phase in the nominal passband. With

$$
\hat{H}(\lambda) = e^{-\hat{W}(\lambda)} \tag{11.38}
$$

$$
\hat{H}(j\Omega) = |\hat{H}(j\Omega)| \, e^{-j\hat{\phi}(\Omega)} \tag{11.39}
$$

we require

$$
\hat{\phi}_1(\Omega) = K \tan^{-1} \Omega \tag{11.40}
$$

where $K$ is a constant. The prototype loss function is defined as

$$
\hat{A}(\Omega^2) = \ln \frac{1}{|\hat{H}(j\Omega)|} \tag{11.41}
$$

and due to the prescribed form of $H(\lambda)$ in (11.29), the amplitude of the prototype is chosen to vary in the stopband according to the relation

$$
\hat{A}_2(\Omega^2) = \frac{1}{2} \ln \left( \frac{(1 + \Omega_0^2)^l (\Omega_0^2 - \Omega_x^2)^{2m} \hat{P}_n(\Omega^2)}{(1 + \Omega^2)^l (\Omega^2 - \Omega_x^2)^{2m}} \right), \qquad \Omega_0 \leqslant \Omega \leqslant \infty \tag{11.42}
$$

where $\Omega_x$ defines the location of the amplitude zeros of transmission. $\hat{P}_n(\Omega^2)$ is an even positive polynomial. For a specific combination $(n, m, l, \Omega_x, \Omega_0)$ the polynomial $\hat{P}_n(\Omega^2)$ is determined such that $\hat{A}_2(\Omega^2)$ fits any prescribed loss curve in the stopband. This can be done numerically using simple least squares fit routine. The degree of $\hat{P}_n(\Omega^2)$ is determined according to the closeness with which the stopband performance is to be approximated. The passband amplitude of the prototype $\hat{A}_1(\Omega^2)$ is determined from (11.37), (11.40), (11.42).

*Stage 2*

Having constructed the prototype function $\hat{H}(\lambda)$ we determine a realizable transfer function $\hat{H}(\lambda)$ satisfying the following conditions.

1. $H(\lambda)$ has the same prescribed numerator $\hat{H}(\lambda)$, since this defines the filter structure. Thus $H(\lambda)$ has the same degree as $\hat{H}(\lambda)$; the same number and locations of the zeros of transmission.

2. The absolute value of the realizable transfer function $H(\lambda)$ is a close approximation to $|\hat{H}(\lambda)|$. This is achieved by means of a fitting algorithm, with the coefficients of $|P_n(j\Omega)|^2$ as the unknowns. Thus we obtain $P_n(\lambda)P_n(-\lambda)$ by analytic continuation, and a Hurwitz factorization allows the construction of a stable transfer function.

Due to the Hilbert transform relation between the amplitude and phase of $H(\lambda)$, it follows that if the amplitude response $|H(j\Omega)|$ is a good approximation to $|\hat{H}(j\Omega)|$ of the prototype, then the phase response of $H(j\Omega)$ will also be a good approximation to the exact linear-phase of the prototype. We now examine the procedure in detail.

### 11.4.3 Construction of the Prototype Function

Let

$$g(\lambda) = \sqrt{1 + \frac{\lambda^2}{\Omega_0^2}} \tag{11.43}$$

so that

$$g(j\Omega) = jv = j\sqrt{\frac{\Omega^2}{\Omega_0^2} - 1} \quad \text{for } \Omega \geqslant \Omega_0 \tag{11.44}$$

and

$$g(j\Omega) = u = \sqrt{1 - \frac{\Omega^2}{\Omega_0^2}} \quad \text{for } \Omega \leqslant \Omega_0 \tag{11.45}$$

Also, let $f_n(\lambda)$ be the Hurwitz part of $\hat{P}_n(\lambda^2)$. Then, given that the prototype function satisfies (11.40) and (11.42) we determine its passband amplitude $\hat{A}_1(\Omega)$ as follows:

$$\hat{A}_1(\Omega^2) = -K\frac{u}{\Omega}\Omega_0\tan^{-1}\Omega + \ln[f_n(u)] - l\ln\left[1 + \frac{\Omega_0 u}{\sqrt{1 + \Omega_0^2}}\right]$$

$$- m\ln\left[1 + \frac{\Omega_0^2 u^2}{\Omega_x^2 - \Omega_0^2}\right], \quad 0 \leqslant \Omega \leqslant \Omega_0. \tag{11.46}$$

The constant $K$ may be chosen to produce a zero loss at $\Omega = 0$, i.e.,

$$K = \frac{\ln[f_n(1)] - l \ln\left[1 + \dfrac{\Omega_0}{\sqrt{1 + \Omega_0}}\right] - m \ln\left[1 + \dfrac{\Omega_0^2}{\Omega_x^2 - \Omega_0^2}\right]}{\Omega_0} \tag{11.47}$$

The polynomial $\hat{P}_n(\Omega^2)$ is chosen to be of the form

$$\hat{P}_n(\Omega^2) = \frac{1}{(1 + \Omega_0^2)^n}[(1 + \Omega^2)^i + (C/\Omega_0^{2i})(\Omega^2 - \Omega_0^2)^l]^k, \quad \Omega_0 \leqslant \Omega \leqslant \infty \tag{11.48}$$

where $n = ik$ and $C$ is a constant. The above expression may be put in the form

$$\hat{P}_n(v^2) = \frac{1}{(1 + \Omega_0^2)^n}[\Omega_0^{2i}(1 + v^2)^i + Cv^{2i}]^k, \quad 0 \leqslant v \leqslant \infty \tag{11.49}$$

For $\hat{P}_n(\Omega^2)$ to be the positive polynomial, we must have

$$C \geqslant -\Omega_0^{2n} \tag{11.50}$$

Now, the choice of $\hat{P}_n(\Omega^2)$ determines how closely an arbitrary stopband loss curve may be approximated. In this regard, choosing $k = 1$ and $i = n$ was found quite satisfactory, while larger values of $k > 1$ made no significant improvement. For this case we have from (11.42)

$$\hat{A}_2(\Omega^2) = \frac{1}{2}\ln\left[\frac{(\Omega_0^2 - \Omega_x^2)^{2m}((1 + \Omega^2)^n + (C/\Omega_0^{2n})(\Omega^2 - \Omega_0^2)^n)}{(1 + \Omega_0^2)^{n-l}(1 + \Omega^2)^l(\Omega^2 - \Omega_x^2)^{2m}}\right], \quad \Omega_0 \leqslant \Omega \leqslant \infty \tag{11.51}$$

For the important special case of no $j\omega$-axis zeros of transmission ($m = 0$), condition (11.50) is both necessary and sufficient for $\hat{A}_2(\Omega^2) \geqslant 0$ for all $\Omega$, and $\hat{A}_2(\Omega^2)$ is a monotonically increasing function in the range $\Omega_0 \leqslant \Omega \leqslant \infty$. However, for $m \geqslant 1$, condition (11.50) is only necessary for $\hat{A}_2(\Omega^2)$ to be non-negative. Rearranging (11.48) and (11.51) we have

$$\hat{P}_n(\Omega^2) = \frac{\exp[2\hat{A}_2(\Omega^2)](1 + \Omega^2)^l(\Omega^2 - \Omega_x^2)^{2m}}{(\Omega_0 - \Omega_x^2)^{2m}(1 + \Omega_0)^l} \tag{11.52}$$

and

$$C = \frac{\Omega_0^{2n}[\hat{P}_n(\Omega^2)(1 + \Omega_0^2)^n - (1 + \Omega^2)^n]}{(\Omega^2 - \Omega_0^2)^n} \tag{11.53}$$

Thus for specific stopband loss function $\hat{A}_2(\Omega^2)$ and a combination of $(\Omega_0, \Omega_x, m, n, l)$, $C$ is calculated using (11.52) and (11.53). It must be observed that the particular choice of $\hat{P}_n(\Omega)$ given here is not unique. However, this form is simple, well behaved, and quite adequate for the class of filters under consideration.

The construction of the prototype function according to the above procedure leads to an exact linear-phase in the passband and a typical amplitude of the form shown in Figure 11.4, where the symbols employed are self explanatory. The phase response is not shown since it is exactly linear for $\Omega \leqslant \Omega_0$. It is convenient to define the linear-phase bandwidth $\beta$ as

$$\beta = \frac{\tan^{-1} \Omega_0}{\tan^{-1} \Omega_p} \tag{11.54}$$

which obviously satisfies

$$0 \leqslant \beta \leqslant 1 \tag{11.55}$$

Now, the prototype function $\hat{H}(\lambda)$ is defined by $(\Omega_x, C, \Omega_0, n, l, m)$. $\Omega_0$ can be used as a bandwidth scaling parameter, while $\Omega_x$ and $C$ are used to shape the stopband and passband, respectively. Consider the typical low-pass amplitude specifications shown in Figure 11.5. Given $\alpha_p$ and $\Omega_p$, then for any combination $(\Omega_0, \Omega_x, n, l, m)$ expressions (11.51) and (11.52) are used to calculate $C$. Fixing $(n, l, m)$ leaves $(\Omega_0, \Omega_x)$ as free parameters to shape the amplitude response of the prototype. It has been found that if $C$ is computed as described, then the passband loss increases with $\Omega_0$. This allows the use of a simple search routine for finding a suitable value of $\Omega_0$. Also, the stopband loss increases with $\Omega_x$. Combining these two results, the prototype can be obtained with the exact required values of passband loss and stopband loss. The prototype construction can be conveniently divided into two stages.

(a) Since $C$ is completely determined by (11.52) and (11.53), then $\Omega_0$ is treated as a free parameter in the process of shaping the passband.

(b) Using $\Omega_x$ as a free parameter, the stopband requirement is met.

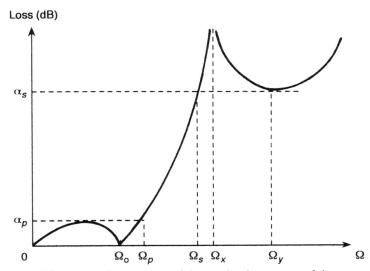

**Figure 11.4**  General features and parameters of the amplitude response of the prototype function.
© 1990 IEEE.

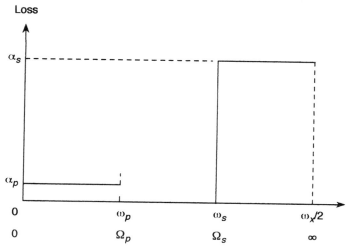

**Figure 11.5** Typical amplitude specifications. © 1990 IEEE.

An algorithm for implementing both stages is easily implemented using a computer program as follows:

(a) To design the passband

1. $[n, m, l, \alpha_p, \Omega_p, \Omega_x]$ are considered the input parameters.

2. $C$ is calculated and if $C < -\Omega_0^{2n}$, then $\Omega_0$ is increased, which results in an increase in $C$, in accordance with (11.52) and (11.53) until a value is found which satisfies (11.50).

3. For certain inputs it may not be possible to find suitable values of $C$ and $\Omega_0$. The result is a continuous loop which is broken by altering $\Omega_x$.

(b) To design the stopband

1. $[n, l, m, \alpha_p, \Omega_p, \alpha_s, \Omega_s, \Omega_{xmax}]$ are the inputs, where $\Omega_{xmax}$ is the maximum value of $\Omega_x$ to be considered and is selected at the outset.

2. For sharp cutoff, $\Omega_x$ is to be minimized. This also results in a smaller value of $\Omega_p$, which in turn maximizes the linear-phase bandwidth $\beta$. Thus $\Omega_x$ is taken as small as possible without violating the specifications.

3. Stage (a) is internal to Stage (b); therefore, the case where Stage (a) enters an infinite loop must be taken into consideration here. This requires a compromise choice of $\Omega_x$ since a very small value causes an infinite loop. This value is easily obtained by several iterations of the algorithm.

We now examine the general issues involved in applying the above procedure and give some typical results.

Let the number of transmission zeros at half the sampling frequency be $q$, then

$$q = n - l - 2m \tag{11.56}$$

Also define the selectivity factor as

$$\zeta = \frac{\alpha_s}{\alpha_p} \tag{11.57}$$

The linear-phase bandwidth is given by

$$\beta = \frac{\omega_0}{\omega_p} \tag{11.58}$$

which is the fraction of the passband over which the phase is linear.

*Example 11.2*   Consider a prototype function of the form

$$\hat{H}(\lambda) = \frac{(1 - \lambda^2)^{n/2}}{\hat{P}_n(\lambda)} \tag{11.59}$$

with the choice of $\hat{P}_n(\lambda)$ in (11.49) with $k = 1$ and $q = 0$. The objective is to meet the following specifications:

1. Passband $0 \leqslant (\omega/\omega_N) \leqslant 0.05$, $\alpha_p \leqslant 1.0$ dB;

2. Stopband $0.1 \leqslant (\omega/\omega_N) \leqslant 0.5$, $\alpha_s \geqslant 38.0$ dB;

3. The phase is linear over 70% of the passband.

Use of the algorithm results in the construction of Table 11.1 containing the necessary data. This shows that the function with $n = 5$ meets the specifications. The prototype responses are shown in Figure 11.6.

**Table 11.1.**   Performance data for a prototype of the form (11.59) with $(\omega_p/\omega_N) = 0.05$, $(\omega_s/\omega_N) = 0.1$, $\alpha_p = 1$ dB. © 1990 IEEE.

| $n$ | $C$ | $K$ | $\beta$ | $\zeta$ |
|-----|---------|-------|--------|-------|
| 4 | 3.1330 | 17.08 | 0.8057 | 32.90 |
| 5 | 1.8540 | 20.06 | 0.7715 | 39.83 |
| 6 | 0.9257 | 24.14 | 0.7422 | 46.60 |
| 7 | 0.4037 | 27.58 | 0.7108 | 53.20 |
| 8 | 0.1629 | 31.03 | 0.6053 | 59.91 |
| 9 | 0.0584 | 34.43 | 0.6758 | 66.45 |
| 10 | 0.01915 | 37.81 | 0.6582 | 72.94 |
| 11 | 0.00590 | 41.18 | 0.6426 | 79.41 |
| 12 | 0.00176 | 44.56 | 0.6289 | 85.89 |
| 13 | 0.000457 | 47.88 | 0.6152 | 92.24 |

*Example 11.3*    Consider a transfer function with a pair of $j\Omega$-axis zeros of transmission whose position is optimized to give a maximum linear phase bandwidth $\beta$. The function is of the form

$$\hat{H}(\lambda) = \frac{(1 - \lambda^2)^{(n-2)/2}(\lambda^2 + \Omega_x^2)}{\hat{P}_n(\lambda)} \tag{11.60}$$

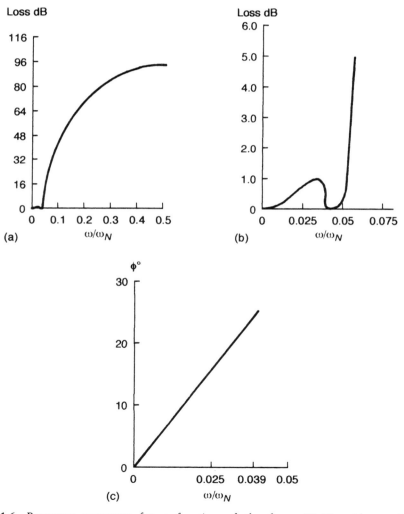

**Figure 11.6**  Prototype responses for a function of the form (11.59) with $n = 5$ and the corresponding parameters in Table 11.1: (a) amplitude response; (b) passband amplitude detail; (c) nominal passband phase. © 1990 IEEE.

Table 11.2. Performance data for a prototype of the form (11.60).

| $n$ | $C$ | $K$ | $\omega_x/\omega_N$ | $\beta$ | $\omega_s/\omega_N$ |
|---|---|---|---|---|---|
| 4 | $7.454 \times 10^{-1}$ | 14.61 | 0.1055 | 0.8184 | 0.09724 |
| 5 | $2.19 \times 10^{-1}$ | 15.50 | 0.09693 | 0.8242 | 0.09145 |
| 6 | $5.91 \times 10^{-2}$ | 16.37 | 0.09424 | 0.8345 | 0.09003 |
| 7 | $1.471 \times 10^{-2}$ | 17.31 | 0.09288 | 0.8413 | 0.08945 |
| 8 | $3.482 \times 10^{-3}$ | 18.34 | 0.9220 | 0.8447 | 0.08929 |
| 9 | $7.911 \times 10^{-4}$ | 19.46 | 0.09207 | 0.8454 | 0.08952 |
| 10 | $1.884 \times 10^{-1}$ | 20.68 | 0.00193 | 0.8461 | 0.08966 |

with the objective of meeting specifications:

1. passband $0 \leqslant (\omega/\omega_N) \leqslant 0.05$, $\alpha_p \leqslant 1.0 \, \text{dB}$;

2. Stopband $0.1 \leqslant (\omega/\omega_N) \leqslant 0.5$, , $\alpha_s \geqslant 40.0 \, \text{dB}$;

3. The phase is linear over 84% of the passband.

Use of the algorithm leads to the construction of Table 11.2 showing that the case with $n = 7$ and $(\omega_x/\omega_N) = 0.092\,88$ meets the specifications.

### 11.4.4 Derivation of the Realizable Transfer Function

The final step in the design is to determine a realizable transfer function whose amplitude approximates that of the prototype arbitrarily closely. Since both functions have the same numerator, the problem reduces to fitting the denominator polynomial $P_n(\Omega^2)$ of $|H(j\Omega)|^2$ to $\hat{P}_n(\Omega^2)$ of $|\hat{H}(j\Omega)|^2$. This can be achieved by using a standard least squares routine, although more elaborate techniques can also be used, however, this was not found to be necessary. Once the polynomial $P_n(\Omega^2)$ has been found, analytic continuation is applied to obtain $P_n(\lambda)P_n(-\lambda)$ and the Hurwitz part is chosen. Naturally, the higher the degree of $P_n(\Omega^2)$, the closer the fit to the prototype characteristics. The problem is formulated as a set of equations with the coefficients of $P_n(\Omega^2)$, as the unknowns. In this regard, since $\hat{P}_n(\Omega^2)$ of the prototype is chosen to be well-behaved polynomial with no zeros in the passband region, it follows that the closer the fit by $P_n(\Omega^2)$ to $\hat{P}_n(\Omega^2)$, the more favourable the result to a Hurwitz factorization of $P_n(\lambda)P_n(-\lambda)$.

### 11.4.5 Design Examples

We now give three examples to illustrate the design technique.

*Example 11.4* Consider the design of a strict LDI ladder filter without finite $j\Omega$-axis zeros of transmission, i.e., its transfer function is of the form (11.15) with the following specifications:

1. passband $0 \leqslant (\omega/\omega_N) \leqslant 0.05, \alpha_p \leqslant 1.0\,\text{dB}$;

2. Stopband $0.1 \leqslant (\omega/\omega_N) \leqslant 0.5, \alpha_s \geqslant 38.0\,\text{dB}$;

3. The phase is linear over 75% of the passband.

To obtain the required filter, we must first construct a prototype which meets the specifications. Using the algorithm, the case with $n = 5$ has $\alpha_p < 1\,\text{dB}$ for $\omega/\omega_N \leqslant 0.05, \alpha_s \geqslant 39.93\,\text{dB}$ for $\omega/\omega_N \geqslant 0.1$, and the phase is linear over 77.15% of the passband. Therefore, this prototype meets the specifications. We next obtain the following realizable transfer function as discussed in the previous section.

$$H(\lambda) = \frac{(1-\lambda^2)^{5/2}}{1 + 26.152\lambda + 294.101\lambda^2 + 2441.87\lambda^3 + 10018.2\lambda^4 + 44783.6\lambda^5} \quad (11.61)$$

which can be realized, using the techniques of Chapter 9 as implemented in **Program LDISCG** of Appendix A in ISICAP. This gives the network of Figure 11.1 with

$$\frac{C_1}{C_s} = \frac{C_2}{C_s} = \frac{C_4}{C_5} = \frac{C_7}{C_8} = \frac{C_{10}}{C_{11}} = \frac{C_l}{C_{13}} = 1.0$$

$$\frac{C_3}{C_1} = 6.946, \quad \frac{C_6}{C_4} = 3.55, \quad \frac{C_9}{C_7} = 7.158, \quad \frac{C_{12}}{C_{10}} = 4.73, \quad \frac{C_{14}}{C_{13}} = 2.76$$

The dashed-line capacitors in Figure 11.1 are not needed in this case, so they are removed from the circuit. Figure 11.7 shows the amplitude and delay responses of the designed filters.

*Example 11.5*  Consider the introduction of a pair of finite $j\omega$-axis transmission zeros to improve the amplitude response and at the same time extend the linear-phase bandwidth beyond that of *Example 4*. Thus, assume a transfer function of the form

$$H(\lambda) = \frac{(1-\lambda^2)^{(n-2)/2}(\lambda^2 + \Omega_x^2)}{P_n(\lambda)} \quad (11.62)$$

obtained to meet the following specifications:

1. Passband $0 \leqslant (\omega/\omega_N) \leqslant 0.05, \alpha_p \leqslant 1.0\,\text{dB}$;

2. Stopband $0.1 \leqslant (\omega/\omega_N) \leqslant 0.5, \alpha_s \geqslant 40.0\,\text{dB}$;

3. The phase is linear over 82% of the passband.

Again, we must first design the prototype with the above specifications. The design algorithm yields the required results with

$$n = 5, \quad C = 0.2192, \quad K = 15.5, \quad \Omega_x = 0.314\,349 \quad (11.63)$$

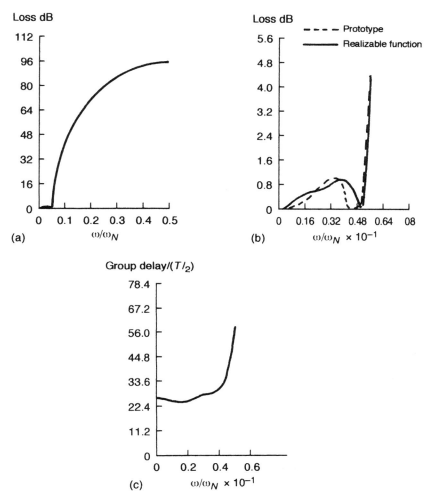

**Figure 11.7** Responses of the filter of Example 11.4: (a) amplitudes; (b) passband amplitude detail; and (c) passband delay. © 1990 IEEE.

corresponding to $(\omega_x/\omega_N) = (\tan^{-1}(0.314\,349)/\pi) = 0.096\,93$. The realizable transfer function is then given by:

$$H(\lambda) = \frac{(1/\Omega_x)^2(1 - \lambda^2)^{3/2}(\lambda^2 + \Omega_x^2)}{1 + 18.773\,8\lambda + 159.662\lambda^2 + 1010.56\lambda^3 + 3428.45\lambda^4 + 10201.4\lambda^5} \qquad (11.64)$$

The realization of the transfer function is accomplished using the techniques of Chapters 9 and 10, which gives the network of Figure 11.1 with capacitor ratios

$$\frac{C_1}{C_s} = \frac{C_2}{C_s} = \frac{C_4}{C_5} = \frac{C_7}{C_8} = \frac{C_{10}}{C_{11}} = \frac{C_l}{C_{13}} = 1.0$$

$$\frac{C_3}{C_1} = 6.4202, \quad \frac{C_6}{C_4} = 3.496, \quad \frac{C_9}{C_8} = 6.25, \quad \frac{C_9}{C_{17}} = 5.1872$$

$$\frac{C_{12}}{C_{10}} = 2.305, \quad \frac{C_{14}}{C_{13}} = 1.707, \quad \frac{C_l}{C_{18}} = 0.823$$

The capacitor values which are not given above are deleted from the circuit. The amplitude and delay responses of the realized filter are shown in Figure 11.8.

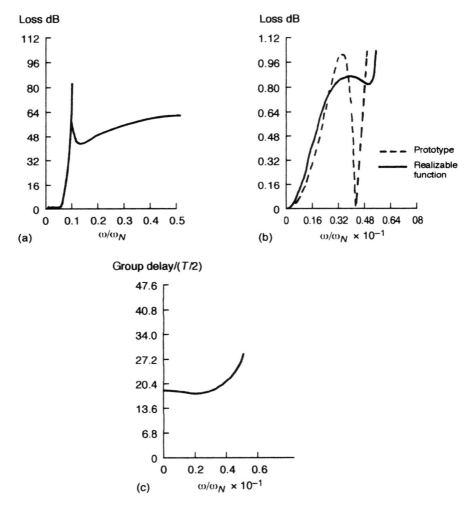

Figure 11.8  Responses for the filter in Example 11.5: (a) amplitude response; (b) passband amplitude detail; (c) passband delay. © 1990 IEEE.

*Example 11.6*   This is given to illustrate the high-pass design with the following specifications:

1. Passband $0.45 \leqslant (\omega/\omega_N) \leqslant 0.5, \alpha_p \leqslant 1.0 \, \mathrm{dB}$;

2. Stopband $0 \leqslant (\omega/\omega_N) \leqslant 0.4, \alpha_s \geqslant 30.0 \, \mathrm{dB}$;

3. The phase is linear over 80% of the passband.

First these high-pass specifications are transformed to low-pass specifications. A low-pass function is then designed with the transformed specifications. This gives

$$H(\lambda) = \frac{(1 - \lambda^2)^{3/2}}{1 + 19.7393\lambda + 186.943\lambda^2 + 930.264\lambda^3 + 4255.46\lambda^4} \tag{11.65}$$

Then we apply the transformation

$$\lambda \to \frac{1}{\lambda}$$

to give the high-pass function

$$\bar{H}(\lambda) = \frac{\lambda(1 - \lambda^2)^{3/2}}{4255.46 + 930.264\lambda + 186.943\lambda^2 + 19.7393\lambda^4} \tag{11.67}$$

which is realized as discussed in Chapters 9 and 10. The amplitude and delay responses of the designed filter are shown in Figure 11.9.

   We finally note that the design technique gives an optimum compromise between phase linearity and amplitude selectivity. This is because the algorithm satisfies the required selectivity, then proceeds by maximizing the portion of the passband over which the phase may be kept linear. This portion is subject to an upper bound which depends on the amplitude selectivity, and a trade-off between the two aspects of the response must be considered. Once the amplitude selectivity is specified, there is an upper bound on the fraction of the passband over which the phase may be kept linear.

## 11.5   OPTIMAL BAND-PASS LADDER DESIGN

This section presents design techniques of switched-capacitor band-pass filters of the lossless-discrete-integrator (LDI) type which exhibit, simultaneously, good amplitude selectivity as well as good approximation to a constant group delay in the passband in an optimum manner [48]. All the available degrees of freedom are exploited to shape both the amplitude and delay responses, by contrast with the available methods of delay equalization which yield sub-optimal designs. As noted earlier, due to the factor $(1 - \lambda^2)$, which multiplies the delay functions, a standard frequency transformation from low-pass to band-pass will not transform a linear phase low-pass filter into a linear phase band-pass characteristic. This means that the design of band-pass filters with prescribed phase response, or on both amplitude and phase bases, should be dealt with independently. The technique presented here is a generalization of the

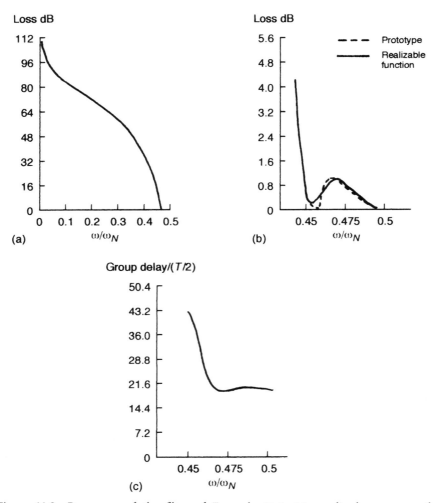

**Figure 11.9**  Responses of the filter of Example 11.6: (a) amplitude response; (b) passband amplitude detail; (c) passband delay. © 1990 IEEE.

low-pass case discussed in the previous section. However, the band-pass problem is a *three-band* approximation problem in the amplitude domain, and when combined with the requirement of phase linearity in the passband, it becomes much more difficult than the low-pass case. Hence, it requires special treatment and examination of the design parameters.

### 11.5.1   Band-pass Transfer Functions

Now consider the use of the lossless discrete integrator (LDI) and the composite building blocks shown in Figures 9.10, 9.11 and 9.14 to construct a leapfrog ladder structure as in Figure 9.9. The transfer functions of the LDI building blocks are given by (9.45)–(9.47). Let the number of Type A building blocks be $k$ and that of Type B be $m$. Both types are used in the leapfrog configuration shown in Figure 9.9 with the conditions that: positive and negative blocks alternate, and the first and last blocks are of the type with subscript B. Then the equivalent network of this SC ladder structure will be in the form shown in Figure 9.16. It has been shown in Chapter 9 that the form of the transfer functions of this equivalent network is

$$H_{21}(\lambda) = \frac{I_L}{V_{in}} = \frac{K\lambda^m(1-\lambda^2)^{\frac{m+k}{2}}}{P_{2m+k}(\lambda)} \qquad \text{for } m \text{ odd} \qquad (11.68)$$

or

$$H_{21}(\lambda) = \frac{V_{out}}{V_{in}} = \frac{K\lambda^m(1-\lambda^2)^{\frac{m+k}{2}}}{P_{2m+k}(\lambda)} \qquad \text{for } m \text{ even} \qquad (11.69)$$

Now our task is to design a prototype transfer function $H_{21}(\lambda)$ which possesses the desired stopband attenuation and passband phase linearity (or equivalently constant passband group delay). But before doing this, a modified Hilbert transform is introduced in the next section. We follow the same lines of treatment as the case of the low-pass filters.

### 11.5.2   The Hilbert Transform

Writing the transfer function $H_{21}(\lambda)$ as

$$H_{21}(\lambda) = e^{-W(\lambda)} \qquad (11.70)$$

then the logarithmic measure of $H_{21}(\lambda)$ is

$$W(\lambda) = \ln\frac{1}{H_{21}(\lambda)} \qquad (11.71)$$

For $\lambda = j\Omega$, (11.71) can be written as

$$W(j\Omega) = \ln\frac{1}{H_{21}(j\Omega)} = A(\Omega^2) + j\phi(\Omega) \qquad (11.72)$$

where the real part $A(\Omega^2)$ and the imaginary part $\phi(\Omega)$ may be called the transfer loss and phase respectively.

Now let $\Omega_{01}$ and $\Omega_{02}$ be the lower cutoff frequency and the upper cutoff frequency respectively and define

$$A(\Omega^2) + j\phi(\Omega) = \begin{cases} A_1(\Omega^2) + j\phi_1(\Omega) & 0 \leqslant \Omega \leqslant \Omega_{01} \\ A_2(\Omega^2) + j\phi_2(\Omega) & \Omega_{01} \leqslant \Omega \leqslant \Omega_{02} \\ A_3(\Omega^2) + j\phi_3(\Omega) & \Omega_{02} \leqslant \Omega \leqslant \infty \end{cases} \qquad (11.73)$$

$A_1(\Omega^2)$, and $\phi_1(\Omega)$ are the loss and phase in the lower stopband;
$A_2(\Omega^2)$, and $\phi_2(\Omega)$ are the loss and phase in the passband and
$A_3(\Omega^2)$ and $\phi_3(\Omega)$ are the loss and phase in the upper stopband

Due to the quasi-minimum-phase nature of the transfer function (11.68) and (11.69), one cannot arbitrarily specify both the transfer loss and phase in the same band, because they are related. Since we require a linear phase response in the passband, only the stopband transfer loss can also be specified. Then the following modified Hilbert transform is a strong non-trivial generalization of the results of the previous section.

$$\frac{2}{\pi\Omega}\left(\frac{\Omega^2}{\Omega_{01}^2 - 1}\right)^{\frac{1}{2}}\left(1 - \frac{\Omega^2}{\Omega_{02}^2}\right)^{\frac{1}{2}}(U + V + W) = \begin{cases} \phi_1(\Omega) & \Omega < \Omega_{01} \\ -A_2(\Omega^2) & \Omega_{01} < \Omega < \Omega_{02} \\ -\phi_3(\Omega) & \Omega > \Omega_{02} \end{cases} \quad (11.74)$$

where

$$U = \int_0^{\Omega_{01}} \frac{x^2 A_1(x^2)}{\sqrt{1 - x^2/\Omega_{01}^2}\sqrt{1 - x^2/\Omega_{02}^2}} \frac{dx}{x^2 - \Omega^2}$$

$$V = \int_{\Omega_{01}}^{\Omega_{02}} \frac{x^2 \phi_2(x)}{\sqrt{x^2/\Omega_{01}^2 - 1}\sqrt{1 - x^2/\Omega_{02}^2}} \frac{dx}{x^2 - \Omega^2}$$

$$W = \int_{\Omega_{02}}^{\infty} \frac{x^2 A_3(x^2)}{\sqrt{x^2/\Omega_{01}^2} - \sqrt{x^2/\Omega_{02}^2 - 1}} \frac{dx}{x^2 - \Omega^2}$$

### 11.5.3   Derivation of the Transfer Function

#### The prototype transfer function

The design begins by constructing a prototype transfer function $\hat{H}(\lambda)$ (which may be unrealizable) such that it possesses an exact linear phase in the nominal passband and the required amplitude selectivity in the stopbands. Therefore, with

$$\hat{H}(\lambda) = e^{-\hat{W}(\lambda)} \quad (11.75)$$

$$\hat{H}(j\Omega) = |(j\Omega)|\, e^{-j\hat{\phi}(\Omega)}$$

we require

$$\hat{\phi}_2(\Omega) = C\tan^{-1}\Omega \quad \Omega_{01} \leqslant \Omega \leqslant \Omega_{02} \quad (11.76)$$

where $C$ is a constant.

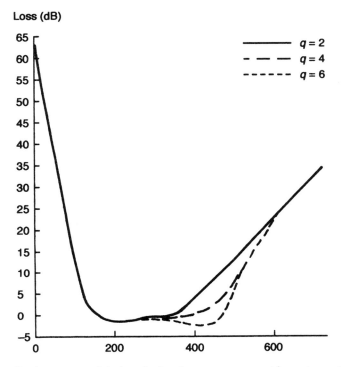

**Figure 11.10** Amplitude response of eigth-order band-pass prototype with varying $q$. © 1990 Kluwer.

The prototype loss function is defined as

$$\hat{A}(\Omega^2) = \ln \frac{1}{|\hat{H}(j\Omega)|} \tag{11.77}$$

Considering the general form of the bandpass LDI filters given in (11.68) and (11.69) we choose the amplitude squared of the prototype to vary in the stopbands according to the following relations

$$\hat{A}_1(\Omega^2) = \frac{1}{2}\ln \frac{\Omega_{01}^{2m}(1+\Omega_{01}^2)^l P_{1N}(\Omega^2)}{\Omega^{2m}(1+\Omega^2)^l} \qquad 0 \leqslant \Omega \leqslant \Omega_{01} \tag{11.78}$$

$$\hat{A}_3(\Omega^2) = \frac{1}{2}\ln \frac{\Omega_{02}^{2m}(1+\Omega_{02}^2)^l P_{2N}(\Omega^2)}{\Omega^{2m}(1+\Omega^2)^l} \qquad \Omega_{02} \leqslant \Omega \leqslant \infty \tag{11.79}$$

where

$$P_{1N}(\Omega^2) = \left(1 + C_1^2\left(1 - \frac{\Omega^2}{\Omega_{01}^2}\right)^N\right) \qquad 0 \leqslant \Omega \leqslant \Omega_{01} \tag{11.80}$$

$$P_{2N}(\Omega^2) = \left(\left(\frac{\Omega}{\Omega_{02}}\right)^{2q} + C_2^2\left(\frac{\Omega^2}{\Omega_{02}^2} - 1\right)^N\right) \qquad \Omega_{02} \leqslant \Omega \leqslant \infty \tag{11.81}$$

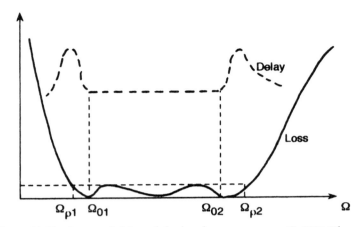

**Figure 11.11**    Loss and delay of the bandpass prototype. © 1993 Kluwer.

Here, although we guarantee that $\hat{A}_1(\Omega_{01}^2) = 0$ (dB) and $\hat{A}_3(\Omega_{02}^2) = 0$ (dB) with $\hat{A}_1(\Omega^2)$ and $\hat{A}_3(\Omega^2)$ being the form described above, for small value of $q$, $\hat{A}_3(\Omega^2)$ will become negative (in dB) for some $\Omega$ near $\Omega_{02}$ because of the behaviour of the term $(\Omega_{02}/\Omega)^{2m}$. To compensate for this the value of $q$ should be properly chosen. Figure 11.10 shows the loss function of a prototype with $N = 8$, $m = 2$, $l = 6$ and different $q$'s.

Then, the parameters $m$, $l$, $q$, $n$, $C_1$ and $C_2$ can be used to adjust the selectivity of the prototype attenuation to meet the specifications.

With $\hat{A}_1(\Omega^2)$, $\hat{A}_3(\Omega^2)$, and $\phi_2(\Omega)$ chosen, the passband attenuation can be computed according to (11.74). However, since the integrals appearing in (11.74) are basically elliptic, numerical integration methods have to be used.

The construction of the prototype function according to the above procedure leads to an exact linear-phase in a significant portion of the passband and a typical amplitude of the form shown in Figure 11.11 Here $\Omega_{p1}$ and $\Omega_{p2}$ are the actual lower passband edge and upper passband edge respectively. From Figure 11.11 we can define the linear-phase bandwidth $\beta$ as

$$\beta = \frac{\tan^{-1}\Omega_{02} - \tan^{-1}\Omega_{01}}{\tan^{-1}\Omega_{p2} - \tan^{-1}\Omega_{p1}} \tag{11.82}$$

which obviously satisfies

$$0 \leqslant \beta \leqslant 1$$

Now, the prototype function $\hat{H}(\lambda)$ is defined by $(\Omega_{01}, \Omega_{02}, C_1, C_2, n, m, l, q)$, in which $(\Omega_{01}, \Omega_{02})$ can be used as bandwidth scaling parameters, while $(C_1, C_2)$ are used to shape the stopbands and the passband respectively.

### The realizable transfer function

Having constructed the prototype transfer function we must determine the realizable transfer function. The realizable transfer function should be consistent with the prototype, i.e., they both must have the same degree $N$, and the same numerator.
Consider the function

$$H_{21}(\lambda)H_{21}(-\lambda) = \frac{K^2\lambda^{2m}(1-\lambda^2)^l}{P_N(\lambda^2)} \tag{11.83}$$

where

$$H_{21}(\lambda) = \frac{K\lambda^m(1-\lambda^2)^l}{D_N(\lambda)} \tag{11.84}$$

and $D_N(\lambda)$ is a strictly Hurwitz polynomial in $\lambda$.
Then obviously we have

$$P_N(\lambda^2) = D_N(\lambda)D_N(-\lambda) \tag{11.85}$$

where $D_N(-\lambda)$ is anti-Hurwitz and has all its roots in the right half-plane.
The amplitude squared of the realizable transfer function is, then, fitted in some way (least squares) to the amplitude squared of the prototype transfer function over the bands of interest with the coefficients of the polynomial $P_N(\lambda^2)$ as the unknown. This is precisely the same procedure as in the low-pass case of the previous section. The synthesis technique of Chapter 9 is used to obtain the switched-capacitor circuit.

## 11.5.4 Design Examples

*Example 11.7* Consider the design of a sixth-order SC-bandpass filter with the following specifications:

Passband ripple $<0.5$ dB. Linear phase over 65% of passband.
Passband width from $f_{p1}/f_N = 0.047$ to $f_{p2}/f_N = 0.104$.
Lower stopband attenuation $\leqslant 20$ dB, for $f/f_N \leqslant 0.025$.
Upper stopband attenuation $\geqslant 30$ dB, for $f/f_N \geqslant 0.2$.

The prototype function with $C = 7.5$, $C_1 = 5$, $C_2 = 5$, $\Omega_{01} = 0.158$, $\Omega_{02} = 0.325$, $n = 6$, $m = 2$, $1 = 4$ and $q = 4$ will meet the specifications.
Using least squares fitting, the realizable transfer function is obtained as

$$H_{21}(\lambda) = \frac{K\lambda^2(1-\lambda^2)^{4/2}}{0.1178 + 0.6173\lambda + 8.709\lambda^2 + 24.66\lambda3 + 137.28\lambda^4 + 160.83\lambda^5 + 511.53\lambda^6}$$

with $K = 1$.

The equivalent network of $H_{21}$ is given in Figure 11.12. Then, using the synthesis algorithm of Chapter 9, the input impedance of the equivalent network of Figure 11.12 is obtained

$$Z_{in} = \frac{\dfrac{1}{\mu}(2.537\gamma^6 + 5.698 \times 10^{-1}\gamma^4 + 3.312 \times 10^{-2}\gamma^2 + 4.605 \times 10^{-4})}{\dfrac{1}{\mu}(1.497 \times 10^{-1}\gamma^5 + 2.368 \times 10^{-2}\gamma^3 + 5.811 \times 10^{-4}\gamma)}$$

$$\frac{+(5.779 \times 10^{-1}\gamma^5 + 7.758 \times 10^{-2}\gamma^3 + 1.832 \times 10^{-3}\gamma)}{+(3.41 \times 10^{-2}\gamma^4 + 2.313\,10 \times 10^{-3}\gamma^2)}$$

The first three steps of the continued fraction expansion of $Z_{in}$ lead to the element values of

$$L_1 = 16.95$$

$$C_2 = 0.8884$$

$$L_3 = 56.055$$

The remainder impedance is given by

$$Z'_{in} = \frac{\dfrac{1}{\mu}(1.362 \times 10^{-2}\gamma^2 + 4.605 \times 10^{-4}) + 1.832 \times 10^{-3}\gamma}{\dfrac{1}{\mu}(3.006 \times 10^{-2}\gamma^3 + 1.721 \times 10^{-4}\gamma) + 6.848 \times 10^4\gamma^2}$$

Extracting the pole at $\gamma = 0$, we have

$$C_3 = 0.3737$$

The remainder admittance is then

$$Y''_{in} = 0.5837\gamma + \frac{1}{32.43\gamma} + 0.1227\mu$$

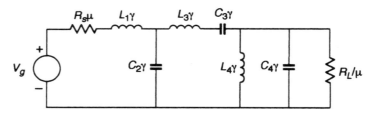

**Figure 11.12**  Equivalent network of the sixth-order band-pass filter of Example 11.7. © 1993 Kluwer.

from which we obtain

$$C_4 = 0.5387$$

$$L_4 = 32.43$$

$$R_l = 8.149.$$

Figure 11.13 shows the SC filter with the capacitor ratios as

$$\frac{C_{c1}}{C_{a1}} = R_s = 1$$

$$\frac{2C_{b1}}{C_{a1}} + \frac{C_{c1}}{C_{a1}} = L_1 = 16.95$$

$$\frac{2C_{b2}}{C_{a2}} = C_2 = 0.8884$$

$$\frac{2C_{b3}}{C_{a3}} = L_3 = 56.05$$

$$\frac{2C_{d3}}{C_{a3}} = C_3 = 0.3737$$

$$\frac{2C_{b4} + C_{c4}}{C_{a4}} = C_4 = 0.5387$$

$$\frac{2C_{d4}}{C_{a4}} = L_4 = 32.43$$

$$\frac{C_{c4}}{C_{a4}} = \frac{1}{R_l} = 0.1227$$

Figure 11.14 shows the amplitude and the delay responses of the realized SC-bandpass filter with the sampling frequency $f_N = 10\,\text{kHz}$. The actual passband is between 45 and 105 Hz with a 0.5 dB passband ripple and the linear phase bandwidth is between 61 and 101 with the delay $-9.75 \pm 0.63$ ms. So

$$\beta = \frac{101 - 62}{105 - 45} = 65\%$$

*Example 11.8*   Consider the design an eighth-order band-pass filter with the specifications as follows:

Passband ripple $\leqslant 1\,\text{dB}$. Linear phase over 80% of passband.

Passband width from $f_{p1}/f_N = 0.047$ to $f_{p2}/f_N = 0.109$

Lower stopband attenuation $\geqslant 20\,\text{dB}$, for $f/f_N \leqslant 0.025$

Upper stopband attenuation $\geqslant 37\,\text{dB}$, for $f/f_N \geqslant 0.2$.

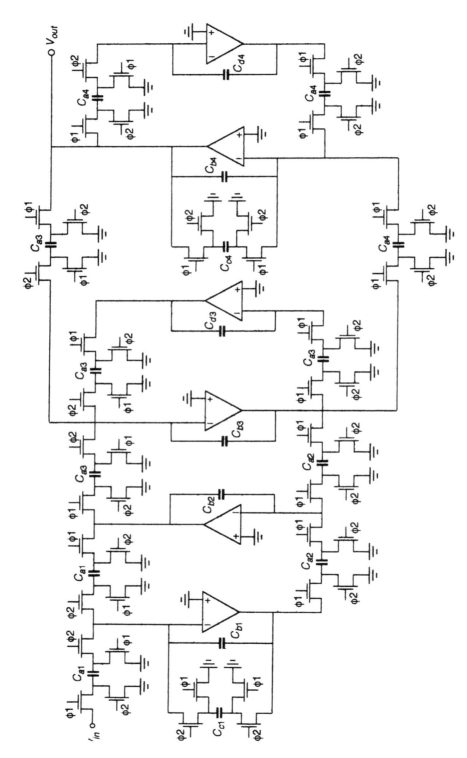

Figure 11.13    The filter of Example 11.7. © 1986 Kluwer.

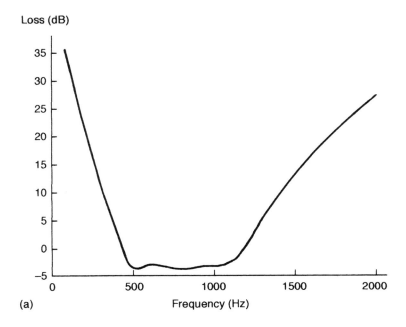

(a)

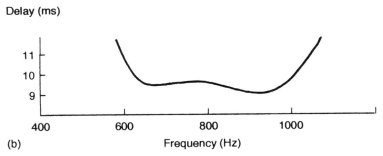

(b)

**Figure 11.14**  Responses of the filter of Example 11.7 with clock frequency of 10 kHz: (a) amplitude, (b) delay. © 1993 Kluwer.

The prototype function with $C = 9.6$, $C_1 = 10$, $C_2 = 10$, $\Omega_{01} = 0.158$, $\Omega_{02} = 0.325$, $n = 8$, $m = 2\ 1 = 6$ and $q = 4$ will meet the specifications.

The realizable transfer function is obtained by least squares fitting as

$$H_{21}(\lambda) = \frac{K\lambda^2(1 - \lambda^2)^{6/2}}{\begin{array}{c} 0.1099 + 0.7452\lambda + 10.947\lambda^2 + 44.246\lambda^3 + 265.17\lambda^4 \\ + 578.53\lambda^5 + 1952.96\lambda^6 + 1910.75\lambda^4 + 4046.55\lambda^8 \end{array}}$$

with $K = 0.951\,26$.

Use of the synthesis algorithm of Chapter 9 results in the input impedance of the equivalent network

$$Z_{in} = \frac{\dfrac{1}{\mu}\displaystyle\sum_{r=0}^{4} a_{2r}\gamma^{2r} + \displaystyle\sum_{r=0}^{4} b_{2r-1}\gamma^{2r-1}}{\dfrac{1}{\mu}\displaystyle\sum_{r=1}^{4} c_{2r-1}\gamma^{2r-1} + \displaystyle\sum_{r=0}^{3} d_{2r}\gamma^{2r}}$$

with

$$a_8 = 3.01892 \qquad\qquad b_7 = 1.00322$$

$$a_6 = 1.145\,62 \qquad\qquad b_5 = 2.528\,19 \times 10^{-1}$$

$$a_4 = 1.317\,76 \times 10^{-1} \quad b_3 = 1.6913 \times 10^{-2}$$

$$a_2 = 5.129\,66 \times 10^{-3} \quad b_1 = 2.669\,92 \times 10^{-4}$$

$$a_0 = 5.429\,87 \times 10^{-5}$$

$$c_7 = 2.493\,36 \times 10^{-1} \quad d_6 = 8.285\,72 \times 10^{-2}$$

$$c_5 = 7.795\,94 \times 10^{-2} \quad d_4 = 1.534\,48 \times 10^{-2}$$

$$c_3 = 6.054\,10 \times 10^{-3} \quad d_2 = 4.981\,69 \times 10^{-4}$$

$$L_1 = 12.1078$$

$$C_2 = 1.236\,18$$

$$L_3 = 35.5406$$

$$C_4 = 0.383\,91$$

$$L_5 = 74.8614$$

$$C_5 = 0.245\,77$$

$$C_6 = 0.230\,45$$

$$L_6 = 66.3390$$

$$R_s = 1 \text{ and } R_L = 13.495$$

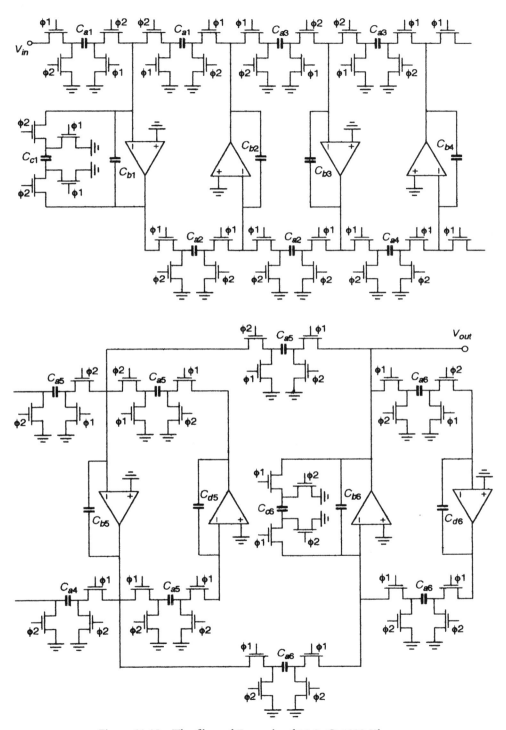

**Figure 11.15** The filter of Example of 11.8. © 1993 Kluwer.

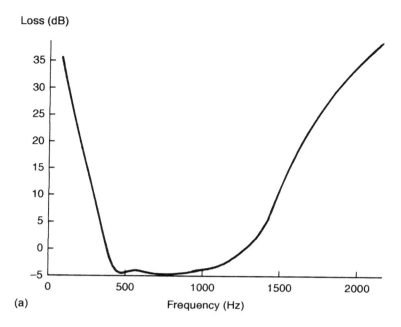

(a)

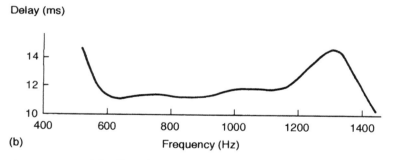

(b)

**Figure 11.16**   Responses of the filter of Example 11.8 with clock frequency of 10 kHz: (a) amplitude, (b) delay. © 1993 Kluwer.

Figure 11.15 shows the SC circuit of this eighth-order bandpass filter. The corresponding capacitor ratios are obtained from

$$\frac{C_{c1}}{C_{a1}} = R_s = 1$$

$$\frac{2C_{b1}}{C_{a1}} + \frac{C_{c1}}{C_{a1}} = L_1 = 12.1078$$

$$\frac{2C_{b2}}{C_{a2}} = C_2 = 1.236\,18$$

$$\frac{2C_{b3}}{C_{a3}} = L_3 = 35.5406$$

$$\frac{2C_{b4}}{C_{a4}} = C_4 = 0.383\,91$$

$$\frac{2C_{b5}}{C_{a5}} = L_5 = 74.8614$$

$$\frac{2C_{d5}}{C_{a5}} = C_5 = 0.245\,77$$

$$\frac{2C_{b6} + C_{c6}}{C_{a6}} = C_6 = 0.230\,45$$

$$\frac{2C_{d6}}{C_{a6}} = L_6 = 66.3390$$

$$\frac{C_{c6}}{C_{a6}} = \frac{1}{R_L} = 0.0741$$

Figure 11.16 gives the amplitude and the delay responses of the realized SC-band-pass filter with the sampling frequency $f_N = 10\,\text{kHz}$. The actual passband is between 44 and 108 Hz, the linear phase bandwidth is between 57 and 108 Hz with the delay $= 11.58 \pm 0.51\,\text{ms}$. Hence,

$$\beta = \frac{108 - 57}{108 - 44} = 80\%$$

## 11.6   GENERALIZED APPROXIMATION

So far in this chapter, we have restricted our discussion of filter transfer functions to those of the minimum-phase or quasi minimum-phase types. This has been done in order to obtain low-sensitivity structures which have passive counterparts. The price to pay for this highly desirable attribute, however, is that there exists an upper bound on the fraction of the passband over which phase linearity can be maintained, once the amplitude selectivity has been specified. For most applications this upper bound is not too restrictive since we have demonstrated that one can achieve an excellent approximation to phase linearity over 85% of the passband with moderate order filters. Nevertheless, it is always desirable to have at our disposal a most general technique for the simultaneous approximation of amplitude and phase which is not subject to these constraints. To this end, we have to drop the minimum-phase nature of the transfer functions, and sacrifice the low-sensitivity property of the structures for increased flexibility. In this section, an approximation technique of considerable generality [45], [49] is given for the derivation of non-minimum-phase transfer functions which satisfy simultaneous conditions on the amplitude and delay responses. These are realizable in cascade form as discussed in Chapter 10. The technique represents the most comprehensive one available for the solution of the simultaneous amplitude and phase approximation problem and leads to a large family of stable transfer functions.

### 11.6.4  Derivation of the Non-minimum-phase Transfer Function

*Theorem 11.1*
Let $G_n(\lambda)$ be an $n$th degree polynomial, which will be termed the *key polynomial*, and write

$$G_n(\lambda) = N(\lambda) + M(\lambda) \tag{11.86}$$

where $N(\lambda)$ and $M(\lambda)$ are the odd and even parts of $G_n(\lambda)$, respectively. Suppose $G_n(\lambda)$ is obtained such that the function

$$\Psi(\lambda) = \tanh^{-1}\left[\frac{N(\lambda)}{M(\lambda)}\right] \tag{11.87}$$

evaluated at $\lambda = j\Omega$, interpolates a desired odd phase function $\phi(\Omega)$ at a number of points $\lambda_i = j\Omega_i$ $(i = 0, 1, 2, \ldots, n)$. Form the function

$$H(\lambda) = \frac{\displaystyle\sum_{r=0}^{m} k_r \lambda^r G_{n-r}(-\lambda)}{\displaystyle\sum_{r=0}^{m} k_r \lambda^r G_{n-r}(\lambda)} \tag{11.88}$$

where $k_r (r = 0, 1, 2, \ldots, m)$ are arbitrary constants, and $m \leqslant n$. Then the following is the case:

(a) The magnitude function $|H(j\Omega)|$ interpolates unity at the points $\lambda = j\Omega_i$ $\{i = 0, 1, \ldots, (n-m)\}$, i.e,

$$|H(j\Omega_i)| = 1, \quad i = 0, 1, 2, \ldots, (n-m) \tag{11.89}$$

(b) The phase of $H(j\Omega)$ interpolates the desired (possibly linear) phase function $\phi(\Omega)$ at the points $\lambda = j\Omega_i$, $i = 0, 1, \ldots, (n-m)$.

(c) The constants $k_r (r = 1, 2, \ldots, m)$ can be chosen to reduce the coefficients of the highest $m$ powers of $\lambda$ in the numerator of $H(\lambda)$ to zero, thus producing $m$ zeros of transmission at $\lambda = \infty$ (i.e., $\omega_N/2$) and a function results which is of the general form

$$H(\lambda) = \frac{f_{n-m}(\lambda)}{g_n(\lambda)} \tag{11.90}$$

*Proof* First note that the determination of the key polynomial $G(\lambda)$ can follow a number of different lines . Indeed, any method for the determination of the key polynomial which satisfies the above conditions is adequate. The validity of the result is, however, independent of the specific method by means of which $G(\lambda)$ is determined. We shall have more to say about this later. Let $G_k(\lambda)$ be a $k$th degree polynomial satisfying

$$G_k(\lambda) = N_k(\lambda) + M_{k-1}(\lambda) \tag{11.91}$$

with

$$\tanh^{-1}\left[\frac{N_k(\lambda_i)}{M_k(\lambda_i)}\right] = \phi(\lambda_i), \quad i = 0, 1, \ldots, k \tag{11.92}$$

where $k$ is taken to be odd for convenience, and $\phi(\lambda)$ is the desired phase function. Thus

$$\frac{N_k(\lambda_i)}{M_{k-1}(\lambda_i)} = \tanh \phi(\lambda_i) \tag{11.93}$$

or

$$N_k(\lambda_i) = M_{k-1}(\lambda_i) \tanh \phi(\lambda_i) \tag{11.94}$$

Rewrite (11.88) as

$$H(\lambda) = \frac{\displaystyle\sum_{r=0}^{m} k_r \lambda^r [N_{n-r}(-\lambda) + M_{n-r-1}(-\lambda)]}{\displaystyle\sum_{r=0}^{m} k_r \lambda^r [N_{n-r}(\lambda) + M_{n-r-1}(\lambda)]} \tag{11.95}$$

where it is assumed that $n$ is odd. Noting that since $N$ is odd and $M$ is even, then $N(-\lambda) = -N(\lambda)$ and $M(-\lambda) = M(\lambda)$. Using this in (11.95), we obtain

$$H(\lambda) = \frac{\displaystyle\sum_{r=0}^{m} k_r \lambda^r M_{n-r-1}(-\lambda)[1 - N_{n-r}(\lambda)/M_{n-r-1}(\lambda)]}{\displaystyle\sum_{r=0}^{m} k_r \lambda^r M_{n-r-1}(-\lambda)[1 + N_{n-r}(\lambda)/M_{n-r-1}(\lambda)]} \tag{11.96}$$

But the key polynomial $G_k(\lambda)$ satisfies(11.93), then use of (11.94) and (11.95) gives

$$H(\lambda) = \frac{\displaystyle\sum_{r=0}^{m} k_r \lambda^r M_{n-r-1}(\lambda)[1 - \tanh \phi(\lambda_i)]}{\displaystyle\sum_{r=0}^{m} k_r \lambda^r M_{n-r-1}(\lambda)[1 + \tanh \phi(\lambda_i)]} \tag{11.97}$$

$$= \frac{1 - \tanh \phi(\lambda_i)}{1 + \tanh \phi(\lambda_i)}$$

Letting $\lambda_i \to j\Omega_i$, and since $\phi(\lambda)$ is assumed odd, we have

$$H(j\Omega_i) = \frac{1 - j \tan \phi(\Omega_i)}{1 + j \tan \phi(\Omega_i)} \tag{11.98}$$

Consequently

$$|H(j\Omega_i)| = 1, \quad i = 0, 1, 2, \ldots, (n - m) \tag{11.99}$$

and the phase of $H(j\Omega_i)$ is such that

$$\Psi(\Omega_i) = -2\phi(\Omega_i), \quad i = 0, 1, \ldots, (n - m) \tag{11.100}$$

i.e., the phase interpolates the desired phase angle $\phi(\Omega)$ at the set of points specified in $\phi(\Omega_i)$.

We now turn to point (c) of Theorem 11.1. We seek a general formulation of the problem of producing an arbitrary number of transmission zeros at $\lambda = \infty$. Let $C_{q,l}$ be the coefficient of $\lambda^l$ in a $q$th-order polynomial satisfying the main theorem. Then, in order to produce $m$ zeros of transmission at $\lambda = \infty$, we use the constants $k_r (r = 1, 2, \ldots, m)$ to force the coefficients of the highest $m$ powers of $\lambda$ in the numerator of $H(\lambda)$ to vanish. So, using these constraints in (11.88), and taking $k_0 = 1$, the following set of linear equations in the $k_m$ unknowns are obtained:

$$\begin{bmatrix} C_{n-1,n-1} & -C_{n-2,n-2} & \cdots & (-1)^{m-1}C_{n-m,n-m} \\ C_{n-1,n-2} & -C_{n-2,n-3} & \cdots & (-1)^{m}C_{n-m,n-m-1} \\ \vdots & \vdots & & \vdots \\ C_{n-1,n-m} & -C_{n-2,n-m} & \cdots & (-1)^{m-1}C_{n-m,n-2m+1} \end{bmatrix} \begin{bmatrix} k_1 \\ k_2 \\ \vdots \\ k_m \end{bmatrix} = \begin{bmatrix} C_{n,n} \\ C_{n,n-1} \\ \vdots \\ C_{n,n-m} \end{bmatrix}$$

$$\tag{11.101}$$

The solution of the above set of linear equations gives the required values of $k_r (r = 1, 2, \ldots, m)$ necessary to produce the desired number of transmission zeros at $\lambda = \infty$. Thus, we generate a function of general form

$$H(\lambda) = \frac{f_{n-m}(\lambda)}{g_n(\lambda)} \tag{11.102}$$

which has the following required properties: its amplitude interpolates unity at a set of $(n - m)$ frequencies, its phase interpolates the desired phase angle at the same set of $(n - m)$ points, and it possesses $m$ zeros of transmission of $\lambda = \infty$, thus allowing arbitrary stopband selectivity to be obtained.

This completes the proof of the procedure for generating the transfer function from a polynomial whose phase angles interpolate the desired characteristics. Next, we give a number of corollaries which follow directly from the Theorem and the properties of the polynomials stated in Section 11.3.

*Corollary 1*   If in (11.88) we let

$$G_k \overset{\Delta}{=} A_k(\lambda, \phi_0, \beta) \tag{11.103}$$

i.e., the equidistant linear-phase polynomial defined by (11.24)–(11.26), then the theorem holds true, but (b) now becomes

(b) The phase function $(\Omega_i)$ interpolates the ideal (linear) phase angles at the points $\Omega_i$, $i = 0, 1, \ldots, (n - m)$.

*Corollary 2*   If in (11.88) we set

$$G_k \overset{\Delta}{=} Q_k(\lambda, \alpha) \tag{11.104}$$

being the maximally flat delay polynomial defined by (11.23), then the function $H(\lambda)$ is such that:

1. The amplitude function $|H(j\Omega)|$ has $(n - m)$ forced zero derivatives at $\Omega = 0$;

2. The delay response of $H(j\Omega)$ has $(n - m)$ forced zero derivatives at $\Omega = 0$;

3. the constants $k_r$ can be chosen, by solving (11.101), to produce $m$ zeros of transmission at $\Omega = \infty$.

*Corollary 3*   The high-pass filter transfer function with the properties corresponding to those of the main theorem can be obtained by letting $\lambda \rightarrow 1/\lambda$ in the low-pass transfer functions.

### 11.6.2   Stability and Performance

#### *The finite-band case*

Consider the function obtained according to Corollary 2 and the main theorem. This produces a finite-band approximation to constant amplitude and linear phase, with an arbitrary number of zeros at $\lambda = \infty$, i.e., at half the sampling frequency $\omega_N/2$. This takes the form

$$H(\lambda) = \frac{\displaystyle\sum_{r=0}^{m} k_r \lambda^r A_{n-r}(-\lambda, \phi_0, \beta)}{\displaystyle\sum_{r=0}^{m} k_r \lambda^r A_{n-r}(\lambda, \phi_0, \beta)} \tag{11.105}$$

in which the key polynomial is the equidistant linear-phase polynomial $A_k(\lambda, \phi_0, \beta)$. Upon solution of (11.101) to determine $k_r(r = 1, 2, \ldots, m)$, $H(\lambda)$ takes the form

$$H(\lambda) = \frac{f_{n-m}(\lambda, \phi_0, \beta)}{g_n(\lambda, \phi_0, \beta)} \tag{11.106}$$

In this connection, the system of linear equations in (11.101) always had a solution, and no case was found in which the square matrix in (11.101) was singular.

As discussed in Section 11.2, frequency scaling in the discrete domain is not possible, and additional parameters are incorporated in the expressions of transfer functions, to allow the bandwidth of the filter to be adjusted according to given specifications. In (11.105) and (11.106), $\beta$ is essentially the bandwidth scaling parameter. The parameter $\phi_0$ determines the points in the passband at which the amplitude interpolates unity, and the phase interpolates the linear-phase function.

A detailed study of the functions under consideration reveals the following dependence of the stability on the design parameters.

(a) For fixed $n, \phi_0$, and $\beta$, low values of $m$ are associated with stable functions, while increasing $n$ beyond a certain value causes the functions to become unstable.

(b) As the product $n\phi_0$ approaches $\pi/2$ the filters become unstable.

(c) For fixed $n$, $m$, and $\phi_0$, the parameter $\beta$ has a range over which the filters remain stable. A necessary condition on $\beta$ is

$$\beta > n - 1 \qquad\qquad (11.107)$$

and the upper limit for stability depends on $m$ and $\phi_0$ for a given degree $n$. As in the lumped case, the above constraints are quite compatible with very stringent specifications.

*Example 11.9*   Consider the following set of specifications:

(Passband edge/sampling frequency) $= 0.2$
with $\leqslant 0.1$ dB ripple in the passband,
(Stopband width/sampling frequency) $= 0.21$,
with attenuation $\geqslant 20$ dB in stopband
Passband delay variation $\leqslant 1\%$.

Note that the passband width is $\omega_{N/2} - \omega_s$. A computer program must be written to obtain the necessary data along the following lines:

(1) Select $(n, m)$ and determine a combination of $(\phi_0, \beta)$ from a set of data such that the stopband amplitude requirements are satisfied;

(2) Calculate the passband delay variation;

(3) Repeat the above steps with increasing $n$.

In this particular example, the function which meets the specifications is defined by (11.88) and (11.19) with

$$n = 13, \quad m = 2, \quad \phi_0 = 0.0571, \quad \beta = 19.9$$

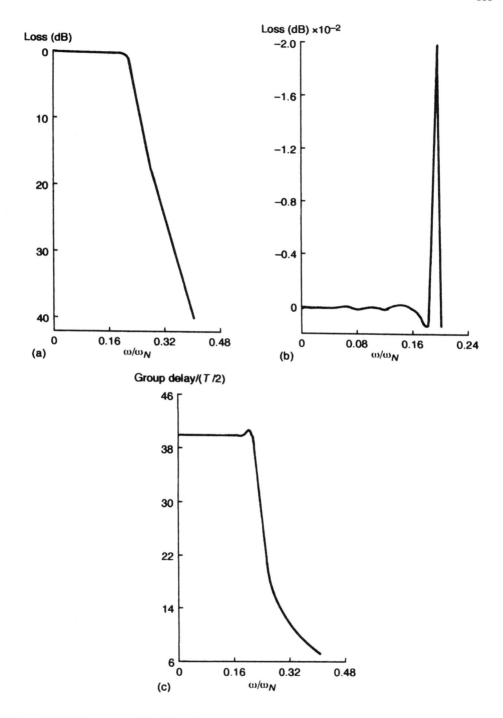

**Figure 11.17**   Responses of the filter of Example 11.9: (a) amplitude; (b) passband amplitude detail; (c) delay. © 1993 IEEE.

Figure 11.17 shows the amplitude and delay responses of the filter with the above degree and parameters.

### The zero-bandwidth case

This can be obtained from Theorem 11.1, Corollary 2, and the maximally-flat delay polynomial $Q(\lambda, b)$ given by (11.23) used as the key polynomial. Thus, we write

$$H(\lambda) = \frac{\displaystyle\sum_{r=0}^{m} k_r \lambda^r Q_{n-r}(-\lambda, b)}{\displaystyle\sum_{r=0}^{m} k_r \lambda^r Q_{n-r}(\lambda, b)} \tag{11.109}$$

and $k_r$ are determined from (11.101) for an arbitrary number of zeros at $\omega_N/2$ (i.e., $\lambda = \infty$). We also note that the expressions for the functions and their characteristics may be obtained from the finite-band cases by taking the limit as $\phi_0 \to 0$. The passband width and stopband width can both be adjusted by a suitable choice of $b$. If, for a combination $(n, m)$, no value of $b$ can be found to meet the specifications, the process is repeated with higher $n$ (and possibly different $m$) until the range of $b$ includes a value defining a passband width and stopband width within the specified values.

## 11.7   DATA TRANSMISSION FILTERS

### 11.7.1   Properties of Data Transmission Filters

A data transmission filter is primarily a pulse (or impulse) shaping network [50]. It has a time domain pulse (or impulse) response which allows the transmission of a train of shifted versions of this pulse without significant intersymbol interference (ISI). In addition, the filter must have a frequency response with sufficient selectivity and attenuation for bandlimiting and suppression of noise and cross-talk. Figure 11.18 illustrates the salient features of the time response and frequency bands of interest of a continuous-time data transmission filter, in which $T$ is the data rate interval and $2\omega_1$ is the highest frequency in the band of interest such that

$$\omega_1 = \frac{\pi}{T} \tag{11.110}$$

Furthermore, the phase response of the filter is important since a deviation from the ideal linear characteristic causes unpredictable distortion of the pulse shape and also such a deviation is incompatible with the minimum ISI requirement.

In this section, we present a technique for the design of switched-capacitor data transmission filters. The filter transfer function is derived entirely in the frequency domain in such a way that the minimum intersymbol interference (ISI), amplitude selectivity and phase characteristics are met simultaneously without the need for subsequent equalization or correction of either the amplitude or delay. The technique

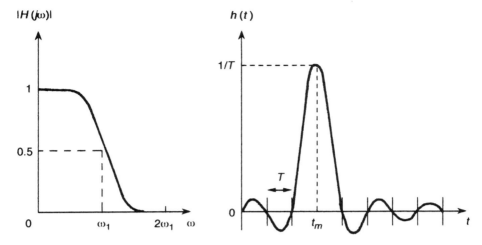

**Figure 11.18**  General features of the expected time and frequency responses of a data transmission filter. © 1993 IEEE.

gives the denominator of the transfer function in analytic closed form while the numerator is determined by solving a set of linear algebraic equations. Furthermore, the cases of impulse and pulse transmission are treated in basically the same manner without the need for external $\sin(x)/x$ corrections.

The conditions on the transfer function are first discussed in the continuous-time domain since this is usually easier to grasp, then the derivation of the transfer function is given for sampled-data filters in general, and switched-capacitor filters in particular.

### 11.7.2  The Conditions in the Continuous-time Domain

Let the continuous-time filter transfer function be written as

$$H(s) = \frac{P_{2m}(s)}{Q_n(s)} \quad 2m < n$$

with

$$P_{2m}(s) = \sum_{r=0}^{m} b_r s^{2r}$$

where $P_{2m}(s)$ is an even polynomial and $Q_n(s)$ is a strictly Hurwitz polynomial.
Define the functions

$$\gamma_1(\omega) = \cos(\beta(\omega_1 + \omega)) \tag{11.113}$$

$$\gamma_2(\omega) = \cos(\beta(\omega_1 - \omega)) \tag{11.114}$$

$$F_1(\omega) = \operatorname{Re} H(j(\omega_1 + \omega)) \tag{11.115}$$

$$F_2(\omega) = \operatorname{Re} H(j(\omega_1 - \omega)) \tag{11.116}$$

where $2\omega_1$ is the highest frequency of the band of interest and $\beta$ is a parameter. Now, suppose $H(j\omega)$ satisfies the following conditions:

(a) $\operatorname{Arg} Q_n(j\omega)$ interpolates the function $\beta\omega$ at $n+1$ equidistant points over the band $0 < |\omega| < 2\omega_1$, i.e

$$\beta\omega_i - \operatorname{Arg} Q_n(j\omega_i) = 0 \text{ for } \omega_i = i\frac{2\omega_1}{n} \qquad i = 0, 1, 2, \ldots, n \qquad (11.117)$$

so that

$$\operatorname{Arg} Q_n(j\omega) \approx \beta\omega \qquad 0 < |\omega| < 2\omega_1 \qquad (11.118)$$

with an approximation error that can be made arbitrarily small by increasing $n$.

(b) $F_2(\omega)\gamma_1(\omega) + F_1(\omega)\gamma_2(\omega)\}$ approximates the function $\gamma_1(\omega)\gamma_2(\omega)$ in the least mean squares sense over the band $0 < |\omega| < \omega_1$ i.e.

$$F_2(\omega)\gamma_1(\omega) + F_1(\omega)\gamma_2(\omega) \approx \gamma_1(\omega)\gamma_2(\omega), \quad 0 < |\omega| < \omega_1 \qquad (11.119)$$

Then, the impulse response $h(t)$ of the filter is such that if we define

$$h_\beta(t) = h(t + \beta) \qquad (11.120)$$

then

$$h_\beta(kT) \approx 0 \quad k = \pm 1, \pm 2, \pm 3, \ldots \qquad (11.121)$$

and it deviates from zero by an error that can be made negligibly small by taking $n$, $m$ and $(n - m)$ sufficiently large. Also,

$$h_\beta(0) \approx \frac{1}{T} \quad \text{with } T = \frac{\pi}{\omega_1} \qquad (11.122)$$

We observe that conditions (11.119) in the *frequency domain* imply a certain symmetry of the real part around the point $\omega_1$. This symmetry, together with the phase response constraints (11.117), lead to the impulse response, in the *time domain*, meeting conditions (11.121) of the minimum intersymbol interference.

## Pulse transmission

Rather than treat the case of arbitrary (or square) pulse transmission as a separate problem, we can reformulate the problem in a manner such that the conditions stated above can be directly applied without modification. To this end, let the transfer function $H_p(s)$ of a filter be formed as the product of two functions as

$$H_p(s) = \Pi(s)H(s) \qquad (11.123)$$

where $\Pi(s)$ is the Laplace transform of a pulse $p(t)$. Then the impulse response of the filter with transfer function $H_p(s)$ is the same as the pulse response of the filter with transfer function $H(s)$. Consequently, if we design the data transmission filter with a transfer function $H(s)$ such that $H_p(s)$ defined by (11.123) satisfies conditions (1.117) and (11.119), then the pulse response of the required filter will have the required properties, those being exactly the same as those of the impulse response corresponding to $H_p(s)$.

### 11.7.3   Switched-capacitor Filters

The derivation of the transfer function of a data transmission switched capacitor filter is accomplished such that the same conditions (11.117) and (11.119) of the are satisfied. Due to the one-to-one correspondence between the $s$-domain and $\lambda$ domain, we use the variable

$$\lambda = \tanh(\tau s) = \frac{1 - z^{-1}}{1 + z^{-1}} \tag{11.124}$$

where

$$\tau = \frac{\pi}{\omega_N} \tag{11.125}$$

and $\omega_N$ is, as usual, the radian sampling frequency.
   Let the transfer function be written as

$$H(\lambda) = \frac{P_{2m}(\lambda)}{Q_n(\alpha|\beta|\lambda)} \tag{11.126}$$

where $Q_n(\alpha|\beta|\lambda)$ is the equidistant linear phase polynomial employed earlier in this chapter and given by (11.25)–(11.26) which are repeated below with a slight change in notation,

$$Q_{n+1}(\alpha|\beta|\lambda) = Q_n(\alpha|\beta|\lambda) + \gamma_n[\lambda^2 + \tan^2(n\alpha)]Q_{n-1}(\alpha|\beta|\lambda) \tag{11.127}$$

$$\gamma_n = \frac{\cos(n-1)\alpha \cos(n+1)\alpha \sin(\beta-n)\alpha \sin(\beta+n)\alpha \cos^2 n\alpha}{\sin(2n-1)\alpha \sin(2n+1)\alpha \cos^2 \beta\alpha} \tag{11.128}$$

$$Q_0(\alpha|\beta|\lambda) = 1 \tag{11.129}$$

$$Q_1(\alpha|\beta|\lambda) = 1 + \frac{\tan(\alpha\beta)}{\tan(\alpha)}\lambda \tag{11.130}$$

and

$$\beta\omega_i - \text{Arg}\{Q_n(\alpha|\beta|j(\tan(\tau\omega_i))\} = 0 \quad \omega_i = i\alpha \quad \text{for } i = 0, 1, \ldots, n \tag{11.131}$$

The necessary and sufficient conditions for $Q_n(\alpha|\beta|\lambda)$ to be strictly Hurwitz are

$$\alpha < \frac{\pi}{2n} \quad \beta > n - 1 \quad (\beta + n - 1)\alpha < \pi \tag{11.132}$$

$H(\lambda)$ will therefore satisfy condition (12.9).

To satisfy condition (11.117), $H(\lambda)$ must be such that $\{\mathrm{Re}\{H(\omega_1 - \omega)\}\gamma_1(\omega) + \mathrm{Re}\{H(\omega_1 + \omega)\}\gamma_2(\omega)\}$ approximates the function $\{\gamma_1(w)\gamma_2(\omega)\}$ in a least mean squares sense over the band $0 < \omega < \omega_1$ and

$$\mathrm{Re}\{H(\omega)\} = \mathrm{Re}\{H(\mathrm{tg}(\omega))\} \tag{11.133}$$

where

$$\mathrm{tg}(\omega) = \tan(\tau\omega) \tag{11.134}$$

The criterion used to define the unknown coefficients $b_r$ is the minimization of the squared error $E_d$ representing the deviation of the real part of $H_n(\omega)$ from the symmetry relation over the band of interest, i.e.

$$\epsilon_{sd}(\omega, r) = -\gamma_1(\omega)\gamma_2(\omega) + \mathrm{Re}\{H(\mathrm{tg}(\omega_1 - \omega))\}\gamma_1(\omega)$$
$$+ \mathrm{Re}\{H(\mathrm{tg}(\omega_1 + \omega))\}\gamma_2(\omega) \tag{11.135}$$

$$E_d = \int_0^{\omega_1} [\epsilon_{sd}(\omega, r)]^2 \, d\omega \tag{11.136}$$

We also add one extra constraint on the coefficients $b_r$ to force the value of the slope of $|H(\omega)|$ at the midpoint $\omega_1$. This will control the selectivity of the filter. Hence

$$|H\omega)| = \frac{|P_{2m}(\mathrm{tg}(\omega))|}{|Q_n(\alpha|\beta|j\,\mathrm{tg}(\omega))|} = F''(\Omega, r)G''(\Omega) \tag{11.137}$$

where

$$\Omega = \mathrm{tg}(\omega)$$

and

$$F''(\Omega, r) = \left| \sum_{r=0}^{m} b_r(-1)^r\Omega^{2r} \right|$$
$$= \sum_{r=0}^{m} b_r(-1)^r\Omega^{2r} \quad \text{for } \Omega \text{ in the vicinity of } \Omega_1 \tag{11.139}$$

with

$$\Omega_1 = \mathrm{tg}(\omega_1)$$

$$G''(\Omega) = \frac{1}{|Q_n(\alpha|\beta|j\Omega)|} \tag{11.141}$$

and

$$\frac{\partial |H(j\omega)|}{\partial \omega}\bigg|_{\omega=\omega_1} = \sum_{r=0}^{m} b_r(-1)^r \{2r\Omega_1^{2r-1}G''(\Omega_1) + \Omega_1^{2r}\}\frac{\partial G''(\Omega)}{\partial \Omega}\bigg|_{\Omega=\Omega_1}\}$$

$$\cdot \left\{\frac{\partial \text{tg}(\omega)}{\partial \omega}\bigg|_{\omega=\omega_1}\right.$$

(11.142)

If we define the constraint as

$$K''(\omega, r) = \frac{\partial |H(j\omega)|}{\partial \omega}\bigg|_{\omega=\omega_1} - \sigma_d = 0$$

(11.143)

where $\sigma_d$ represents the desired slope of $|H(j\omega)|$ at the midpoint $\omega_1$.
The corresponding system of equations will be

$$\sum_{r=0}^{m} b_r A''(k, r) + \xi L''(k) = B''(k) \quad \text{for } k = 0, 1, 2, \ldots, m$$

(11.144)

$$\sum_{r=0}^{m} b_r(-1)^r \{2r\Omega_1^{2r-1}G''(\Omega_1) + \Omega_1^{2r}\frac{\delta G''(\Omega)}{\delta \Omega}\bigg|_{\Omega=\Omega_1}\} \cdot \left\{\frac{\delta \text{tg}(\omega)}{\delta \omega}\bigg|_{\omega=\omega_1}\right\} = \sigma_d$$

(11.145)

with

$$L''(k) = (-1)^k \left\{2k\Omega_1^{2k-1}G''(\Omega_1) + \Omega_1^{2k}\frac{\delta G''(\Omega)}{\delta \Omega}\bigg|_{\Omega=\Omega_1}\right\}\left\{\frac{\delta \text{tg}(\omega)}{\delta \omega}\bigg|_{\omega=\omega_1}\right\}$$

(11.146)

$$A''(k, r) = \int_0^{\omega_1} \{f_k''(\omega_1 - w)\gamma_1(\omega) + f_k''(\omega_1 + w)\gamma_2(\omega)\}$$

$$\cdot \{f_r''(\omega_1 - w)\gamma_1(\omega) + f_r''(\omega_1 + w)\gamma_2(\omega)\}\, dw$$

(11.147)

$$B''(k, r) = \int_0^{\omega_1} \{f_k''(\omega_1 - w)\gamma_1(\omega) + f_k''(\omega_1 + w)\gamma_2(\omega)\}\{\gamma_1(\omega) + \gamma_2(\omega)\}\, dw$$

(11.148)

$$f_r''(\omega) = \frac{(-1)^r \text{tg}(\omega)^{2r} \text{Re}\{Q_n(\alpha|\beta|\text{tg}(\omega))\}}{|Q_n(\alpha|\beta|\text{tg}(\omega))|^2}$$

(11.149)

and where $\xi$ is a Lagrange multiplier.

*Example 11.9* Figure 11.19 gives the impulse and frequency responses for a 10th degree switched-capacitor impulse shaping filter with $m = 4$, $\omega_1 = \pi/8$, $\alpha = \pi/40 = 0.0785$, $\beta = 16$ and the slope $\sigma_d = 3.328$.
The transfer function is given by

$$H(z) = \frac{N(z)}{D(z)}$$

with

$$N(z) = 4.619\,637 - 281.0891z + 296.0831z^2 + 113.5679z^3$$

$$+ 44.702\,72z^4 + 847.0423z^5 - 44.702\,72z^6 + 113.5679z^7$$

$$+ 296.0831z^8 - 281.0891z^9 + 4.619\,637z^{10}$$

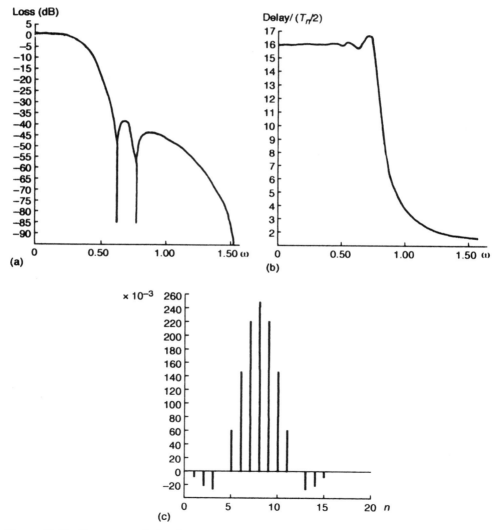

**Figure 11.19**  Responses of the data-transmission filter of Example 12.9 designed for impulse transmission. (a) amplitude, (b) delay, (c) impulse response. © 1993 IEEE.

$$D(z) = 1489.422 - 12\,945.68z + 55\,727.74z^2 - 15\,6511.2z^3$$

$$+ 318\,069.5z^4 - 490\,121.3z^5 + 582\,466.0z^6 - 530\,424.3z^7$$

$$+ 357\,297.9z^8 - 162\,687.2z^9 + 38\,663.07z^{10}$$

In this example, the maximum overshoot is 0.077 dB, the minimum attenuation 38.05 dBs, the sampling error has a value of $1.86\ 10^{-5}$ and the first side lobe size is equal to 10.04% of the main lobe.

## CONCLUSION

In this chapter, a high degree of sophistication has been reached in the design techniques of switched-capacitor filters. Not only are we able to meet stringent amplitude specifications but we can also design filters with excellent phase as well as high amplitude selectivity. Furthermore, the time domain response of filters used in data and image transmission is of paramount importance. We have seen that such data transmission filters are possible using switched-capacitor techniques which meet the requirements of minimizing intersymbol interference in the time-domain, while providing high amplitude selectivity and good approximation to phase linearity over the band of interest in the frequency domain.

## PROBLEMS

11.1 Design a low-pass LDI ladder filter with maximally-flat delay response meeting the following specifications:

    Passband 0 to 3.4 kHz with 1 dB maximum attenuation
    Passband delay variation 10 ns
    Stopband 4.6–15 kHz with 25 dB minimum attenuation.

    [Use Program MXLP in Appendix D for the generation of the transfer function, followed by Program LDISCG in Appendix A for the realization].

11.2 Design a filter with the same specifications as those in Problem 11.2 but using an equidistant linear phase approximation.

    [Use Program ELPP in Appendix E, followed by Program LDISCG for the realization].

11.3 Design an optimum low-pass strict LDI filter with the following specifications:

    Passband 0 to 3 kHz, attenuation $\leqslant 0.8$ dB
    Stopband 10 kHz to 20 kHz, attenuation $\geqslant 30$ dB
    The phase approximates linearity over 75% of the passband.

11.4 Design a filter to meet the same amplitude specifications as those in Problem 11.3 but extending the linear phase approximation bandwidth to 85% of the passband.

11.5 Design a band-pass filter with the following specifications:

Passband 800 Hz–1600 Hz with 0.5 dB maximum attenuation
Phase linearity is maintained over 80% of the passband
Lower stopband edge at 400 Hz with attenuation $\geqslant 20$ dB
Upper stopband edge at 2000 Hz with attenuation $\geqslant 30$ dB
Sampling frequency is 10 kHz.

11.6 Write a computer program to implement the evaluation of the transfer function in (11.85) using the formulation in (11.98) to obtain the general form in (11.99).

11.7 Use the program obtained in Problem 11.6 to design the filters of Problems 11.1, 11.2, 11.3, and 11.4. Compare the degrees of the filters obtained in these cases.

11.8 Realize the three filters of Problem 11.6 in cascade form using the technique of Section 10.4. Check the results using Program ISICAPC on the diskette.

# 12

# Practical Considerations

## 12.1 INTRODUCTION

In Chapters 8–11, the design techniques of switched-capacitor filters were delineated assuming idealized components. On the other hand, Chapters 2 to 6 dealt with the integrated circuit building blocks together with their non-ideal behaviour. In this chapter, we combine the results of these chapters to examine the effect of the non-ideal behaviour of the integrated circuit building blocks on the overall response of a switched-capacitor filter. We also examine other practical issues which may be of interest to the designer.

## 12.2 EFFECT OF FINITE OP AMP GAIN

The building blocks of switched-capacitor filters are first or second order sections containing Op Amps, switches and capacitors. In the design methods presented in the previous chapters, it was assumed that the Op Amps are ideal, having infinite gain values. However, as we know from Chapters 4 and 5, real CMOS Op Amps have large, but finite gain values. This factor has to be taken into account in the final simulation of the switched-capacitor filter to determine its exact response, particularly for high frequency designs where the operating frequencies approach the bandwidth of the Op Amp. The finite gain of the Op Amps results in a distortion of the transfer functions of the building blocks, and hence the overall filter transfer function. Furthermore, the strays-insensitive property of building blocks used throughout, depends on the establishment of a virtual ground, which in turn is only closely approximated for very high gain Op Amps. It is necessary, therefore to model the finite gain effect and incorporate the model in the filter, then perform the analysis on the modified circuit to find out the response. This is illustrated here [51] by the typical first order transfer function of the building block shown in Figure 12.1 which is that of Figure 9.10(a). We assume that the Op Amp has a finite gain of $A$, and write

$$
\begin{aligned}
C_A[ - V_x(n) &- V_1(n-1)] \\
&= -C_B([V_o(n) - V_x(n)] - [V_o(n-1) - V_x(n-1)])
\end{aligned}
\tag{12.1}
$$

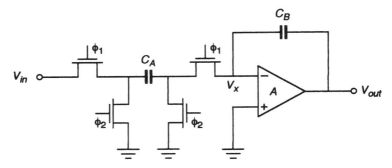

**Figure 12.1**    A first-order switched-capacitor circuit.

Substituting for $V_x = -V_o/A$ we have

$$C_A[(1/A)V_o(n) - V_1(n-1)] = -C_B([V_o(n) - V_o(n)n - 1])$$
$$+ (1/A)[V_o(n) - V_o(n-1)])$$

$$(12.2)$$

Letting $C_B/C_A = \alpha$ and $1/A = \eta$ then taking the z-transform of both sides of the above equation, we can form the transfer function of the building block as

$$H(z) = \frac{z^{-1}}{\alpha(1+\eta)(1-z^{-1}) + \eta} \qquad (12.3)$$

which approaches (9.37) as $A \to \infty$ i.e. $\eta \to 0$. It is easy to see that $H$ is related to the ideal transfer function of (9.37) by

$$H = H_{ideal}/[\alpha(1+\eta) + \eta/2 + (\eta/2)\coth(Ts/2)] \qquad (12.4)$$

Thus, with $E$ denoting the ratio $H/H_{ideal}$ and neglecting terms containing $\eta^2$ we have on the $j\omega$ axis

$$|E|^2 \cong \frac{\alpha}{\alpha + \eta} \qquad (12.5)$$

and

$$\text{Arg}(E) \cong \frac{\eta}{2\alpha\tan(\omega T/2)} \qquad (12.6)$$

From the above expressions, the error in the magnitude is negligible for reasonably large values of $A(>1000)$, but is easily taken into consideration by noting that

$$|H|^2 = |H_{ideal}|^2(1 + \eta/\alpha) \qquad (12.7)$$

which is equivalent to an error in the capacitor ratio by the factor $\eta$. However, the phase error is frequency dependent and can be quite large.

Obviously, the only 'cure' for the finite gain effects, is to design Op Amps with sufficiently high gain as discussed in Chapter 4. So, the purpose of the above discussion is purely informative.

## 12.3   EFFECT OF FINITE BANDWIDTH
## AND SLEW RATE OF OP AMPS

Each Op Amp has a unity gain frequency which limits the frequency range over which the switched-capacitor filter is used. In particular, the unity gain frequency affects the settling time of each Op Amp as may be clearly seen by representing the Op Amp by its frequency dependent transfer function, inserting this model in the switched-capacitor section, then finding the corresponding time response. For a first order section of the type used in the previous analyses, the output voltage will settle to

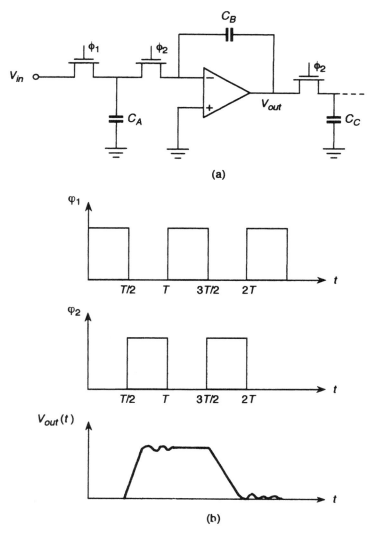

**Figure 12.2**   Pertinent to the discussion of settling time and slew rate effects.

its final value either exponentially or passing through a phase of damped oscillations. Thus the time $T_{on}$ corresponding to the clock phase during which the switch is ON, must be large enough to allow the Op Amp to settle to its final value (within certain error). This implies that the settling time of each Op Amp must be less than half the sampling period, $T$ or,

$$T_{settling} < 0.5T < 1/2f_N \qquad (12.8)$$

Thus, the real solution to this problem, is to design Op Amps with fast settling time. Calculations on both the positive and negative first order sections have revealed that the unity gain frequency $\omega_t$ of the Op Amp must satisfy

$$\omega_t \gg \omega_N/\pi \qquad (12.9)$$

and a value of $\omega_t = 5\omega_N$ is found satisfactory in practice to result in negligible errors. In addition to the settling time, there is a delay caused by the finite slew rate of the Op Amp as shown in Figure 12.2. Thus, instead of (12.8) we have

$$T_{settling} + T_{slew} < 0.5T \qquad (12.10)$$

## 12.4   EFFECT OF FINITE OP AMP OUTPUT RESISTANCE

The Op Amp must charge a load capacitance $C_L$ through its output resistance $R_{out}$. For unbuffered Op Amps, this is of the order of a few hundred $k\Omega$. The charging time constant can be found to be of the order of $2R_{out}C_L$, which must be less than $0.5T$ to give acceptable charge transfer. Naturally, buffered Op Amps can be used to reduce this effect.

## 12.5   SCALING FOR MAXIMUM DYNAMIC RANGE

Consider a typical Op Amp in a switched-capacitor filter [8] as shown in Figure 12.3. Let all the transfer functions $\Delta Q/V$ of the branches connected to the output of the $k$th Op Amp be multiplied by a factor $\beta_k$. In Figure 12.3 these are branches $B_4, B_5$ and $B_6$. This leads to scaling of the output voltage of the OP Amp by $1/\beta_k$. The input branches are left unchanged and their charges remain the same which results in the charge in $B_4$ retaining its original value. The same is true for branches $B_5$ and $B_6$ since their capacitances are scaled by $\beta_k$ while their voltages are scaled by the inverse of this factor. We conclude that multiplying all capacitor values which are connected ,or switched, to the output of an Op Amp by a factor, scales its output voltage by the inverse of this factor, while leaving all charges flowing between this Op Amp and the rest of the circuit unchanged. This process allows the improvement of the dynamic range of the filter according to the following procedure.

(a)  Set $V_{in}(\omega)$ to the largest value for which the output Op Amp does not saturate.

(b)  Calculate the maximum values $V_{pk}$ for all internal Op Amp output voltages. These values usually occur near the passband edge of the filter.

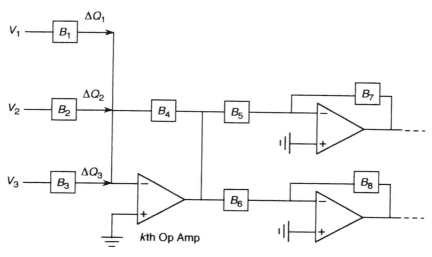

**Figure 12.3** Pertinent to the discussion of maximum dynamic range and minimum capacitance scaling.

(c) Multiply all capacitors connected or switched to the output terminal of Op Amp $k$ by $\beta_k = V_{pk}/V_{k,max}$ where $V_{k,max}$ is the saturation voltage of Op Amp $k$.

(d) Repeat for all internal OP Amps.

The programs in **ISICAP**, which have been employed throughout the book, include an option for scaling the filter for maximum dynamic range.

## 12.6   SCALING FOR MINIMUM CAPACITANCE

This scaling operation can be used to reduce the capacitance spread and hence the total capacitance used by the filter [8]. This relies on the very simple principle that if all capacitors connected to the input terminal of an Op Amp are multiplied by the same number, then the output voltage remains the same. The procedure is as follows:

(a) Divide all capacitors in the filter into non-overlapping sets. Those in the $i$th set $S_i$ are connected or switched to the input terminal of Op Amp $i$.

(b) Multiply all capacitors in $S_i$ by $m_i = C_{min}/C_{i,min}$, where $C_{min}$ is the smallest capacitor which the fabrication technology allows and $C_{i,min}$ is the smallest capacitor in $S_i$.

(c) Repeat for all sets, including that associated with the output Op Amps.

Finally, we note that scaling for maximum dynamic range should be performed before scaling for minimum capacitance, since the former changes the voltages of the circuit nodes, while the latter does not.

Again, the programs used for the design of the filters, which are available in ISICAP, include an option for scaling the filter for minimum capacitance.

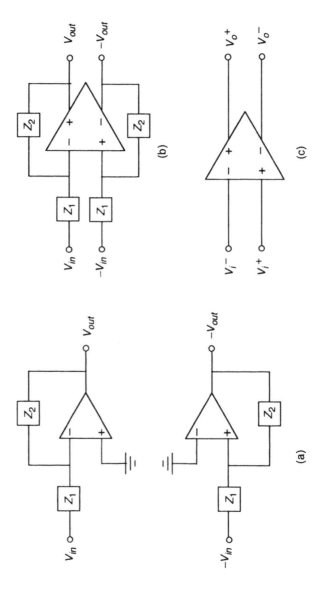

**Figure 12.4**   (a) A general Op Amp circuit and its mirror image, (b) conversion into fully differential equivalent, (c) symbol of the fully differential Op Amp.

## 12.7  FULLY DIFFERENTIAL BALANCED DESIGNS

It was shown in Chapter 4 that fully differential balanced MOS Op Amp designs offer several advantages over single-ended ones. We now justify and extend this concept to the entire switched-capacitor filter structure. The dynamic range of a switched-capacitor filter is determined by the ratio of the maximum signal swing giving acceptable distortion to the noise level. It follows that an improvement in the signal swing, results in improvement in the dynamic range. Since fully differential designs double the signal swing, they also improve the dynamic range. Moreover, since the signal paths are balanced, noise due to power supply variation and clock charge injection are also reduced. This is in addition to the noise reduction inherent in the Op Amp design itself. The principle of conversion from a structure using single-ended output to fully-differential balanced structure is as follows [34]:

(a) Sketch the single-ended circuit and identify the ground node(s).

(b) Duplicate the entire circuit by mirroring it at ground.

(c) Divide the gain of all active devices by two.

(d) Reverse the signs of all duplicated active elements and merge each resulting pair into one balanced differential-input differential output device.

(e) Simplify the circuit, if possible by replacing any device which merely achieves sign inversion by crossed wires.

Figure 12.4 shows the application of the procedure for a general Op Amp circuit. Figure 12.5 shows a fully differential first-order switched-capacitor section [34]. In this circuit $v_{i1}^+$ and $v_{i1}^-$ constitute one of the differential input signals while $v_{i2}^+$ and $v_{i2}^-$

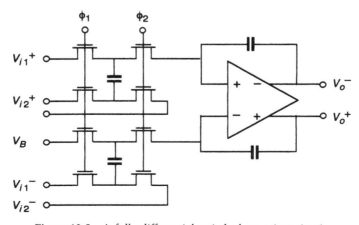

**Figure 12.5**  A fully differential switched-capacitor circuit.

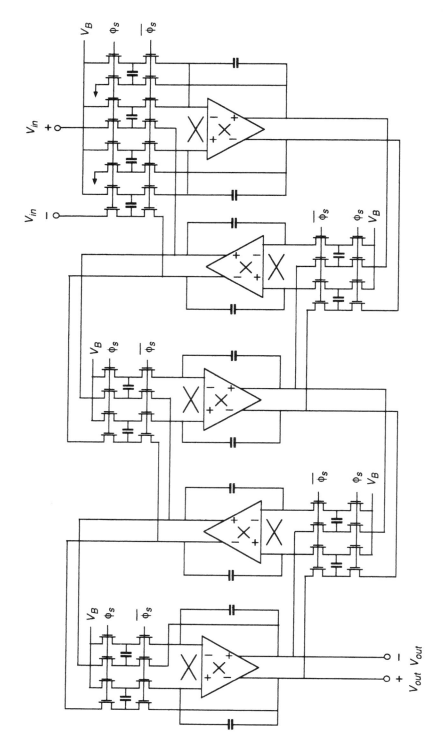

**Figure 12.6** A fifth-order filter using chopper stabilization and fully-differential topology.

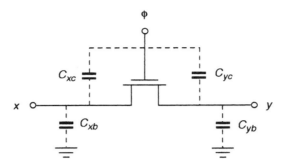

**Figure 12.7**  A switch with the associated parasitic capacitances.

form the other. With the assumption of zero common-mode signal, the common mode voltage of the Op Amp is $V_B$ which can be set to a suitable value by an on-chip circuit.

With this technique , a filter can be designed with a dynamic range exceeding 100 dB. An example [34] of a fifth-order filter structure using both chopper stabiliza-tion and fully-differential techniques is shown in Figure 12.6

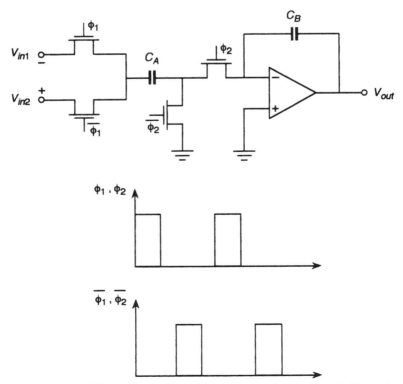

**Figure 12.8**  Typical first-order section with conventional two-phase locking scheme.

## 12.8   MORE ON PARASITIC CAPACITANCES
## AND SWITCH NOISE

It has been emphasized that the use of parasitic-insensitive structures is essential for good performance of switched-capacitor filters. The adoption of switched-capacitor filter circuits which are parasitic insensitive has been a major factor in the development of high quality integrated filters. These circuits have the desirable property that their transfer functions are unaffected by additional capacitance from node to ground, assuming ideal Op Amps. However, this immunity does not extend to the effect of the ungrounded parasitic capacitances $C_{xc}$ and $C_{yc}$ associated with the control terminals of the switches as shown in Figure 12.7. These parasitic capacitances can be shown to cause dc offset voltages, modification of the transfer function and distortion. Figure 12.8 shows a typical first-order section together with the conventional two-phase clocking scheme. A new four-phase clocking scheme has been proposed [52] as shown in Figure 12.9 which, theoretically, eliminates the remaining parasitic effects and in practice reduces them drastically. This is achieved by slightly delaying the switching waveforms of two of the switches. Figure 12.10. shows the improvement in the performance due to the various waveform arrangements.

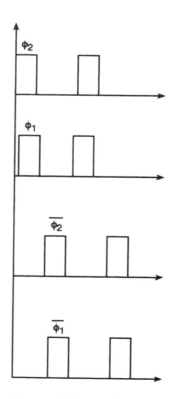

**Figure 12.9**   A more elaborate clocking scheme for the circuit of Fig. 12.8.

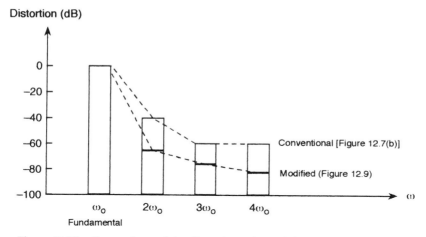

**Figure 12.10**   Comparison of the distortion values of the two clocking schemes.

## 12.9   PRE-FILTERING AND POST-FILTERING REQUIREMENTS

As pointed out in Chapter 6, a switched-capacitor filter, working in continuous-time environment, should be preceded by an analog continuous-time filter which ensures that aliasing does not occur. This is called an *antialiasing filter* (AAF) and is realized on the same chip using an active RC circuit. A possible realization is the Sallen–Key second-order section shown in Figure 12.11. It is designed to have a typical response as shown in Figure 12.12 in relation to the response of the switched-capacitor filter. Also, due to the (sin x/x) effect present at the output of the filter, as explained in Chapter 6, an amplitude equalizer circuit with response approximating the inverse function may be needed if the passband edge of the filter is not small compared with the sampling frequency. If the sampling frequency is greater than ten times the passband edge, then a simpler continuous-time filter similar to the prefilter can be used. If the sampling frequency is not very large compared with the pass-band edge of the SC filter, then the required degree of the AAF could be large resulting in an increased area on the chip. To ease the requirement on the selectivity of the AAF and hence reduce the degree, a decimator [53] can be used as shown in Figure

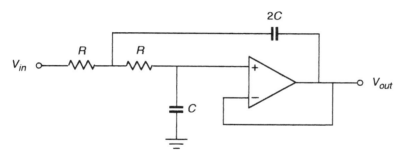

**Figure 12.11**   Sallen–Key second-order section.

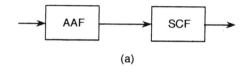

(a)

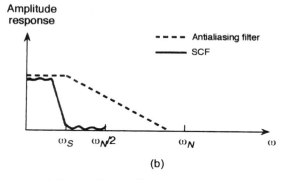

(b)

**Figure 12.12**   Response of the antialiasing filter relative to that of the switched-capacitor filter.

12.13(a). This introduces zeros at multiples of the sampling frequency as shown in Figure 12.13(b). The required degree of the AAF in this case is obviously lower than that which would be required for the direct implementation of Figure 12.12. Thus, in Figure 12.13 aliasing is eliminated in two stages. A circuit which implements the scheme of Figure 12.13 is shown in Figure 12.14 together with the clock forms and

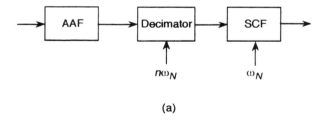

(a)

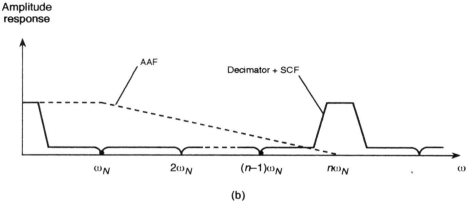

(b)

**Figure 12.13**   The use of a decimator and AAF for prefiltering. (a) Scheme, (b) amplitude response of AAF relative to that of decimator + SCF.

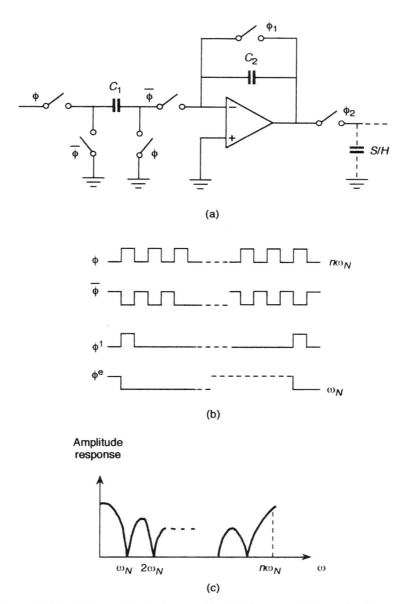

**Figure 12.14**  (a) A possible decimator, (b) clock forms, and (c) amplitude response.

amplitude response. The latter is given by

$$|H(\omega)| = \frac{C_1}{C_2} \left| \frac{\sin(\pi\omega/\omega_N)}{\sin(\pi\omega/n\omega_N)} \right| \tag{12.11}$$

These techniques are only successful at relatively low frequencies (a few kHz) and for a complete non-contrived solution to the antialiasing and smoothing filters problem, good continuous-time filtering techniques seem to be the only real answer. For a good treatment of MOS integrated active RC filters the reader is referred to reference [6].


## 12.10   PROGRAMMABLE FILTERS

A switched-capacitor filter can be made programmable by simply varying the clock frequency. As we have seen the response of the filter contains the key points such as the passband edge and stopband edge. These are basically determined as ratios of the actual frequencies to the clock frequency. Thus, if the clock frequency is multiplied by a factor $k$, the filter key frequencies and hence the entire frequency axis will be multiplied by the same factor. The clock can be programmed digitally.

Another method is to replace the capacitors by capacitor arrays which are switched into place in various combinations; thus controlling the response of the filter. Such arrays can also be programmed digitally.

A third approach is mask programmability. In this case, the components such as operational amplifiers and switches are provided on the chip without interconnection, and a separate portion is dedicated to the capacitors. Other ancilliary components such as resistors and clock generators are also provided. At the final mask stages, the appropriate connections are made, together with the implementation of the capacitors.


## 12.11   LAYOUT CONSIDERATIONS

Analog integrated circuits are particularly sensitive to the geometrical and physical arrangements of their components. This layout of the integrated circuit affects the performance of the switched-capacitor filter by influencing a number of design parameters. Among these are: the noise injection from power lines, clock lines, ground lines and the substrate, clock feed-through noise, accuracy of matching components, and the high-frequency response. This is particularly the case when both digital and analog circuits exist on the same chip, which is a very common situation in communication systems.

We have seen how to obtain designs which minimize noise and clock feed-through, so that care in the layout should be excercised in order not to lose these attributes. So the following, rather common-sense precautions should be observed in addition to the more fundamental design techniques discussed before, such as the use of fully-differential balanced topology, chopper stabilization and delayed clocking schemes for reduction of clock feed-through.

First, power lines, clock lines and ground lines should be kept free from noise and the noise coupling between lines must be minimized. Secondly, separate power lines for digital and analog signals should be used if possible. Also separate bonding pads for these circuits can be used, together with separate pins which can be connected externally together. Finally, external decoupling capacitors can be used between the pins, thus reducing the impedance coupling to the external supply. With these precautions, the bias voltage lines for the subtrates and wells can be connected to the analog supply pads without introducing digital noise into the substrate or wells. If a certain signal or clock line is particularly troublesome, it can be shielded as shown in Figure 12.15. The shielding is formed from two grounded metal lines and a grounded polysilicon layer. Obviously, this arrangement can also be used to isolate the analog and digital circuits and to prevent noise coupling from and to the substrate.

Noise coupling into the substrate can be reduced by using a clean power supply for biasing, by shielding the substrate from all capacitors, by placing grounded wells below them , and by establishing a good bond using gold if possible between the back surface of the substrate and the package header. A good method for reducing substrate noise is to use epitaxy, since the resulting $n^+$ layer would act as a ground plane.

For the reduction of noise from the substrate, the fully differential balanced principle, discussed before, is used together with shielding.

The most significant capacitances coupling the substrate to the circuit are those between the substrate and the bottom plates. Each of these could be as large as 20% of the nominal top-to-bottom plate value. Therefore, the bottom plate should not be connected or switched to the inverting input of an Op Amp since this terminal has a high noise gain to the output. In addition, the lines connecting the input nodes to any capacitors or switches should be as short as possible, and made of polysilicon or metal; diffusion-type lines should be avoided since they are too strongly coupled capacitively to the substrate.

The crossing of input-node lines with other signal lines should be avoided. The lines should be shielded and guard rings used to shield the input devices of an Op Amp. Also the number of components connected to an input node should be kept to a minimum. Only one switch should be used at the input terminal and implemented with minimum area transistors.

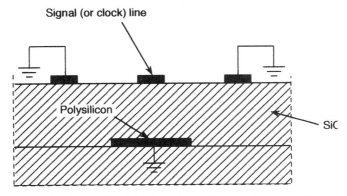

**Figure 12.15**  A possible shielding scheme for signal or clock lines.

## CONCLUSION

In this chapter, we considered many practical issues characteristic of the integrated circuit realization of switched-capacitor filters. These included high frequency designs and fully differential circuits, as well as the effects of the non-ideal behaviour of Op Amps and switches on the overall performance of the designed filter. Scaling for maximum dynamic range and minimum capacitance, prefiltering, postfiltering and layout considerations were also discussed.

In conclusion, we note that several design techniques were not discussed in this book but they can be useful in some situations. These include the use of coupled biquads [54], voltage-inverter-switches [55] and N-path designs [8]. At any rate, the reader who has mastered the design techniques discussed in this book should be able to follow these off-stream methods with no difficulty, if the need arises. Moreover, the reader who is interested in the design of switched-capacitor filters using the current-mode approach may consult reference [56].

# References

[1]  H. Baher, *Synthesis of Electrical Networks*, J. Wiley and Sons, 1984.

[2]  J. D. Rhodes, *Theory of Electrical Filters*, J. Wiley and Sons, 1976.

[3]  L. Huelsman and P. Allen, *Introduction to the Theory and Design of Active Filters*, McGraw-Hill, 1980.

[4]  G. Daryanani, *Principles of Active Network Synthesis and Design*, J. Wiley and Sons, 1976.

[5]  G. Moschytz and P. Horn, *Active Filter Design Handbook*, J. Wiley and Sons, 1981.

[6]  M. Banu and Y. Tsvidis, 'Fully integrated active RC filters in MOS technology', *IEEE J. Solid State Circuits*, vol. SC-18, No. 6, pp. 644–651, Dec. 1983.

[7]  G. M. Jacobs *et al*, 'Design techniques for MOS switched-capacitor ladder filters', *IEEE Trans. Circuits and Systems*, vol CAS-25, No. 12, pp. 1014–1021, Dec. 1978.

[8]  R. Gregorian and G. Temes, *Analog MOS Integrated Circuits for Signal Processing*, J. Wiley and Sons, 1986.

[9]  R. Castello and P. Gray, 'A high performance micropower switched-capacitor filter', *IEEE J. Solid State Circuits*, vol. SC-20, No 6, pp. 1122, 1132, December 1985.

[10]  D. Haigh and J. Taylor, 'High frequency switched-capacitor filters for CMOS technology', *Proc. Int Synposium on Circuits and Systems*, pp. 1469–1472, 1988.

[11]  R. Unbehauen and A Cichcki, *MOS Switched-capacitor and Continuous-time Integrated Circuits and Systems*, Springer Verlag, Berlin 1989.

[12]  T. Choi *et al*, 'High frequency CMOS switched-capacitor filters for communications application', *IEEE J Solid State Circuits*, vol. SC-18 No. 6 pp. 652–663, Dec. 1983.

[13]  B. Song, 'A 10.7-MHz Switched-capacitor band-pass filter', *IEEE J. Solid State Circuits*, vol. 24, No. 2, pp. 320–324, April 1989.

[14]  K. Matsui, 'CMOS video filters using switched-capcitor 14 MHz circuits', *Digest of Tech Papers, IEEE Int Solid State Conference*, pp. 282–283, and 364, 1985.

[15]  H. Baher and E. Afifi, 'A fourth order switched-capacitor cascade strucure for sigma-delta converters', *Int. J. Circuit Theory and Applications*, vol. 23, pp. 3–21, Feb. 1995.

[16]  D. Haigh and J. Everard (Eds.), *GaAs Technology and its Impact on Circuits and Systems*, P. Peregrinus, 1990.

[17]  Y. Tsvidis and P. Antognetti (Eds.), *Design of MOS VLSI Circuits for Telecommunications*, Prentice Hall, 1985.

[18]  R. Gregorian and W. Nicholson, 'CMOS switched-capacitor filters for a PCM voice codec', *IEEE J. Solid State Circuits*, vol. SC-14 pp. 970–980, Dec. 1979.

[19]  Y. Kuraishi *et al*, 'A single chip NMOS analog front-end for modems', *IEEE J. Solid State Circuits*, vol. SC-17, No. 6, pp. 1039–1044, Dec. 1982.

[20] L. Lin and H. Tseng, 'Monolithic filters for 1200 baud modems', *ISSCC Dig. Tech. Papers*, pp. 148–149, Feb. 1982.

[21] B. White, G. Jacobs and G. Landsburg, 'A monolithic dual tone multifrequency receiver', *IEEE J. Solid State Circuits*, vol. SC-14, No. 6, pp. 991–997, Dec. 1979.

[22] P Fleisher *et al*, 'A single chip dual-tone and dial-pulse signalling receiver', *ISSCC Dig. Tech. Papers*, pp. 212–213, Feb. 1982.

[23] R. Brodersen,, P. Hurst, and D. Allstot, 'Switched-capacitor applications in speech processing', *Proc. IEEE Int Symposium on Circuits and Systems*, pp. 732–737, April 1980.

[24] L. Lin *et al*, 'A monolithic audio spectrum analyzer for speech recognition systems', *Int. Solid State Conf. Digest of Tech. Papers*, pp. 272–273, 1983.

[25] MAGIC version 5, *User Manual*, Computer Science Division, Electrical Engineering and Computer Sciences, University of California, Berkeley.

[26] CAzM (Circuit Analyzer with Macromodelling), *User Manual*, Microelectronics Centre of North Carolina.

[27] C. Fang, Y. Tsvidis, and O. Wing, 'SWITCAP: A switched-capacitor network analysis program Part I: Basic features', *IEEE Circuits and Systems Magazine*, pp. 4–10, Sep. 1983.

[28] C. Fang, Y. Tsvidis, and O. Wing, 'SWITCAP: A switched-capacitor network analysis program Part II: Advanced applications', *IEEE Circuits and Systems Magazine*, pp. 41–46, Sep. 1983.

[29] R. Plodeck *et al*, 'SCANAL – A program for the computer-aided analysis of switched-capacitor networks', *Proc. Inst. Elec. Eng.* vol. 128, Pt. G, No. 6, pp. 277–285, Dec. 1981.

[30] Y. Tsvidis, *Operation and Modelling of the MOS Transistor*, McGraw-Hill, 1987.

[31] P. Gray and R. Meyer, *Analysis and Design of Analog Integrated Circuits*, J. Wiley and Sons, third ed., 1993.

[32] P. Allen and D. Holberg, *CMOS Analog Circuit Design*, Holt, Reinhart and Winston, 1987.

[33] R. Jolly and R. McCharles, 'A low-noise amplifier for switched-capacitor filters', *IEEE J. Solid State Circuits*, vol. SC-17, No. 6, pp. 1192-1194, Dec. 1982.

[34] K. Hsieh, P. Gray, D. Senderowitcz and D. Messerschmitt, 'A low noise chopper stabilized differential switched-capacitor filter technique', *IEEE J. Solid State Circuits*, vol. SC-16, No. 6, pp. 708–715, December 1981.

[35] K. Bult and G. Geleen, 'A fast settling CMOS Op Amp with 90 dB dc Gain and 116 MHz unity-gain frequency', *IEEE Int. Solid State Conf. Tech. Papers Digest*, pp. 108–109, 1990.

[36] H. Baher, *Analog and Digital Signal Processing*, J.Wiley and Sons, 1990.

[37] M. Abramowitz and I. Stegun (Eds.), *Handbook of Mathematical Functions*, Dover Publ., NY, 1970.

[38] R. Daniels, *Approximation Methods for Electronic Filter Design*, McGraw-Hill, 1974.

[39] R. Saal, *Handbook of Filter Design*, AEG Telefunken, 1979.

[40] H. Baher and S.O. Scanlan, 'Stability and exact synthesis of low-pass switched-capacitor filters', *IEEE Trans. Circuits and Systems*, vol. CAS-29, No. 7, pp. 488–492, July. 1982.

[41] H. Baher, 'Transfer functions for switched-capacitor and wave digital filters', *IEEE Trans. Circuits and Systems*, vol. CAS-33, No. 11, pp. 1138–1143, Nov. 1986.

[42] H. Baher and S.O. Scanlan, 'Exact synthesis of band-pass switched-capacitor ladder filters', *IEEE Trans. Circuits and Systems*, vol. CAS-31, No. 4, pp. 342–349, April 1984.

[43] R. B. Datar and A.S. Sedra, 'Exact design of strays-insensitive switched-capacitor ladder filters' *IEEE Trans. Circuits and Systems*, vol. CAS-30, pp. 888–898, Dec. 1983.

[44] P. E. Fleischer and K.R. Laker, 'A family of active switched-capacitor biquad building blocks' *Bell System Tech. J*, pp. 2235–2289, Dec. 1979.

[45] H. Baher, 'Selective linear-phase switched-capacitor filters modelled on classical structures', *IEEE Trans. Circuits and Systems*, vol. CAS-33, No. 2, pp. 141–149, Feb. 1986.

[46] H. Baher, *Selective Linear-phase Switched-capacitor and Digital Filters*, Kluwer Academic Publishers, 1993.

[47] H. Baher and M. O'Malley, 'Design of switched-capacitor and wave digital filters with linear phase and high amplitude selectivity' *IEEE Trans. Circuits and Systems*, vol. CAS-37, No. 5, pp. 614–622, May 1990.

[48] H. Baher and S. Zhuang, 'Switched-capacitor band-pass filters with simultaneous amplitude selectivity and pass-band phase linearity' *Journal of Analog Integrated Circuits and Signal Processing*, vol. 2, No. 1: pp. 5–17, Feb. 1992.

[49] H. Baher and M. O'Malley, 'Generalized appoximation techniques for selective linear-phase digital and non-reciprocal lumped filters' *IEEE Trans. Circuits and Systems*, vol. CAS-33, No. 12: pp. 1159–1169, Dec. 1986.

[50] H. Baher and J. Beneat, 'Design of analog and digital data transmission filters', *IEEE Trans. Cicuits and Systems*, vol CAS-40 No. 7 pp. 449–460, July 1993.

[51] K. Martin and A. Sedra, 'Effects of the Op Amp finite gain and bandwidth on the performance of switched-capaitor filters', *IEEE Trans. Circuits and Systems*, vol. CAS-28, pp. 822–829, August 1981.

[52] D. G. Haigh and B. Singh, 'A switching scheme for switched-capcitor filters which reduces the effect of parasitic capacitances associated with switch control terminals', *Proc. IEEE Int. Symp. Circuits and Systems*, pp. 586–589, 1983.

[53] D. Grunigen *et al*, 'Integrated switched-capacitor low-pass filter with combined anti-aliasing decimation filter for low frequencies', *IEEE J. Solid State Circuits*, vol. SC-17, No.6, pp. 1024–1029, December 1982.

[54] K. Martin and A. Sedra, 'Exact design of switched-capacitor band-pass filters using coupled-biquad structures', *IEEE Trans. Circuits and Systems*, vol. CAS-27, No. 6, pp. 469–475, June 1980.

[55] A. Fettweis, 'Basic principles of switched-capcitor filters using voltage inverter switches' *Arch. Elektr.Ubertrag.* vol. 33, pp. 13–19, Jan 1979.

[56] C. Toumazou, F. Lidgey and D. Haigh, '*Analogue IC Design: the current-mode approach*', Peter Peregrinus, London, 1990.

# Index